The Fukushima and Tohoku Disaster

The Fukushima and Tohoku Disaster

A Review of the Five-Year Reconstruction Efforts

Faculty of Societal Safety Sciences
Kansai University

ELSEVIER

Butterworth-Heinemann
An imprint of Elsevier

Butterworth-Heinemann is an imprint of Elsevier
The Boulevard, Langford Lane, Kidlington, Oxford OX5 1GB, United Kingdom
50 Hampshire Street, 5th Floor, Cambridge, MA 02139, United States

Library of Congress Cataloging-in-Publication Data
A catalog record for this book is available from the Library of Congress

British Library Cataloguing-in-Publication Data
A catalogue record for this book is available from the British Library

ISBN: 978-0-12-814078-9

For information on all Butterworth-Heinemann publications
visit our website at https://www.elsevier.com/books-and-journals

Publisher: Candice Janco
Acquisition Editor: Amy Shapiro
Editorial Project Manager: Tasha Frank
Production Project Manager: Anitha Sivaraj
Cover Designer: Miles Hitchen

Typeset by SPi Global, India

Contents

Chapter authors

CHAPTER 1 – The Great East Japan Earthquake: Catastrophic

Yoshiaki Kawata
Faculty of Societal Safety Sciences, Kansai University

CHAPTER 2 – The Great East Japan Earthquake: Current Status and Problems of Recovery

Shingo Nagamatsu
Faculty of Societal Safety Sciences, Kansai University

CHAPTER 3 – Current Status of Rebuilding Houses and Reconstructing Local Regions After the Great East Japan Earthquake

Kenji Koshiyama
Faculty of Societal Safety Sciences, Kansai University

CHAPTER 4 – Problems With Lifeline Recovery and Public Transportation

Hiroshi Nishimura and Seiji Abe
Faculty of Societal Safety Sciences, Kansai University

CHAPTER 5 – Supporting Health of the Great East Japan Earthquake Evacuees

Toshio Takatorige
Faculty of Societal Safety Sciences, Kansai University

CHAPTER 6 – Seismological Research and Earthquake Risk Assessment After the Great East Japan Earthquake

Yoshinari Hayashi
Faculty of Societal Safety Sciences, Kansai University

CHAPTER 7 – Concerns and New Measures for Tsunami Disaster Mitigation in Preparation for the Next Major Earthquake

Tomoyuki Takahashi
Faculty of Societal Safety Sciences, Kansai University

CHAPTER 8 – Liquefaction With the Great East Japan Earthquake

Tomofumi Koyama
Faculty of Societal Safety Sciences, Kansai University

CHAPTER 9 – Legal System for Supporting Disaster Victims

Eiichi Yamasaki
Faculty of Societal Safety Sciences, Kansai University

CHAPTER 10 – The Great East Japan Earthquake and Insurance

Kinzou Kuwana
Faculty of Societal Safety Sciences, Kansai University

CHAPTER 11 – Problems With "Disaster Relief Volunteers"

Mashiho Suga
Faculty of Societal Safety Sciences, Kansai University

CHAPTER 12 – How the Great East Japan Earthquake Affected Corporate Risk Management

Katsuyuki Kamei
Faculty of Societal Safety Sciences, Kansai University

CHAPTER 13 – Problems and Future of Post 3.11 Disaster Journalism

Seiji Kondo
Faculty of Societal Safety Sciences, Kansai University

CHAPTER 14 – Nuclear Plant Disaster and Safety

Emiko Kanoshima
Faculty of Societal Safety Sciences, Kansai University

CHAPTER 15 – Socioeconomic Issues, Reluctance in Buying Produce and Dealing With Highly Radioactive Waste

Yukio Hirose
Faculty of Societal Safety Sciences, Kansai University

CHAPTER 16 – The Government and TEPCO Problems in Communicating Information With the Public During the Fukushima-1 NPP Accident

Shoji Tsuchida
Faculty of Societal Safety Sciences, Kansai University

CHAPTER 17 – The Great East Japan Earthquake and Nuclear Power Safety Regulation

Mamoru Ozawa and Seiji Abe
Faculty of Societal Safety Sciences, Kansai University

Preface

Kansai University is a private institution, with 130 years of history since its foundation in 1886. The university has 13 faculties, 14 graduate schools, Faculty of Law and School of Accountancy, with about 30,000 undergraduate and graduate students enrolled each year. The main campus is in the City of Suita, Osaka Prefecture, located at the geographic center of Japan. The university also has campuses in the City of Takatsuki, and City of Sakai in Osaka Prefecture.

The Faculty of Societal Safety Sciences opened in April 2010, after 5 years of preparation, as the 13th school. It is conveniently located in the new campus in the City of Takatsuki. The campus was named Kansai University MUSE (myooz) Campus, after the Greek goddesses of art.

The Faculty of Societal Safety Sciences and the Graduate School of Societal Safety Sciences are dedicated to promoting research and educational activities that contribute to creating safe and secure societies. Concerns that threaten the safety and security of societies today take many forms, including natural disasters such as earthquakes and tsunamis, accidents in transportation and engineering plants, global environmental issues, food safety, pandemics such as new types of flu, crimes and international terrorism, or threats to information security. Taking proper actions against these concerns to keep citizens safe and secure is one of the most important challenges societies in the 21st century face. Advanced educational institutions are obligated to play major roles to meet these societal needs through developing theories, methodologies and tools, educating the human resources to take charge in designing new policies, and putting them into practice. Europe and the United States have already established universities and graduate schools with such missions; however, none has yet been present in Japan. Our school and graduate school are the first institutions in Japan that systematically aim at societal safety sciences.

Interdisciplinary research that traverses existing academic fields needs deep digging for integration and synthesis of "safety knowledge," theoretical and political research for accident prevention and reduction, and making practical proposals for disaster prevention and mitigation. Our 28 faculty members, for the promotion of joint research that steps over the boundaries of art and science, consist of 16 with a background in arts and 12 in scientific fields. The breakdown of their specialties is as follows: four in law, four in psychology, four in economics and business administration, two in sociology, one in education, one in philosophy, two in science, seven in engineering, one in information science, and two in social medicine.

The magnitude 9.0 mega-earthquake on March 11, 2011, with the tsunami and aftershocks that followed, devastated the Pacific coastal area of Tohoku over a length of about 500 km. Tokyo Electric Power Company's Fukushima Daiichi Nuclear Power Plant suffered a series of severe accidents because the tsunami destroyed the emergency electricity supply systems.

The Great East Japan Earthquake was a huge compound disaster of earthquake, tsunami, and nuclear plant accidents. The number of casualties, still not finalized, consists of 15,893 dead and 2556 missing (announcement by National Police Agency). In the approximately 100 years of modern Japanese history, it was the third largest disaster, following the 1923 Great Kanto Earthquake with over 100,000 deaths and the 1896 Sanriku tsunami that counted over 20,000 deaths.

Immediately after the strike, our schools and research centers faced this megadisaster and carried out research and analyses. Our work was compiled into the book *Review of the Great East Japan Earthquake* published in March 2012 by Minerva Shobo in Kyoto. The book attracted a large reader base and made the second edition. It was translated into Korean by The Center for Japanese Studies at Korea University and published in Seoul. The book had an international reputation.

March 2016 marks 5 full years since the breakout of the disaster. The Japanese government had announced these 5 years as "time for concentrated reconstruction" and have spent a total recovery and reconstruction budget of 26.3 trillion JPY. For partial finance of this budget, the government set a new special reconstruction tax. These tax increases are projected to provide a total amount of 10.5 trillion JPY.

The people of Japan, in general, have strong reactions against raising taxes. This time, however, the people did not show such a reaction, because the use of the money is limited to reconstruction support for the damaged areas. Most people understood the need and importance in supporting the devastated regions.

Five years from the disaster, reconstruction of lifelines and social infrastructures is almost complete. Rebuilding residents' lives and towns, however, is still only halfway done. The Prefecture of Fukushima especially is faced with a pile of difficult problems, including residents who evacuated out of the prefecture, still close to 100,000, or processing contaminated water in Fukushima Daiichi Nuclear Power Plant and its decommission. This book reviews the 5-year recovery and reconstruction efforts after the Great East Japan Earthquake disasters and clarifies what problems lie ahead.

This book is a sequel to the mentioned *Review of the Great East Japan Earthquake*. It is said that Japan will be attacked by a Nankai Trough Megathrust Earthquake or a Tokyo Metropolitan Epicentric Earthquake in the near future, and these predictions are given fairly high odds. The death toll forecast in the worst case, especially for a Nankai Trough Megathrust Earthquake, is 320,000 or more. This book is not limited to reviewing the reconstruction processes of the Great East Japan Earthquake, but it also summarizes the issues we have to solve and makes political suggestions in preparation for the Nankai Trough Megathrust Earthquake and Tokyo Metropolitan Epicentric Earthquake.

Finally, in publishing this book, Dr. Kenji Iino worked hard in translating the Japanese manuscripts. Translating this type of specialized book requires not just superb skills in English, but it also requires ample knowledge over a variety of specialized fields. This book would not have been published without his dedicated efforts. We would like to express our deep gratitude for his work. We would also like to thank Ms. Amy M. Shapiro and Ms. Tasha Frank at Elsevier for their timely advice in the process of publishing this book.

Seiji Abe
Kansai University, Osaka, Japan

Chapter 1

The Great East Japan Earthquake: Catastrophic

CHAPTER OUTLINE

I was assigned to be a member of two important committees set by the Japanese government immediately following the Great East Japan Earthquake. One was "The Reconstruction Design Council in response to the

The Fukushima and Tohoku Disaster. https://doi.org/10.1016/B978-0-12-814078-9.00001-7

Great East Japan Earthquake" and the other, "Committee for Technical Investigation on Countermeasures for Earthquakes and Tsunamis Based on the Lessons Learned from the 2011 off the Pacific coast of Tohoku Earthquake." Upon finishing the tasks of these committees, I then served as a committee member of the "Committee for Policy Planning on Disaster Management of the Central Disaster Management Council," and then for the "Disaster Management Implementation Committee." Through this service, I carefully looked after the reconstruction and recovery after the earthquake and tsunami disaster. My work was based on the experience and knowledge gained from the 1995 Great Hanshin-Awaji Earthquake and the 2004 Niigata Prefecture Chuetsu Earthquake. We strongly hope that we will reach our common goal of "Build back better than before," first accepted as the Hyogo Framework for Action during the Second United Nations (UN) World Conference on Disaster Risk Reduction in 2005 and practically adopted in the Third Sendai Framework Action (SFA) in 2015.

1.1 TROUBLE WITH POLICIES IN RECOVERY AND RECONSTRUCTION

1.1.1 Problems in Response at the Local Community Level

Let's first list the clear reasons for the problems:

(1) *Belief and confirmation bias*: Everyone believed that if there was an earthquake, it had to be offshore of Miyagi Prefecture. The prediction back then called for 99% occurrence there within the next 30 years, and this was believed to be the next big one to hit Japan in future. The Japan Meteorological Agency was no different and its tsunami forecast had not calculated an earthquake of magnitude 8 or larger off the shore of Sanriku. The administration and local residents all shared the same belief.
Miyagi Prefecture had the highest rate of earthquake insurance coverage, in excess of 1230 billion yen. The coverage was about 16 times that of the amount, roughly 78 billion, for the 1995 Great Hanshin-Awaji Earthquake. Due to this good coverage, rebuilding of homes was not a major social issue. Rather, new problems surfaced for relocating houses to higher-level areas.

(2) *Illusion that countermeasures were sufficient*: Tsunami hazard maps had been prepared with almost all local administration offices in preparation for Miyagi Prefecture offshore earthquakes. Communities of Iwate Prefecture had seawater gates and walls against hypothetical 1933 Showa Sanriku offshore earthquakes and tsunamis, and with the assumption that no tsunami would go over these walls, had no tsunami hazard maps. Tsunami evacuation drills were held, but the number of local residents participating in them kept dropping every year. The administration had some fear that the lack of preparedness within the residents could lead to serious damages. The number of residents in the submerged area during the Great East Japan Earthquake was about 510,000 and the death toll reached about 4.2% of that. On the other hand, the Taro area of Miyako City was protected by a seawater wall 10 m high, and the death rate there was also about the same, 4%. The percentage of residents that did not run for safety was the same everywhere.

(3) *Unexpected damage to the communities*: No one thought the community administration buildings would collapse and be underwater, killing mayors and administrative workers. The local disaster prevention plans had no conditions of such damage to administrative bodies except perhaps some damage to the buildings. This lack of preparation caused delays in the rescue activities, with no proper damage information sent out after the disaster struck.

(4) *Neglecting the power of tsunami*: A tsunami is one of the most dangerous natural disasters and it was not well understood that fleeing from it is the only way of surviving. The last damaging tsunami dates back to that caused by the Great Chilean Earthquake in 1960. Along the coastline of Sanriku, resonation amplified the tsunami size when it reached the end of the Ofunato Bay, causing concentrated damage there. Over the 50 years since the disaster, the threat had faded from people's memories.

(5) *Misunderstanding past experience*: On February 27, 2010, a year before the Great East Japan Earthquake, when a magnitude 8.8 quake shook Chile, the Japan Meteorological Agency announced a large tsunami warning along the coast of Iwate and Miyagi. The actual tsunami heights observed, however, reached less than 1 m at most observation spots and the warning ended up appearing to be "crying wolf." The Japan Meteorological Agency never explained its false alarms and the residents gradually lost faith in its alerts. Moreover, with 5–6 m high walls right in front of their eyes, the residents had the misconception that the walls would protect them from tsunamis, which could reach only 3–6 m: another misjudgment. Their past experience directed the residents into underestimating the size of tsunamis that could hit the area.

(6) *Difficulty in taking actions without proper information*: With total lack of disaster information, the communities that suffered damage set up their disaster response headquarters, but had no idea where to start the discussion. In the era of Internet and cellular phones, there was a misconception that they would operate even during times of disaster, and the administration had not made any a priori evaluation of how to estimate the degree of damage and what to report to the prefecture and central government.

(7) *Tsunami damage to the designated shelters of schools and community halls was unexpected*: The people did not understand that going to the designated shelters was insufficient and that they had to climb to higher locations. As people age, they tend to be self-centered and base their decisions on experience only. The elderly population in the areas did not consider walking long distances to escape the possible disaster.

(8) *The affected prefectures ignored collaboration with the communities during the disaster*: The Basic Act for Disaster Countermeasures assigns responsibility for residential aid to local communities, and if the local communities fail, to the prefecture. The affected prefectures, however, made no cooperative efforts to help the communities. The lack of collaboration was the same as during the 2003 Sanriku Minami Earthquake, the 2005 Miyagi Earthquake, and the Iwate-Miyagi Nairiku Earthquake in 2008. Rather the local Diet members would run to the local area that elected them with the arrangement of The Cabinet Office, and the local communities had no expectations regarding the prefectures, and the prefectures in turn had the same idea.

1.1.2 **Problems in Response at the Prefecture Level**

This set of problems was the biggest cause of the delay in reconstruction and recovery from the Great East Japan Earthquake.

(1) *No collaboration with affected local communities*: As I stated in item (8) in the previous section, the lack of collaboration was a crucial issue. The prefectures of Iwate, Miyagi, and Fukushima had no procedures or disaster prevention plans stating what countermeasures to take in the event of disasters and had no idea how to organize administrative actions. Especially in the early stage of the disaster, with the lack of disaster information, many of the prefectural government employees were seen leaving for home when the end-of-the-day bell rang. All three government capital hall buildings survived with only minor damage, causing misunderstandings among the employees about the severity of the damage. This fact shows that they learned nothing from the lessons of the Great Hanshin-Awaji Earthquake.

(2) *Administration offices without field experience*: The prefectural civil engineering and land administration departments had daily discussions with local communities and communications with residents about land acquisition and civil planning. The administration and planning department workers, however, were constrained to their desks and had hardly any communication with local communities. The same lesson taken from the Great Hanshin-Awaji Earthquake, that "You can only expect to do what you do daily," applies here.

(3) *Huge disaster area*: The number of communities covered by the Disaster Relief Act was up to 241. Faced with this huge number, authorities in the prefecture were lost as to how to proceed with the recovery and reconstruction. They did not even think about dispatching prefectural officials to areas that had not made reports to the prefecture. The prefecture kept its stance that it had to rely on the central government.

(4) *Governors of Iwate, Miyagi, and Fukushima Prefectures lacked leadership*: As a committee member of The Reconstruction Design Council in response to the Great East Japan Earthquake, I came across a number of opportunities to meet with these three governors in the Prime Minister's Office starting in May. I then made three proposals: the prefecture to build a recovery foundation, to distribute the monetary donations, and to carry out the rubble disposition. They all ignored my proposals, saying they were not for the prefecture. In particular, the idea for a recovery foundation was turned down because they did not want to "borrow money." These three governors were all self-centered, lacking an attitude of supplementing their lack of knowledge with that from outside. The lesson for the affected prefecture during the 1991 eruption of Unzen-Fugen Mountain had not made its way to teach these chiefs.

(5) *Recovery and reconstruction were almost fully funded at national expense*: The prefectures judged that any plan approved by the central government, especially the reconstruction plans, could start, and a number of almost identical businesses made their way through the negotiations. The lack of personnel resources probably contributed to the affected prefectures not having unique ideas. The worst was Fukushima Prefecture, with the attitude of trying to have as many plans approved as possible during discussions in the Prime Minister's Office.

(6) *Lack of personnel resources*: The affected prefecture administration should have dispatched a large number of employees to the affected communities, but it did not. Cooperative disaster management was difficult due to the physical distance between the administration offices. As the unaffected Tohno City did, the prefecture could also have built a temporary office. After all, over 2000 local community employees from around the nation have been dispatched to the affected communities, even at this point, i.e., about 6 years after the disaster.

(7) *Poor recognition of information*: Even with the advanced information technologies used today, e.g., satellite information, geographic information systems (GIS), or global positioning systems (GPS), the prefecture lacked recognition of their needs. The lack of knowledge caused delays in collecting information for rebuilding towns and issuing certificates of disaster affliction.

1.1.3 Problems in Response by the Central Government

The Prime Minister and his cabinet, having learned from the Great Hanshin-Awaji Earthquake, initially responded successfully. The actions that followed, however, kept opening up new problems, one after another.

(1) *Problems with gathering information*: They could have used geospatial intelligence to analyze images from satellites to better understand real-time situations over the wide disaster-hit areas. The government did not take the initiative and left the information gathering to Japan Aerospace Exploration Agency (JAXA) and other private companies. Disaster information reached the Cabinet for the first time when the Japan Ground Self-Defense Force (JGSDF) entered the affected area.

(2) *Poor leadership by the Cabinet*: The Democratic Party of Japan, at the time in power, dispatched Vice-Ministers and Parliamentary Secretaries to disaster-struck prefectures to take government-driven leadership in response. It was, however, its first time to take such measures and those headed into the fields had no delegated power, e.g., to commit securing recovery funds or dispatch national public officers, and they ended up merely transferring correspondence between the Cabinet and the local prefectures. The action had no advantages and it only delayed the decision-making process.

(3) *Undervalue of administration*: Before the Great East Japan Earthquake hit, the Democratic Party of Japan had led the Administrative Reform Council, which undervalued the work of all government ministries and agencies. The reform had led to national public officers' loss of morale and weak efforts in taking action for projects to restore the disaster-struck areas. Disposing of the disaster debris was the responsibility of the Ministry of the Environment; however, with the added work of responding to the nuclear accident, the debris was left untouched for an extended time without clear assignment of responsibilities.

(4) *Disrespect for and ignoring of tacit knowledge*: The Emergency Response Headquarters was established for the first time, headed by the Prime Minister; however, members of the Cabinet Secretariat and the Cabinet Office had shared hardly any tacit knowledge and could not take proper actions at the time. These organizations lacked the necessary knowledge and power to respond to large-scale disasters, and the administration offices had not taken measures to document and turn the officers' tacit knowledge into organizational knowledge. Without proper knowledge, most officers were forced through their first experience of countering a large-scale disaster. It is regretful that the system was not prepared to take advantage of past experience. Even with the experience of the Great Hanshin-Awaji Earthquake, no thorough efforts had been made to review mistakes in responses at the time.

1.1.4 **Causes of Delays in Incomplete Recovery and Reconstruction Projects**

A number of intertwined factors led to this problem, and it is hard to point at what were the exact causes. From the aforementioned (1)–(3), however, we can at least point to the following problems:

(1) Did not learn from past lessons:
As we mentioned previously, none of the following actions for the victim prefectures took place: to establish recovery funds, to distribute the relief funds, to make debris disposition plans or, with consultation with the central government, to guide local administration offices to follow the plans. Those that suffered the most were the afflicted residents. The lessons from the Great Hanshin-Awaji Earthquake, that we have to make the residents central and discuss recovery and reconstruction plans around them, did not have any effect on the administration. The prefectural administration should have asked for advice from Hyogo Prefecture, which went through the 1995 Hanshin-Awaji Earthquake, or Niigata Prefecture which went through the Niigata Prefecture Chuetsu (2004) and Chuetsu Offshore (2007) earthquakes. Hardly any action took place in seeking guidance.

(2) There were no wide-area collaborations:
Hardly any collaboration took place among victim prefectures and their local communities, or among the central administration and victim prefectures. The prefectures barely communicated among themselves, and merely had governors' signatures listed together on their joint petition to the government. One of the problems proceeded like this: The government set up the Reconstruction Design Council and the Special Research Group in the Central Disaster Management Council to discuss technical aspects of the recovery efforts. Each council published its report in late September to October in 2011. Similar actions took place in victim prefectures and local communities, but the

reconstruction efforts progressed without any consistency among different organizations. The disorganization resulted in the mistake of setting the levee height to the maximum height of the Leven-1 tsunami. In the future, when the residents are faced with evacuation orders or advice, they will again not flee the area, to repeat the same tragedy. The cause was probably the council members not understanding the purpose of regulations for local tsunami disaster prevention. Misunderstanding by the prefectural council technical group of the guidelines from the central government led it to build its own theory, another factor in this mistake.

(3) Reconstruction Charter was incomplete:
Experts had agreed that the reason for delays in the reconstruction after the Great Hanshin-Awaji Earthquake was the lack of a Reconstruction Charter. In 2008, the Japan Society for Disaster Recovery and Revitalization was established mainly to form this charter, with its administration office, naturally, set inside the Institute for the Research of Disaster Area Reconstruction of Kwansei Gakuin University. Later, when the Great East Japan Earthquake hit, experts feared that not completing the Reconstruction Charter at the time could lead the government to set erroneous reconstruction guidelines, a fear that turned into reality. The Reconstruction Agency spent no time committing the mistake. One of the biggest flaws was relying on the central government budget to support all funding for reconstruction. It may seem natural under the financial situations with local governments, but the national budget turned away any ideas from local communities for their reconstruction. All reconstruction projects were approved on the desks at the central government. In fact, few proposals for reconstruction projects from victim communities had any relevance to reconstructing the communities. In particular, almost all proposals from the Fukushima Prefecture were approved, possibly with consideration for the nuclear disaster it had to suffer from. Nevertheless, as the projects proceeded, the planned project budget ran out in the middle, a totally irresponsible management of the development.

(4) Actions after the disaster and emergency responses are not reviewed
Six years have passed since the earthquake, but disaster reviews by the central and local governments are not in process. We must fear that we would suffer the same mistakes again. The purpose is not to point fingers at persons or groups. Japan has a hard time understanding this concept. Improvements can only be made after a thorough investigation. The United States suffered a fatality count of about 1800 and 125 billion dollars in damage, the worst in their history, with the 2005 Hurricane Katrina. The disaster statistics worsened with faulty

action after the disaster struck. The thorough investigation that followed is called an After Action Review (AAR). Every ministry was instructed to turn in its investigation report within 2 years of the disaster. The report had 15 assignments of emergency support functions (ESFs). The United States publicizes what each ministry should do, before a disaster takes place.

The biggest faults after the Great East Japan Earthquake were marked by the Japan Meteorological Agency. They were: (i) Undervaluing the earthquake magnitude Ms to 7.9. This surface wave magnitude, when it exceeded 8, was known to read smaller than the actual among seismologists. Almost all seismographs in Japan scaled out and when England announced a moment magnitude Mn of 9 from its readings, most of the damage had already been done. (ii) Why did it announce deterministic tsunami warnings of magnitudes 3 m in Iwate, 6 m in Miyagi, and 3 m in Fukushima? These were clearly underestimations that should have been avoided, "a looking strikeout," by any means. Its fear of making "a swung strikeout" alarm ended up with having to make corrections that warned of tsunami sizes of about twice the first alarms, 30 minutes after the first set. By that time, most of the regions had lost electricity and the new alerts only rarely made their way to the residents. Moreover, some say that all wave-height observation stations were destroyed (there is no record of wave shapes and the tsunami characteristics cannot be evaluated), and the wave heights were also underestimated after the Great Chilean Earthquake in February of 2010, but what improvements had been made after that are still not clear.

Without a complete review of actions after the event, the Japan Agency for Marine-Earth Science and Technology (JAMSTEC) and the National Research Institute for Earth Science and Disaster Prevention (NIED) have not explained why they need huge funding for developing seismographs and tsunami surveillance stations (called DONET or S-net) to be installed on the ocean bottom. The fact that their installation is about to be completed while a thorough discussion of disaster prevention and reduction is not even on the table suggests that research topics for natural scientists of mechanisms for earthquake tsunami generation were given priority.

1.2 QUICK OVERVIEW OF THE 3.11 TSUNAMI GENERATION MECHANISM

Why did the tsunami grow so large? JAMSTEC has been gradually unveiling the mechanism [1]. The explanation of the phenomenon is that it was triggered by an almost simultaneous occurrence of plate boundary earthquake and tsunami earthquake. The mechanism that produces tsunami

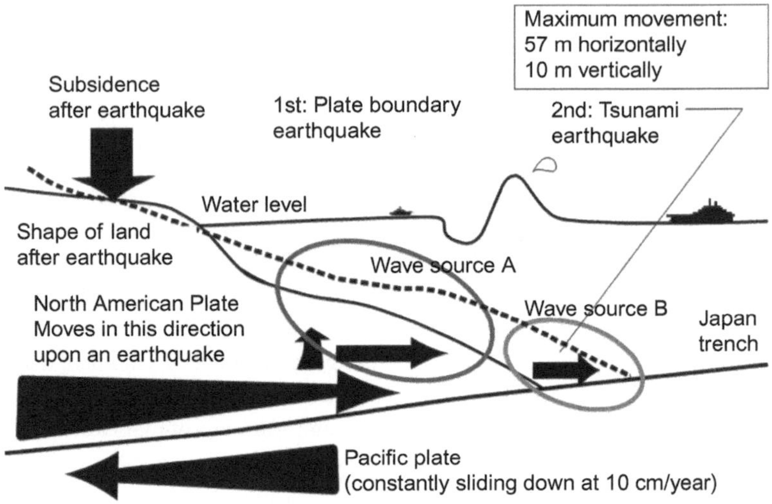

■ **FIG. 1.1** Two types of mechanisms for earthquake and tsunami to occur offshore Sanriku coast.

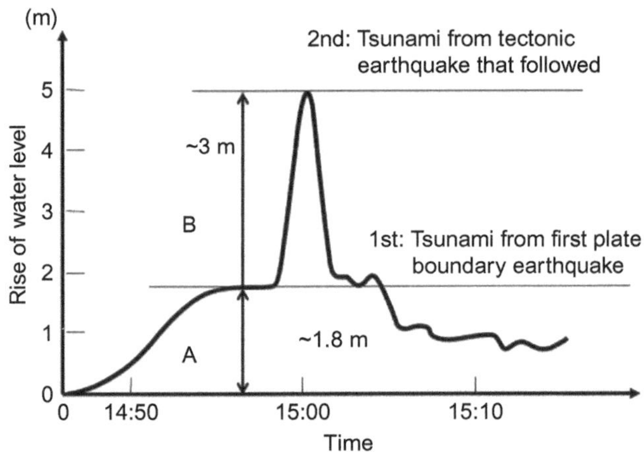

■ **FIG. 1.2** Two tsunamis superimposed to reach a huge size.

earthquakes has so far not been clear. Fig. 1.1 shows the mechanisms that produce the two types of tsunami. Fig. 1.2 shows the wave formation when the two tsunami waves superimpose.

The tsunami earthquake is now known to have caused a huge landslide at the bottom of the sea that moved a massive amount of soil towards the Japan trench axis, about 8000 m deep, about 50 m horizontally and 10 m vertically. The new knowledge we obtained recently is that the soil did not fall freely (freefall of soil under water is slow) but it accelerated as it approached the trench axis. The huge volcanic soil layer was subject to compression caused by submergence of the plates and when the North American plate and Pacific plate binding was released, the resulting friction heat caused high enough temperature to turn the compression into expansion, which pushed the soil towards the trench axis. This phenomenon explained the disturbance of soil layers after the earthquake.

These models were verified by the results of drilling and observations by the deep-sea drilling vessel Chikyu in the area near the earthquake center and the following analyses.

In the case of tsunami earthquakes, we still do not have a model that produces a tsunami. We just manipulate numerical eight-fault parameters to match the records. We need to continue efforts in developing our dynamic models to cover the mechanisms of tsunami earthquake generation.

1.3 TSUNAMI PREPARATION BEFORE AND AFTER THE 3.11 DISASTER

It took about half a year to obtain a clear picture of the actual tsunami evacuation that took place. Other papers and journals discuss this in detail, so we will point out problems with the tsunami evacuation warnings, and recommendations. Coastal engineering has clear definitions for terms such as tsunami levee, storm surge barrier, coastal revetment, and sea embankment. The people of Japan, however, are not so interested in the terminology and we will collectively call these *tsunami levee* to mean installments to prevent tsunamis at banks or coastlines. The following content will discuss conventional guidelines for tsunami levee design, Level-1 and 2 tsunamis, and as a real example, the tsunami levee construction in the City of Rikuzen-Takata.

1.3.1 Tsunami Evacuation Warnings and Advices

During the Great East Japan Earthquake, about 30% of the residents did not start immediate evacuation, as Fig. 1.3 shows [2]. Adding this percentage to those that fled to elevations after seeing the tsunami approaching, as many as 40% of the residents were exposed to the danger. If we apply the same percentage to the people of Osaka Prefecture, who will be attacked by a 3.8 m tsunami during the hypothesized Nankai Trough Megathrust Earthquake (M9.0), we reach an estimate of over 110,000 victims [3]. Furthermore, during the recent typhoon season, results show a national average of only 1% of residents taking refuge in evacuation camps after heavy rain and flood warnings.

When heavy rain hit east and northern Japan in 2015, the Kinugawa River opened a gap in its bank and about 4000 residents were left isolated in the

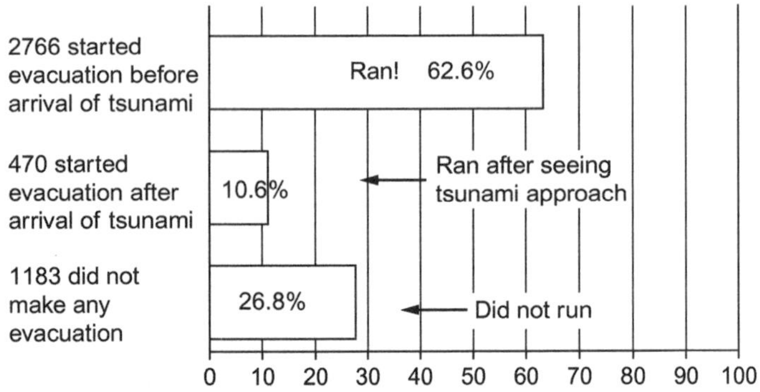

FIG. 1.3 Start of evacuation (4421 people).

flooded area, and about 1300 of them were rescued by helicopters. When the media asked them why they did not evacuate the area, many of them answered that the evacuation order was late or was not issued. Those words are not true. Residents that were left in the flooded area were aware that they themselves were to blame for the disaster and were trying to push the responsibility onto the administration. This is a typical reaction with many victims of disasters. For example, when residents suffer earthquake damage to their houses, when they come to the city hall to receive damage certificates, almost all of them, at the clerk's desk, blame the local government for not strongly recommending aseismic reinforcement, and some even go to the extreme of blaming the damage on the local government.

There is no medicine to cure such misunderstandings and misinterpretations. No tsunami warnings were announced during the 1896 Great Meiji Sanriku Tsunami or the 1933 Showa Sanriku Tsunami. Japan Meteorological Agency started its tsunami warnings in 1952. Residents, up to that time, had to make their own judgments. After the 1960 Great Chilean Earthquake, the tsunami warning was announced after the first tsunami. Far-field tsunamis caused by earthquakes with centers distant by over 500 km that did not cause tremors was not in the scope of the Japan Meteorological Agency. The 1983 Sea of Japan Earthquake caused tsunamis that attacked many areas, some with late tsunami warnings and a number of fatalities. Thinking about these events tells us the importance of information.

In discussing this issue of ineffective tsunami warnings and recommendations, more people are concerned about natural scientific problems, e.g., with how measurements are taken; however, taking views from sociological science angles is more important. Why weren't such concerns raised from the devastated areas? Once the memories of this earthquake fade away from people's memories, residents will again not flee to safety in future disasters.

Moreover, when several decades have past after the Great East Japan Earthquake, people will be living with a tsunami levee made of concrete and over 10 m tall. Who can blame them from thinking that they will be safe? We have to continue efforts to carry on the lesson that "running is for living" throughout the regions even when such times arrive.

1.3.2 Tsunami Levee Design Before the Quake

During the Great East Japan Earthquake, the tsunami levee built on land was destroyed over a length of 190 km. The damage was mostly caused by scouring power at the base when the tsunami waves spilled over the levee. Waves exceeding the levee height were not expected to begin with and, even if the designers did expect them, the budget to build such a levee would not have been approved. In other words, there were no designers or builders that would have explained the necessity from a technical standpoint to win a budget to build one. There would be no exaggeration in saying that we lacked profound insight into natural phenomena.

When building coastal structures, we work safety factors into the design, e.g., we increase the expected external load by 10%–20%. Many think that safety factors account for variations in the external force, but that is not what they are for. Safety factors are to take into consideration internal error with the design methodology. Therein lies a big misunderstanding. In fact, computational errors in expected height of tsunami waves that hit the coast of Japan are within 3%, thanks to the precise seabed maps that surround our country. Earthquake characteristics, on the other hand, that define the initial waveform are described using eight parameters, and if we think of the total variations that change in these parameters can cause, they can quickly multiply rather than increasing only by 10%–20%. The 2011 Off the Pacific Coast of Tohoku Earthquake was exactly one of such cases.

When the Great East Japan Earthquake hit, with a magnitude of 9, it instantly caused "unexpected phenomena." However, not many of us know that even if it had been of magnitude 8, it would have caused a tsunami that "exceeded" our expectations. The margin of design was that small, probably because structures were designed to fit the budget with cost (C) benefit (B) analysis (in public works, usually B/C is larger than 2). There we can see the poor safety philosophy of the designers and builders.

1.3.3 Level-1 (L1) Tsunami and Level-2 (L2) Tsunami

I chaired the Committee for Technical Investigation on Countermeasures for Earthquakes and Tsunamis Based on the Lessons Learned from the 2011 off the Pacific coast of Tohoku Earthquake established in May of 2011 in the Central Disaster Management Council. During my service, I proposed defining L1 and L2 tsunamis. The proposal was approved by the Ministry of Land, Infrastructure, Transport and Tourism.

The biggest reason for approving the proposal was the "huge damage to the people of Japan that far exceeded expectations." It was not because huge areas of farmland were submerged under seawater, nor was it because of the enormous damage to the social capital. We must not have this misunderstanding.

Our misunderstanding is much like the one we have about aseismic retrofitting of housing to counter earthquakes. The Great Hanshin-Awaji Earthquake led to so many fatalities caused by the collapse of residential housing. Then, reinforcing houses to counter earthquakes was strongly encouraged to prevent collapsing houses from "killing people." As time passed, the purpose switched to "reinforcing houses." The original purpose of protecting lives transformed into just reinforcing houses. This led the government to disapprove partial reinforcement of each room of a house. Alas, the aseismic rate of housing rises by about 1% annually. The increase, in fact, owes to the demolishing of old houses when the residents pass away. This natural increase with new housing buildings let the government escape without taking a real interest in aseismic reform of houses. Such policy is not humanistic at all.

So, in fact, it is misleading to think that there is no need to evacuate because the tsunami levee would protect people from an L1 tsunami. When a tsunami warning is issued and an evacuation warning or recommendation is announced, we must evacuate. Officers in local government, especially those in prefectural administration offices, often have this misconception. Furthermore, in the regions hit with this earthquake, an L1 tsunami will not arrive in 5 minutes after the shaking. The area is different from Shizuoka or Wakayama Prefectures that will see a 10 m or so tsunami 5 minutes after a Nankai Trough Megathrust Earthquake attacks.

When I visited the Taro area in Miyako-city 35 years ago, I asked the mayor the question,

> *If a 15 m tsunami like the one with Great Meiji Sanriku Earthquake hits, with the tsunami levee being 10 m tall, wouldn't the town turn into a 10 m deep lake killing a large number of residents?*

He then answered,

> *The town has prepared twelve evacuation routes and the residents will be safe because they will evacuate.*

During the Great East Japan Earthquake, 4.2% of the residents in the tsunami-hit areas died. The number for the Taro area was about the same, 4%. The 10 m tall tsunami levee had no effect in protecting the residents. In fact, there were quite a few people standing on the levee to watch the tsunami approach.

In Japan, designers are possessed by the old idea that they cannot start structural design unless the tsunami height design is set to a certain value. First, L1 tsunamis are caused by simple plate boundary earthquakes and, along the coast of Sanriku, they have hit about once every 37 years since the Edo era. L2 tsunamis, on the other hand, are now known to be caused by plate boundary earthquakes followed by a huge landslide of sea bottom soil that is dynamically pumped towards a trench axis (compression). The mechanism is explained with a rise in the water pressure in the soil grain gaps caused by high heat upon shift of the plate boundary. Given these circumstances, we would like to set the L1 tsunami height to one that can possibly occur. Numerical simulation will allow us to do so; in other words, for the extreme case, we can calculate a tsunami height as high as we want by manipulating the fault parameters.

1.3.4 Problems With Constructing Tsunami Levees

The Iwate Prefecture tsunami disaster prevention technical committee, for example, set the expected tsunami height at the City of Rikuzen-Takata to 12.5 m upon Miyagi offshore earthquakes. This is about twice the height of the highest on record. This setting is wrong. The reason is that the Takata pine forest with over 70,000 pine trees has a history of 400 years. If there was a tsunami of 11.5 m, all the pine trees would have been washed away. (Experience from the Great East Japan Earthquake taught us that a pine tree will be washed away with a tsunami that reaches half its height.) The Technical Investigation Committee is not saying to set the L1 height at the biggest of all L1 tsunamis. If the setting is at the biggest, the residents will again not evacuate when the time comes. Common sense tells us that maybe it should be set to half the biggest. Table 1.1 shows regions planning to construct tsunami levees that exceed 10 m. Some communities in the same region, however, have settled on lower levees. An L1 tsunami is caused by plate boundary earthquakes. On the other hand, regions with 10 m or taller levees are thinking of arming against the Great Meiji Sanriku Earthquake. This was a tsunami earthquake and not an L1 tsunami. We could perhaps call it a Level-1.5 tsunami. We can probably say that the technical evaluation committees of the disaster-struck prefectures intently misinterpreted the report by the Technical Investigation Committee of the Central Disaster Management Council.

Table 1.1 List of Tsunami Levees 10 m or Higher Planned for Construction After the Great East Japan Earthquake

Municipality	Tsunami Levee Height (m)
Iwate Prefecture	
Hirono	12.0
Kuji	8.0–12.0
Noda	14.0
Fudai	15.5
Tanohata	14.3
Iwaizumi	14.3–14.7
Miyako	10.4–14.7
Yamada	9.7–12.8
Otsuchi	14.5
Kamaishi	6.1–14.5
Ofunato	7.2–14.3
Rikuzentakata	12.5–12.8
Miyagi Prefecture	
Kesennuma	5.0–14.7

The design tsunami height is just an indicator. What is important is to share the common knowledge that tsunamis can exceed this value. In other words, it is important that facilities and structures should perform their functions to certain levels and we should not bear the illusion that they provide absolute safety.

Why did such short-sighted conclusions come into play? First, we must acknowledge poor knowledge of the technical members of the prefectural committees that set the heights of L1 and L2 tsunamis. It is common sense not to rely just on natural scientific knowledge in setting such heights. Even if the committee members understand it properly, once set, the numbers start walking out of control. I talked about this concern during my speech at the Foreign Correspondents' Club in Tokyo, but the media also is hesitant to make any noise. That is because they know the residents think the higher the tsunami levees are the better. Local governments have no financial burden in building tsunami levees. Not even 1%. Not having any burden is against the spirit of the principle of self-responsibility and the basics of self-help in disaster prevention.

The German city of Hamburg, frequently hit by storm surges, for example, has regulations to keep the ground floor a parking space or waterproof the windows and doorways on the ground floor. Apartment complexes have to surround the buildings with watertight walls to obtain a building permit. In the United States, in the suburbs of New Orleans, devastated by the 2005 Hurricane Katrina, all houses have to be built with pilotis and the owner has to cover 30% of the construction cost. Advanced countries in the west completely follow the principle of self-responsibility in disaster prevention and reduction.

Given this background in reconstructing areas afflicted by the Great East Japan Earthquake, the "Act on Development of Areas Resilient to Tsunami Disasters" that covers tsunami levee construction could easily be misunderstood as well. What I will discuss here about revitalizing local communities in the next section is making progress, but with a number of misunderstandings as well.

1.4 TROUBLES IN THE EFFORTS TOWARD REVITALIZING LOCAL COMMUNITIES

1.4.1 Lack of Coordination From the Start

The rough picture of reconstructing offered to the afflicted regions includes the aforementioned tsunami levee construction and additionally relocation to higher ground and lifting the ground level. The formal names of the latter two projects are: "Promoting Group Relocation for Disaster Mitigation" and "Land Readjustment Project." The old town area is marked as disaster prone and can only accommodate commercial and touring business facilities. Tens of such areas are about to be born along the Sanriku coastline. These may work for current residents of the areas, but it is questionable whether local industries will come back to uninhabited old towns protected by mere concrete levees, not to mention the difficulty in luring new businesses. What tourist attraction can develop there? We can only picture big towns with isolated buildings here and there, with hardly anyone walking on the streets, and lots of open space.

There is a predecessor, the Kobe Biomedical Innovation Cluster built in Port Island in the City of Kobe in 1998, which is a number of large facilities built on a huge landfill with wide roads but with no one living there. People working there commute on a monorail called Port Liner that runs through the middle of the landfill. If the monorail was not there, the only way to commute would be by car. We have never heard plans for building public transportation through the Sanriku afflicted area. When we turn our eyes to the high land relocation area of the City of Rikuzen Takata, no one lives in the old town and there are plans to lure commercial facilities around the high land. The open land may very well be left untouched. The scene is similar to the distribution of ruins and shell mounds left along Tokyo Bay when the coastline receded about 6000 years ago.

The causes are clear. They were the tsunami levee construction and areal revitalization made independently. It was a serious mistake that the former took off first and the latter followed much later, on its own. The new law, the "Act on Development of Areas Resilient to Tsunami Disasters," was built with the assumption that the two developments would proceed in a unified manner, i.e., the levee height is supposed to be set in relation to how the land it protects will be used. The reality, however, is that community reconstruction is proceeding with eyes on the neighboring community so that they are at about the same level. There is no mindset to target the optimum for the overall area. The communities lacked proper personnel resources, but the prefectures are also responsible for not providing proper guidance and advice while they were aware of the manpower problem. This is another negative effect of the afflicted prefectures not establishing their own reconstruction funds.

1.4.2 A Case Study of Rikuzen-Takata Pine Field

The City of Rikuzen-Takata will not see a tsunami within 5–10 minutes after an earthquake. It takes longer for the waves to arrive. As long as there are tsunami evacuation routes, most lives will be saved. Each local community has to more carefully build its community revitalization plan. As we discussed earlier, the landmark Takata Pine Field had a history of about 400 years. In other words, over the past 400 years, the L1 tsunamis that hit the area every 37 years on average did not destroy the pine forest. Then why was the L1 tsunami set to T.P. + 12.5 m? (T.P.: Tokyo Peil, the average water elevation at Tokyo Bay.) Was this value agreed to among a number of experts? If there were such high L1 tsunamis, the pine forest would not have survived. (For the pine trees to survive such an L1 tsunami, their average height would need to exceed about 25 m.) Why were such questions never raised? Tsunami simulations can produce different results in any way wanted, with different initial conditions.

Let us introduce reconstruction plans of the City of Rikuzen-Takata, a symbolic community of The Great East Japan Earthquake, and point out their problems. The total budget for the project is 7.1 billion yen. Fig. 1.4 shows the planar and sectional views. Levee 2 being too high is out of the question but we cannot understand the overall design, for these reasons:

(1) Views of the memorial pine forest are totally blocked by Levee 2. Its presence has no meaning as a symbol. The entire pine forest will be washed away by a tsunami smaller than L1. The original expectation for

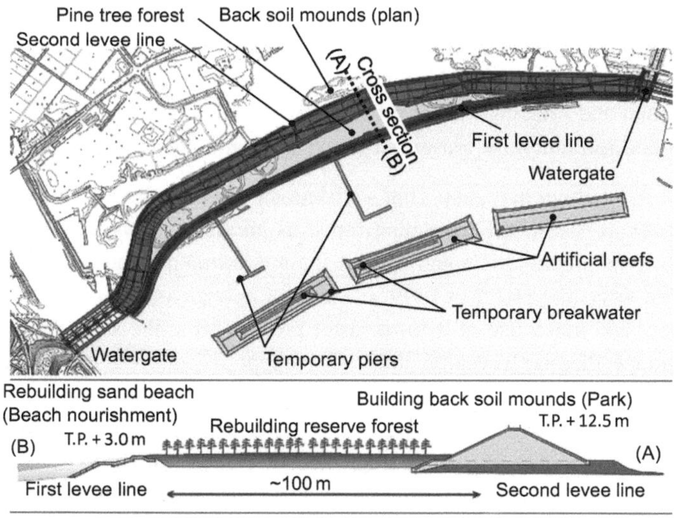

■ **FIG. 1.4** Reconstruction plans along the coast of Rikuzen Takata.

the Takata pine forest was probably to have it block stormy waves. Before the disaster, there were reasons for building a pine forest in the back of the sand beach, which we can find from another case: The world's first soil-filled tsunami levee in Hiromura, Wakayama Prefecture was built after the Great 1854 Ansei Nankai Earthquake. The base is about 3 m high with a rock wall where the base rises. The pine trees on the lifted soil have big trunks, a sign that shows caretaking by thinning. The levee protected the residential area behind it when the 1946 Showa Nankai Earthquake attacked the area 88 years after the levee was built.

The Hiromura levee bank, instead of simple concrete lining on three faces, was formed by laying out stacked rocks, each weighing 2–3 tons, covering them with soil to hide them, and then planting pine trees on the top. Even the largest pine tree has roots only 10%–15% of its height in the ground. Thus, building a 2 m tall rock bed levee, covering it with soil, and planting pine trees on the top would block tsunamis up to the height of the rock bed levee plus half the height of the pine trees. This type of natural levee would provide a good scenic view and protection to the environment.

(2) The source of drift sand for the Matsubara beach is the Kesen River and proper design of the floodgate where the water hits could guide the drift sand to the beach. Such a design would secure the continuity of the sediment environment. The plan seems to be to build three artificial leafs for nourishing the beach but there is no reason to build them at the same time. Selecting sand with grain size similar to that before the tsunami for the bottom material would stabilize the beach. Pumping up bottom material from the ocean bed offshore, placing the sand on the beach and leveling it with a bulldozer would do the job. No need to spend too much by bringing in sand from another area.

As we pointed out here, applying well-known coastal engineering practices could revive the lost Takata pine forest as an attractive beach. This plan would have practical use and protect the environment. The revised Coast Act that went into effect in 1999 was a set of new ideas to protect the environment and make use of it for disaster prevention at the same time. This earthquake disaster revealed that the concept is not well understood in the design field. The locals in the area raised their voices from a number of fields, requesting construction of a well-balanced tsunami levee [4]. However, they were almost all neglected. K-center, which carried out the basic design of the new construction, had to show an attractive design, only possible because the coast was destroyed by the earthquake, to demonstrate the importance of the earthquake recovery project. Building tsunami levees with the same look everywhere cannot win the respect of the affected residents and will never produce a tourist attraction. The precious value of the public project is being lost. The think tank, technical skills of consultants, experience of the designers and governance are all under question.

1.4.3 **House Rebuilding Out-of-Sync**

As of August 13, 2015, the number of evacuees was 198,513 and since its peak in June, 2012, the number is reducing by about 40,000 a year (about 3500 a month). The resident housing at temporary shelters is also dropping from the same year at the rate of about 19,000 houses a year (about 1600 monthly). With the fact that the amount of public housing for disaster victims remains unchanged over the years with each afflicted prefecture, the residents have been rebuilding their homes at the rate of about 20,000 a year.

On the other hand, the system for supporting disaster victims for their reconstruction of their livelihood supports the victims with base fund payment and additional fund payment. The base fund payment applies to the following four cases:

(1) Households whose house was totally destroyed
(2) Households whose house was half destroyed totally
(3) Households whose house was half destroyed and they had to break the house down
(4) The land was damaged and the house had to be broken down

In contrast, additional fund payment applies to those receiving base payment when they rebuild their houses. Currently, 191,352 households (23,000 in Iwate, 122,000 in Miyagi, and 29,000 in Fukushima) are receiving the base payment and half of them are receiving the additional fund payment. Tada investigated the annual change of additional fund payment and reported [5] the reducing number of evacuees is partly due to their own rebuilding of houses. He then pointed out the drop in the number of civil housing area development plans for group relocations in Miyagi and Iwate and concluded that many evacuees have selected to rebuild their own houses.

This fact shows that public housing for disaster victims cannot be the center of housing reconstruction. On the other hand, however, the number of totally demolished houses in the urban area readjustment project for Promoting Group Relocation for Disaster Mitigation and in the area for the Land Readjustment Project compose as many as about one-fourth of all the houses that were totally destroyed. Rebuilding such a large number of totally destroyed houses would result in patchy house rebuilding in the afflicted area. Especially for the City of Ofunato where the tsunami swelled inside the bay, causing different levels of destructions in patches, the city area will be a mixture of new and old households.

1.5 PROMOTING DISASTER REDUCTION POLICIES

1.5.1 Newly Developed Disaster Reduction Policies

People used to believe that disaster prevention and reduction policies improved the level of safety and security. After World War II, for example, storm and flood disasters kept annual victim counts of over 1000 for about 15 years. They were peculiar years of disaster. Building flood-control dams and floodways for class A rivers played a big role in stopping those disasters. The change, however, also caused some serious negative effects of destroying the natural environment and interrupting the preservation of continuity of the ecological system. Since then, we have changed our ideas of "flood control is the only way" to modify the River Act in 1997 and the Coast Act in 1999.

On the other hand, the Basic Act for Disaster Countermeasures that went into effect in 1961 was born in the old years of poor concepts and under the idea "not to repeat the same suffering again," which kept revising the design standards after each damage from a disaster. This law is based on the idea that "measures are not taken until a disaster happen," with the only exception being the Act on Special Measures Concerning Countermeasures for Large-scale Earthquakes that went into effect in 1978. The logic was that the predicted Tokai Earthquake would certainly take place and preparatory measures had to be implemented. The underlying idea of the Basic Act on Disaster Countermeasures did not change even after the 1995 Great Hanshin-Awaji Earthquake and continued until the 2011 Great East Japan Earthquake.

1.5.2 "The Worst Disaster Scenario" and "Disaster Prevention and Reduction in the Mainstream"

The Great East Japan Earthquake made a huge impact on our disaster prevention and reduction policies. This is because it took away over 22,000 lives and caused over 20 trillion yen in damage. Furthermore, the Tokyo Metropolitan Earthquake and Nankai Trough Megathrust Earthquake, which would cause damage that would far exceed the one in 2011, are urgent topics. The 2011 quake also caused a nuclear power plant accident. It was a "Level 7" accident, the worst on the International Nuclear and Radiological Event Scale (INES), which melted the reactor core. Our nation, for the first time, experienced a compound disaster of earthquake, tsunami, and nuclear power plant accident and still has not been able to come up with a stable formation of effective countermeasures.

If we humbly reflect on what forced us into this situation, we realize that we never assumed "the worst disaster scenario" beforehand. What we thought of as "something we do not want" had turned into "something that will never happen" inside our minds. If only we had had the slightest hint of how we could destroy that power plant, we would have easily realized that a station blackout (SBO) was one way to cause the Fukushima Daiichi Nuclear Power Plant accident.

Also, if we had had an idea that a tsunami could flow over a tsunami levee, we could have prevented its destruction over a total length of 190 km. If we decided that the functional requirement had to be met even in the case of excessive external force over the design value, the financial department would not have declared it an excessive design. It was a fatal wound that we never reached a fundamental agreement in planning disaster prevention and reduction when we were dealing with natural phenomena that are full of uncertainties.

The Committee for Technical Investigation on Countermeasures for Earthquakes and Tsunamis Based on the Lessons Learned from the 2011 off the Pacific coast of Tohoku Earthquake set a guideline that we must assume the worst possible earthquake and tsunami without hesitation. This change, therefore, was fundamental in the way we countered and prepared for disasters.

The Disaster Management Implementation Committee, formed afterwards, upon its establishment claimed in its concept of "mainstream disaster prevention" that all projects forming social foundations shall start with disaster prevention and reduction measures at the very first stage. It was a breakthrough for us.

1.5.3 Disasters at Their New Stage

The Great East Japan Earthquake was the first disaster described with the words "beyond expectation." During the years that followed, however, disasters beyond expectation have constantly been taking place. We now have to recognize that disasters have entered a new stage. That is, we must recognize the fact that rain has started to fall in manners never observed before. A typical example is the typhoon No. 18 in 2015. Here are some details: On September 9th, the National Graduate Institute for Policy Studies hosted the Symposium of Policy Research on Crisis Management and Disaster Prevention Practices and I was scheduled to give the keynote lecture and participate in the panel discussion. I did not want to trouble the organizers, so departed Shin-Osaka station on the Nozomi bullet train to arrive 40 minutes earlier at Tokyo station than my schedule called for. The typhoon had a central pressure of 985 hPa so I judged that the wind was not so severe and the rain would not affect the bullet train schedule.

I arrived at Tokyo station 40 minutes early, as I had planned. It was an example of my daily practices not to make any embarrassing mistakes, since I am someone that specializes in crisis management. Tokyo, however, was facing a severe rainstorm. After the symposium closed, when I was about to board the bullet train at 8:30 pm, the typhoon had traveled away to the Japan Sea and had turned into an extratropical cyclone. The rain in Tokyo was still very hard.

At this time, heavy rain was falling on Tochigi and Ibaraki Prefectures, which had both issued emergency warnings. This extratropical cyclone, another typhoon No. 17 at about 2000 km east-southeast in the Pacific with a central pressure of 975 hPa, and the wiggling westerly jetstream combined to form a precipitation band in the south-north direction to cause severe rainfall over the Kinugawa River area. The rainfall poured about 650 million tons of water over an area of 1700 km^2 and the river levee broke at multiple locations around 20 km upstream of where the river meets the Tone River, to flood the area. We have to acknowledge from this flooding disaster that rainfall has entered a new stage of characteristics which we have to cope with.

The flooding in the Kinugawa area was caused not by a direct hit of a typhoon, but by a change in the streams above the islands of Japan. To counter heavy rainfalls, the former basics were to control flooding that fell over the upper and middle regions of rivers with facilities such as dams to prevent flooding in the downstream area. When we review the Automated Meteorological Data Acquisition System (AMeDAS) records with this incident, heavy rain hit the downstream area first and adding the pumped discharge water from the urban area, water level in the downstream area was higher. This caused stagnation to the floodwater from upstream. The four dams in the upstream area, on the other hand, had already discharged water following the past practices and when the river water overspilled and destroyed the levee, they still had about 30% margin to their flood control capacity. The rained dropped on land in a pattern that we had never experienced before and that was why we were not able to manage the situation.

Global warming has changed the weather patterns in Hokkaido Prefecture, carrying typhoons in their full power all the way up there, and turning the rainy season front in the north-south direction. The change with heavy rainfalls is demanding us to conduct an overall review of our plans of controlling river water. It will be too late after a major flooding.

1.6 DISASTER RESILIENCE AND NATIONAL RESILIENCE

1.6.1 Pictures of New Disasters to Plan Against With National Grand Design 2050 and National Spatial Planning

In July of 2014, the Ministry of Land, Infrastructure, Transport and Tourism (MLIT) announced the "Grand Design of National Spatial Development towards 2050" and National Spatial Planning in August of 2015. Also, on June 3, 2014, the cabinet council approved Japan for National Resilience Plans. The council, at the same time, approved the "Japan for National Resilience Action Plans 2014" that listed key performance indicator (KPI) targets and primary policies. The plans call for annual review of progress and promotion of effective policies. It also expects active involvement of local governments in establishing local plans and of private businesses. Additionally, National Grand Design 2050 and National Spatial Planning are required to be consistent with Japan for National Resilience Plans.

These plans, however, point out two major crises we are faced with from the beginning. They are rapid loss of population and major disasters. The former crisis in local regions leads to poverty such that people cannot lead fruitful lives. The government, however, did not explicitly state "poverty." The fear of not being able to reconstruct the east Japan area after the earthquake comes from the local financial crisis caused by loss of population to support the pension funds.

The plans, however, do not explain the seriousness of the Tokyo Metropolitan Earthquake or Nankai Trouph Megathrust Earthquake and do not put the highest priority on disaster reduction plans. We voice our concerns not because disaster researchers are sensitive, but because when disaster do happen, our concerns will turn into reality. We shall not underestimate the possible risk of the Tokyo Metropolitan Earthquake. If such a risk was pointed out to the capital city of a western leading country, its government would proceed with the highest priority on disaster reduction. Leading countries other than Japan would develop their countries under the assumption that such risks do take place. Tokyo is the only metropolitan area in the world that continues to grow and "people, things, information, and resources" keep accumulating. And, instead of regulating the trend, the national government and the local government of Tokyo Metropolis both approve the growth; somewhere, something must be out of its mind.

An earthquake that can devastate our nation is constantly targeting us. That is the Tokyo Metropolitan Earthquake, but once a damage estimate was announced, superficial disaster reduction plans were proposed and people tended to relax all over our country. If the Tokyo Metropolitan Earthquake breaks out, our capital will be devastated and the country will be exhausted. Our nation will not prosper in the future without facing these odds. Before we worry about the drop in population or the lowered birth rate and aging pollution, we should worry about a disaster that is a ticking time bomb and will abolish our country. The sense of risk that this nation holds is far too small.

Aren't we falling into the trap of our "virtual enemy?" As more things are concentrating to a single location, just one blow can finish the job. Concentration into Tokyo continues, and such a trend is only in Tokyo, not anywhere else around the world. To make things worse, looking towards the 2020 Tokyo Olympics, investigation and development are accelerated around the four districts of Minato, Chuo, Shinjuku, and Chiyoda. These are the areas where most of the damage will take place. We do not have the understanding that we are building a most dangerous situation ourselves. To prepare against new types of disasters, we have to place the highest priority on dispersion of the capital function in preparation against the Tokyo Megaquake. If we continue the current trend, we will surely face a situation that we will regret.

1.6.2 **Disaster Resilience**

Disaster resilience had not been on the table of those involved with disaster prevention and reduction among the central and local governments. What triggered the movement was plans for having the Third United Nations

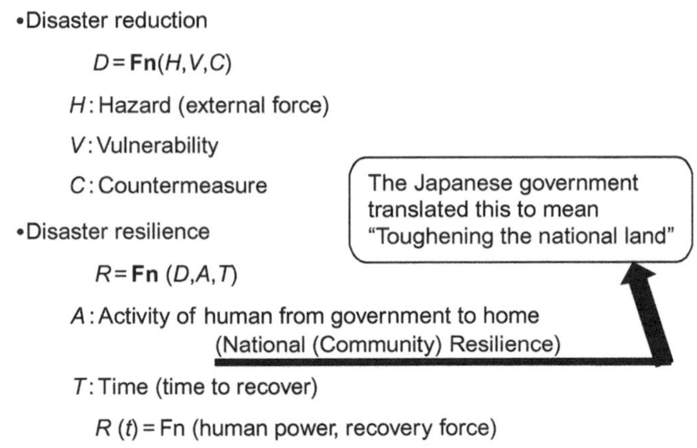

FIG. 1.5 Concepts of disaster mitigation and disaster resilience.

World Conference on Disaster Reduction in March of 2015. Upon deciding on this hosting, it was discussed that the 2005 hosting of the second conference in Kobe had set the goal for the Hyogo framework to realize a resilient society. Then the people involved found they had to report the progress in Sendai. A resilient society quickly recovers from disasters and realizes disaster reduction through people's actions (Fig. 1.5). Thus, national resilience, unlike how the government interpreted it, to mean building a stronger homeland, actually means "disaster reduction by the people." That is why it is a movement by the people of Japan.

To accomplish disaster resilience, there are two new tools that our organizations must learn. They are timeline and after action review (AAR).

(1) Timeline:
The 2005 Hurricane Katrina caused a wide area disaster around New Orleans in the United States. The fatality count was about 1800 and the total damage summed to 125 billion dollars. It was the worst in US history. This system of timeline was born from lessons learned from this disaster. It sets 0 hour to the time the hurricane makes landfall and, as Fig. 1.6 shows, what actions to take before and after the landfall. The system is found useful even for sudden disasters such as earthquakes in addition to disasters with lead times, like flooding that allows a certain time margin after the rain starts falling to when the water overflows. In short, it is a way to decide what to do beforehand, assuming a situation in which the head of a local community lacks sufficient information to make proper judgements, and to share plans among those that need to make actions. If a group makes actions on its own without such planning,

Timeline		Activity	Emergency support function (setting roles of each organization ahead of time)					
Base time (When)		Disaster management actions (What)	Forecast, warning	ESF#4 Water and fire management	ESF#2 Transportation	ESF#5 Crisis management	ESF#3 Social infrastructure	ESF#6 Evacuation, resident support
		(Who)	Meteorological observatory; National river manager; Prefecture river manager	Water management administration; Regional development bureau; Prefecture river management; Cities; Water management	Road management; Japan Coast Guard; Ports and Harbors Bureau; Police; Railway business; Transportation business	Regional development bureau; Prefecture disaster management; City disaster management; Fire department	Regional development bureau; Regional transport bureau; Prefecture construction department; City hall construction department	City disaster management; Fire department; Self-defense force; Police; Volunteer
Before typhoon landing	120–96 h	Preparation for typhoon landing, shelters		○	○	○	○	○
	72–48 h	Expert advice	○			○		
	72–48 h	Alerting possible subway stoppage			○			
	48 h	Building shelters				○		○
	36 h	Weather alerts Water hazard alerts River flood warnings	○	○		○	○	
	24–36 h	Water management announcement Evacuation advisory from Mayor		○	○	○		○
	24–12 h	Subway stoppage Underground arcade closure			○	○		
	12 h	Special warning announcements Evacuation advisory to highland by Mayor	○	○		○		○
	12 h	Flooding caused by high waves	○			○		○
	6–0 h	Evacuation advisory by water management, police, and fire department		○				
Typhoon landing ➡								
After landing	+3–12	Rescue and emergency supplies		○	○	○	○	
	24	Water drainage and emergency recovery		○	○	○	○	

■ **FIG. 1.6** Sample timeline in Japan.

many of the other organizations and individuals end up with difficulties. When typhoon No. 19 hit Japan in October of 2014, JR West announced, on the day before, that it would stop all trains and this announcement caused great confusion among the people and other train companies. JR West needs to review its action, which was based on its sole logic and which caused great social confusion.

To make effective use of timeline, those involved should practice it in workshops and have plenty of preparation to minimize trouble for the people. In Japan, MLIT started with timeline planning in late March, 2015 for the 109 Class A rivers and their watersheds. The hazard maps and timelines are expected to allow providing proper and timely service to the people at times of disasters.

(2) After Action Review:

This system shows how to handle cases when actions failed. It is important to reveal why actions against disasters failed and to learn from them so the system can make modifications so as not to repeat the mistakes. The federal government of the United States made mistakes during the Hurricane Katrina disaster. It then took 2 years to review what went wrong. AAR is one of the outcomes of these reviews. In Japan, on the other hand, when the Great East Japan Earthquake hit, a number of questionable actions were made, e.g., smaller than actual tsunami height estimates by Japan Meteorological Agency, but no information about reviews and lessons of these incidents have been made public. Such an attitude poses a risk for us to repeat the same mistakes in future. We shall start an AAR system soon.

REFERENCES

[1] Japan Agency for Marine-Earth Science and Technology (JAMSTEC), Kanagawa Prefecture, Mysterious mechanism of great earthquakes and huge tsunami revealed, JAMSTEC, Blue Earth 130 (2014) 1–21.

[2] Handout: Committee for Technical Investigation on Countermeasures for Earthquakes and Tsunamis Based on the Lessons Learned from the 2011 off the Pacific Coast of Tohoku Earthquake, Cabinet Office, 2011. http://www.bousai.go.jp/kaigirep/chousakai/tohokukyokun/pdf/sankou.pdf.

[3] Workgroup report about Great Nankai Trough Earthquake disaster countermeasures, Osaka prefecture, 2014. http://www.pref.osaka.lg.jp/attach/20357/00000000/bukaihoukoku.pdf.

[4] Opinion: Requesting review of coastal levee construction along the coastline of Miyagi prefecture after the Great East Japan Earthquake, 2014, Sendai Attorney Association (addressed to Governor of Miyagi prefecture, Chairman of prefectural assembly, Mayor of the City of Sendai, Chairman of the city assembly).

[5] T. Tada, Reconstructing houses after the Great East Japan Earthquake, Res. Inf. 43 (2014) 8–9.

Chapter 2

The Great East Japan Earthquake: Current Status and Problems of Recovery

CHAPTER OUTLINE

2.1 INTRODUCTION: TAKING OVERVIEWS OF DISASTERS

Five years have passed since the Great East Japan Earthquake. When we discuss huge disasters, not only this earthquake, looking only at what happened on the spot can lead us to wrong conclusions. To grasp the overall picture of disasters, we need to obtain overviews of them through quantitative data.

The larger a disaster is, the more difficult is the collection of data for overall statistics. On the other hand, for huge disasters like the Great East Japan Earthquake, which caused serious damage over multiple prefectures, just reviewing general disaster statistics can tell the disaster impact on the

The Fukushima and Tohoku Disaster. https://doi.org/10.1016/B978-0-12-814078-9.00002-9

society and the progress of recovery to a certain extent. This chapter, thus, provides quantitative data for overviewing the damage and the progress of recovery from the Great East Japan Earthquake. The data give input and background information to the discussions throughout this book.

We would like to point out before we open the data that this work had serious obstacles. Statistical data from seriously afflicted areas may contain errors, due to the severity of the damage, or moreover, some numbers may end up not reported or excluded from reports. Socioeconomic statistics are often quoted for political purposes or reported as results of policies and such factors may affect the meaning of statistics when there have been special policies applied to certain regions. The reader shall not simply believe in the numbers displayed here and read through the pages while reviewing situations at the actual disaster hit areas and how recovery policies are working there.

2.2 CHARACTERISTICS OF THE GREAT EAST JAPAN EARTHQUAKE

Let's first review the numbers associated with the Great East Japan Earthquake. The Great East Japan Earthquake is not just the magnitude 9.1 earthquake that hit eastern Japan at 14:46 on March 11, 2011, but it also refers to the series of aftershocks and damage that followed over about a month afterward. It also includes the huge tsunami waves that the earthquake caused, the Fukushima Daiichi Nuclear Power Plant Accident, and damage that the power plant accident caused. As of September 1, 2015, National Police Agency reported the number of deaths at 15,893, whereas, Fire and Disaster Management Agency 19,335. This discrepancy is due to the latter including deaths related to the earthquake disaster. 99% of the deaths and lost bodies were concentrated in the three prefectures of Iwate, Miyagi, and Fukushima. The three prefectures combined were also where 95% of the totally destroyed houses stood. The analyses of this chapter are thus based on numbers from these three prefectures.

We will now point out the level of damage in each community in Iwate and Miyagi Prefectures. Fig. 2.1 graphs relative damage to housing (totally destroyed) and human (death), plotting the count in the vertical axes verses the number of households in the community in the horizontal axes. The left graph is for communities hit by tsunami and the right for those that were not. The graphs clearly show tsunami hit communities with greater numbers, meaning both the counts and relative damages concentrated in the tsunami hit communities rather than those inland.

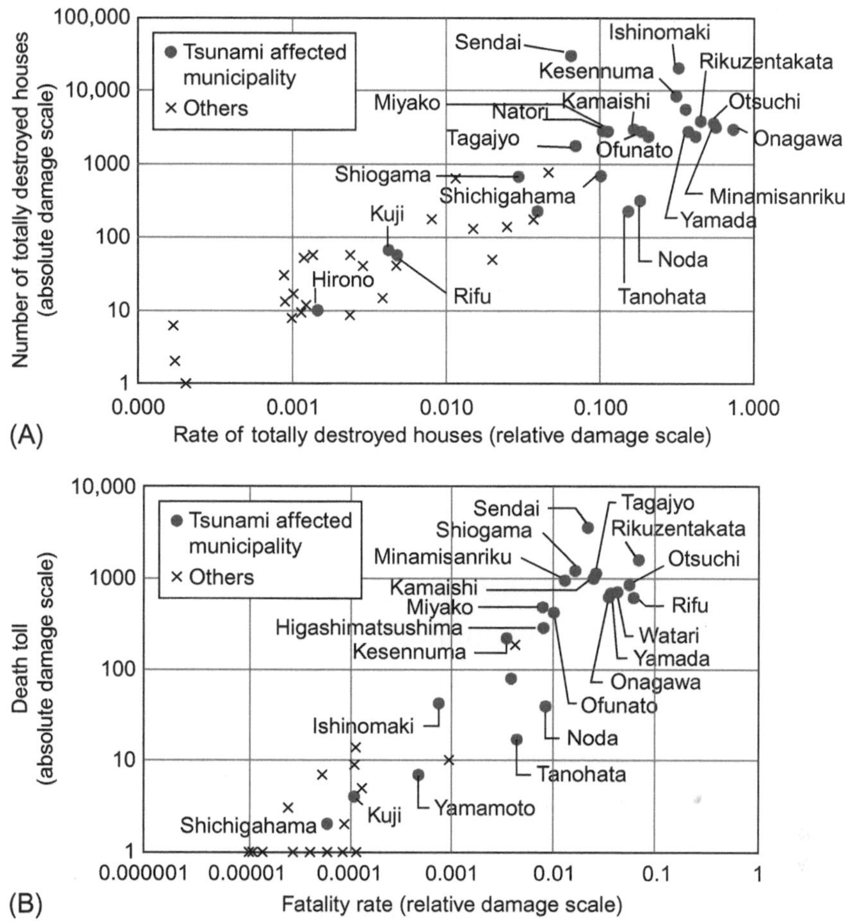

■ **FIG. 2.1** Absolute and relative damages in Iwate and Miyagi. (A) Damage to houses and (B) damage to people. *(Data on number of dead from "About Tohoku Area Pacific Offshore Earthquake (The Great East Japan Earthquake)" Report #152, September 9, 2015, Ministry of Internal Affairs and Communications, Fire and Disaster Management Agency. Population and count of houses are from author's review of Basic Resident Registry of February, 2011.)*

The figure points to the fact about the Great East Japan Earthquake that we tend to look at the wide area hit by the quake; however, within the three prefectures with severe damages, the level of damage was quite different between the inland communities with hardly any damage, and those along the coastline. This gap between the communities was made even larger throughout the recovery process. We will discuss this point later.

Another fact we would like to point out about the Great East Japan Earthquake is the long-term designated evacuation areas within Fukushima prefecture with the nuclear accident. Immediately following the accident, evacuation orders were placed around the nuclear power plant and with a number of later changes, e.g., a 20 km radius area designated as a caution zone, and as of April 1, 2014, evacuation orders continued for about 80,000 people centered around Futaba-gun. The situation where tens of thousands of people were under evacuation orders for such an extended time period was unprecedented in our history of disaster management. Huge compensation is being made to the people and companies that suffered damage from this accident. These facts all have their reasons. However, a huge amount of public funds is being poured into the nuclear plant accident management and its decommissioning, as well as decontamination of the surrounding areas, lifting of the evacuation orders and recovery. These actions are having effects that may not be desirable for the recovery of the regions and have caused a number of side effects, including steep rises in the price of labor and building material due to shortage of bidding contractors.

2.3 CURRENT STATE OF RECOVERY

2.3.1 Population

With these characteristics of the disaster in mind, let's review the current state of recovery within these disaster-afflicted regions. Fig. 2.2 plots the population growth rate within a year from the disaster (March 1, 2011 to March 1, 2012) in the horizontal axis versus that in the second year and after (March 1, 2012 to August 1, 2015) in the vertical axis for each community. The figure shows, in the year that immediately followed the earthquake, the communities affected by the tsunami faced a large drop in population compared to the other communities.

These communities, however, are not necessarily recovering their population. Among those that were hit by the tsunami, communities with growing population are limited to those around Sendai, i.e., the cities of Sendai, Natori, and Tagajo and the town of Rifu. In many of the tsunami-hit communities, people still tend to move out. It may be a general trend over the prefectures of Iwate and Miyagi because the figure shows a population drop at the same time in areas not affected by tsunami. The drops are, nevertheless, rapid for the communities of Minamisanriku, Yamamoto, and Onagawa. The population drop in the town of Onagawa is noteworthy with a rate equivalent to that immediately after the earthquake, which still continued during the following recovery period.

When we show the same graph for Fukushima Prefecture, we can see that the situation is different from Iwate and Miyagi Prefectures. Nuclear-affected

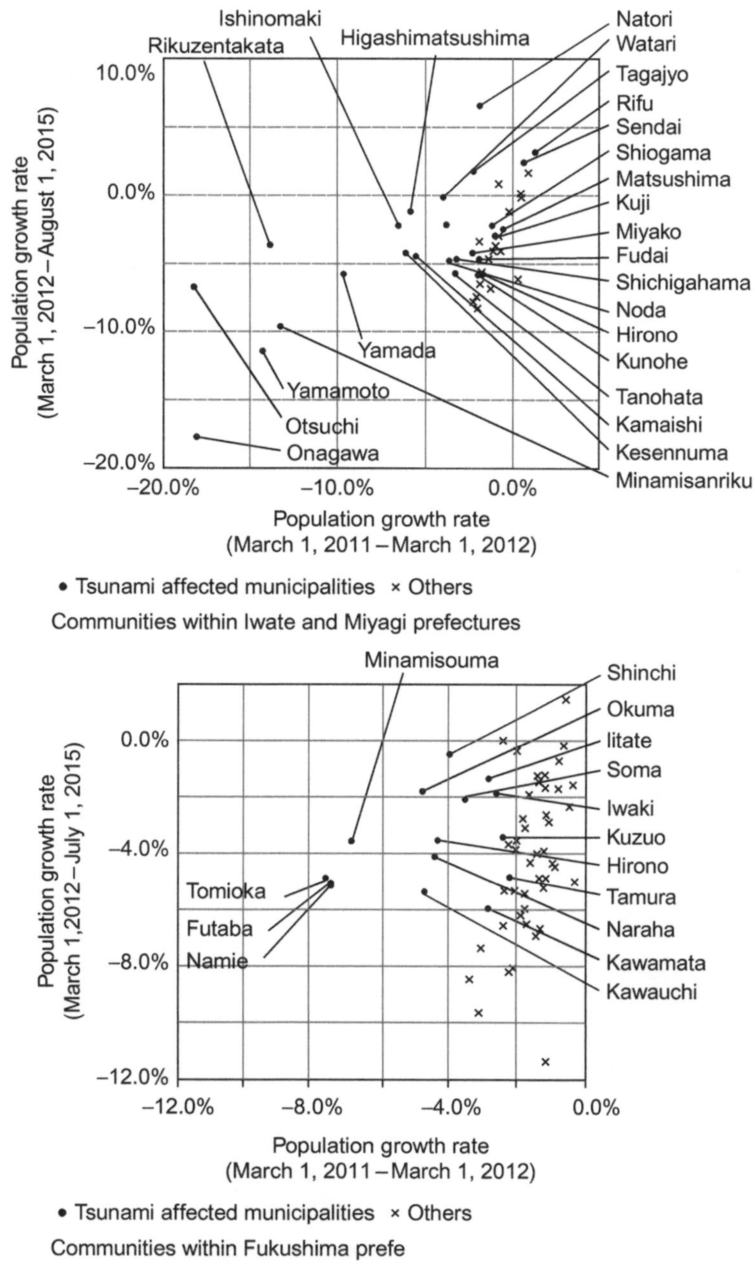

FIG. 2.2 Population change within a year from the earthquake and then after. *(Data from the Basic Resident Registry.)*

communities in the diagram mean those with evacuation orders within their borders on and after April 1, 2012. First of all, there was hardly any community with a growing population after the earthquake. The tsunami caused people to move inland and to urban areas, whereas the nuclear power disaster caused the population to drop over the entire prefecture of Fukushima.

Secondly, we see a paradoxical trend that the population drop in the second and following years in the nuclear power- and tsunami-affected regions were not that bad when compared to the number in the overall Fukushima Prefecture. The population here is based on the Basic Resident Registry, which does not mean the actual number of people living in the area, because many of the nuclear-afflicted communities are still under evacuation. The reason for the relatively small population drop here is probably due to benefits from special measures received from local governments at the destinations of evacuation, receivable without having to relocate their family registrations. On the other hand, we must be aware that when the evacuation orders are lifted, the population in these areas will not necessarily maintain their sizes.

2.3.2 **Recovery of the Industries**

Now let's turn our eyes to the recovery of the industries. Fig. 2.3 shows how commercial and manufacturing production changed before and after the disaster for each community. Note that data are missing for nuclear-affected areas. Some tsunami-afflicted communities, especially Yamada, Otsuchi,

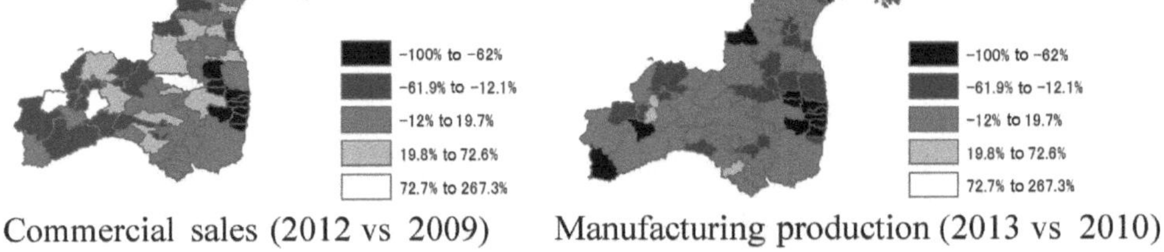

■ **FIG. 2.3** Growth rates of commercial sales and manufacturing production in local communities. *(Data from METI, Census of Commerce and Census of Manufactures.)*

Rikuzentakata, Minamisanriku, and Onagawa, show huge drops. Inland communities, on the other hand, like Miyako, Tono, Ichinoseki, and Kitakami, show large growth in contrast. These facts are strong indications that disaster-afflicted residents along the coast moved their consuming activities inland.

Manufacturing production also shows drops along the coast and increases inland, but the trends are not as obvious as those for commercial activities. Commercial activities in the figure target the needs in the disaster-affected area, so a drop along the coastal regions naturally leads to an increase in neighboring areas, whereas manufacturers target national and even overseas markets. Shifts of manufacturing activities in the neighboring areas do not necessarily take place.

Fig. 2.4 shows growth rates of commercial sales and manufacturing production on the horizontal and vertical axes. The figure shows a large variation even within the tsunami-afflicted communities. When we omit data that miss either commercial or manufacturing results, to compute the average and variance values, commercial sales give 0.165%–14.5% and manufacturing, 0.147%–7.0%. In other words, manufacturing activities recover quicker

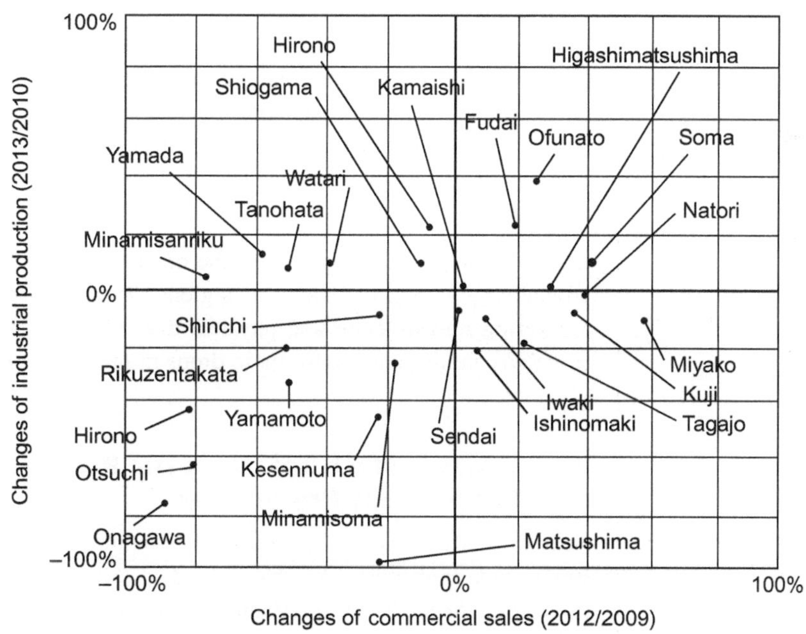

■ **FIG. 2.4** Comparison of growth rates of manufacturing production and commercial sales. *(Data from METI, Census of Commerce and Census of Manufactures.)*

than commercial with smaller variance. Although different sampling periods may have caused the difference, however, manufacturing with markets outside the disaster area is said to recover quickly compared to commercial activities that are affected by a drop in population. Recovery after this earthquake showed this trend. When manufacturing recovers to gain an increase in labor population, commercial sales can expect its recovery that follows.

The diagram shows, however, communities with major damage from the tsunami are late in recovering in both industries. Onagawa and Otsuchi, especially, show serious delays in both commercial and manufacturing activities. If left untouched, these regions will face severe decline of local economies. Realistic strategies for accumulating population and industries are needed.

One of such strategies is growing industries with high added value. If productivity increases, people's earnings will go up and more labor from other regions will be attracted to the area in the future. Let's look at how manufacturing production per employee changed before and after the earthquake. Fig. 2.5 shows the values for 2013 normalized to those of 2010 at 1.

Among the disaster-afflicted communities, a number of them raised their labor productivity. What is intriguing is that some communities that were badly hit by the tsunami, like Ofunato, Yamada, Rikuzentakata, and Minamisanriku, are among those that are ranked high. It is possible that during their recoveries, capital equipment was renewed to raise their labor productivities. We cannot say for sure whether these results are just for now, or if they are sustainable effects; however, the facts do show some hope for these regions.

Otsuchi, Natori, and Onagawa, on the other hand, have notably low labor productivity among the badly hit regions. We can suspect that delays in recovering infrastructure and physical difficulties with large area reconstructions are responsible for these drops in productivity.

2.3.3 Gross Prefectural Domestic Product of the Three Disaster-Hit Prefectures

Next we study the overall recovery of economic activities in the disaster-struck three prefectures. We base our discussions on Annual Reports on Prefectural Accounts by the Cabinet Office of the Government of Japan. Unfortunately, the newest data as of now are the results of 2012, so we cannot look at more recent results. Our analyses, however, provide important

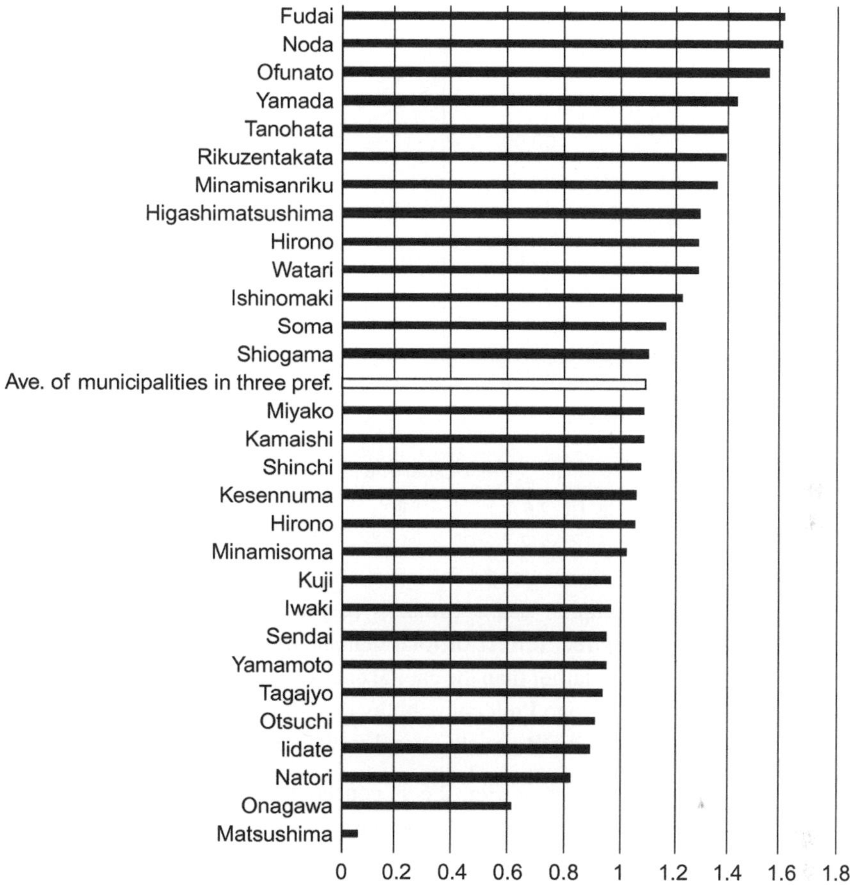

■ **FIG. 2.5** 2013 Production per employee (2010 = 1.00). *(Data from METI, Census on Manufactures.)*

information about how the disaster and recovery from the disaster affected the economies in these three prefectures.

First, we look at the Gross Prefectural Domestic Product (GPDP). Here we normalized the 2010 GPDP to 100 and show the changes of the three prefectural nominal GPPDs and the nominal gross national products (GDPs) over the years. The 2010 GDP includes, strictly speaking, the earthquake effect during the 21 days of March 11–31. However, we set the year 2010 to the base reference because this period was too short to have significant impact on the results.

Outbreak of disasters, in general, lead to an increase in total demand with the special demand for recovery to push up the GPDP. Damages to production

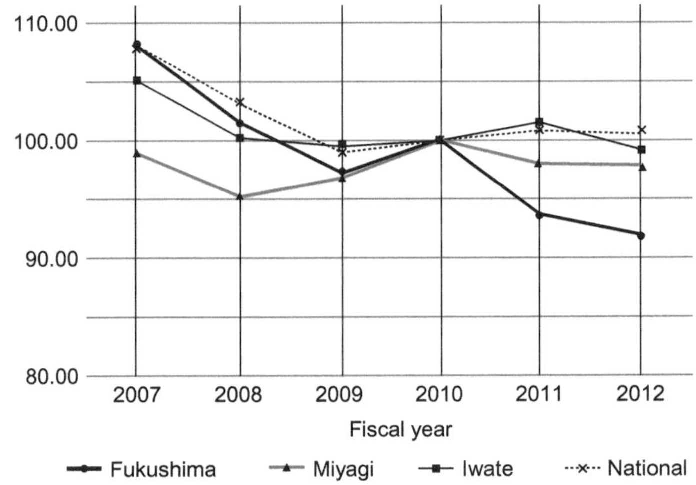

FIG. 2.6 Nominal GPDP in the disaster-struck three prefectures (2010 = 100). *(Data from Annual Reports on Prefectural Accounts by the Cabinet Office of the Government of Japan.)*

capital cause cutbacks in production activities; however, the multiplier effect (effect of total demand growth due to increase in 1 unit of capital) is larger than the capital coefficient (production per 1 unit of capital). In fact, Hyogo Prefecture, hit by the Great Hanshin-Awaji Earthquake in 1996, saw a GPDP growth over 7%, which was an effect of recovery demand.

As far as Fig. 2.6 is concerned, we cannot see clear marks of recovery demand in 2011 because Iwate was the only prefecture whose GPDP increased. However, we can see an obvious increase in the GPDP of Iwate and Miyagi after 2012. All three prefectures shown in this figure show an increase of GPDP from 2011 to 2013, which is an interesting contrast with the case of the Great Hanshin-Awaji earthquake disaster whose recovery demand had disappeared in only 2 years after the disaster. We should note, however, that it was in 2013 when Fukushima Prefecture had recovered its GPDP to the predisaster level.

To find the cause, let's look at the total expenditure by the prefectures (Fig. 2.7). Common trends with the three prefectures were: (1) drop in private-sector final consumption expenditure, and (2) drop in statistical difference between import and export of goods and services. The former is likely to be a result of population drop due mainly to the earthquake. The latter was also observed after the Great Hanshin-Awaji Earthquake and was caused by drop of productivity in the prefectures and import of goods and services from outside. These facts have the following two meanings: First, businesses (primarily manufacturing) that were

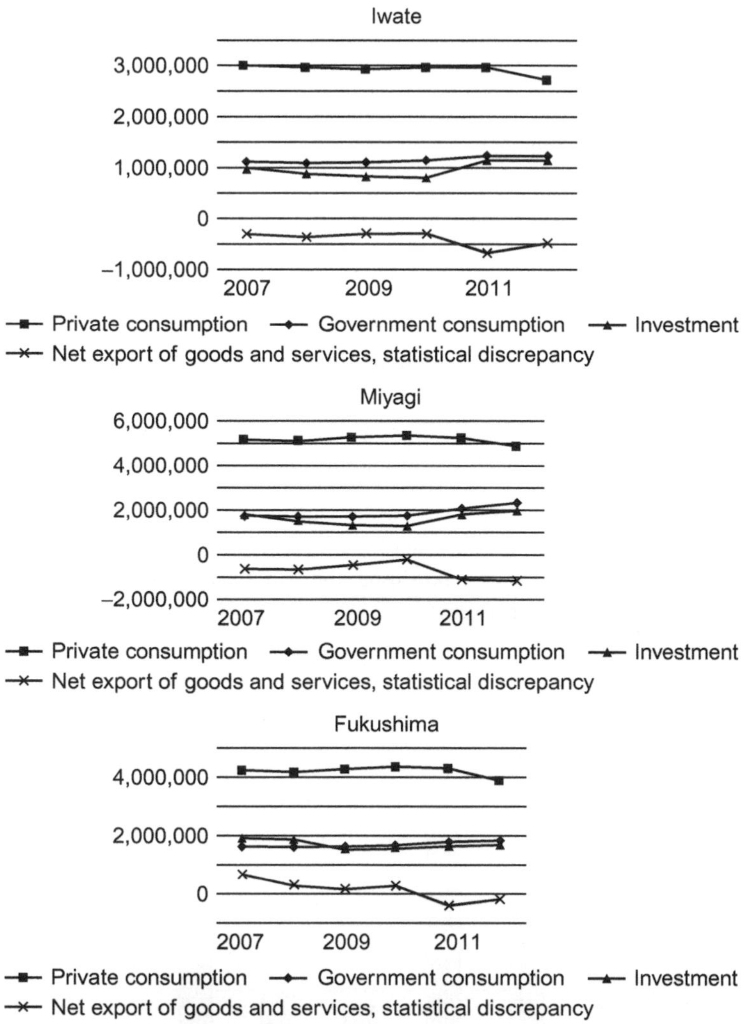

FIG. 2.7 Prefectural total expenditure with the disaster-hit three prefectures. *(Data from Annual Reports on Prefectural Accounts by the Cabinet Office of the Government of Japan.)*

providing goods and services to outside the prefectures could no longer operate, and secondly, demands for recovery businesses were provided by outside companies instead of locals. Government expenditure for a major amount of decontamination work in the evacuation-ordered areas in Fukushima Prefecture were directly managed by the national administration. The jobs were given to out-of-prefecture businesses. This was probably another cause.

Why do we not see a rise in the recovery demand? Demand for recovery primarily includes reconstructing infrastructure, corporate facilities, and rebuilding houses; thus growth in prefectural total capital is expected. In fact, Iwate and Miyagi Prefectures saw increases in prefectural total capital from before the earthquake. Fukushima, however, did not show such a clear increase.

Let's look at the details of the actual recovery demand in the afflicted areas. Fig. 2.8 shows contributing factors for the increase in prefectural total capital after the earthquake. For all three prefectures, contributions by the central government were outstanding and the main factor of recovery demand was recovery investigation by the government. On the other hand, for Iwate and Fukushima, we can point out that capital investments in the private sector were negative. In other words, hardly any company made capital investment for recovery in these prefectures.

The reason for this in Iwate is that the tsunami-hit regions are along the rough shoreline that limited the land use until recovery was complete over a wide area. In the case of Fukushima, the still active evacuation order is limiting the full-scale recovery.

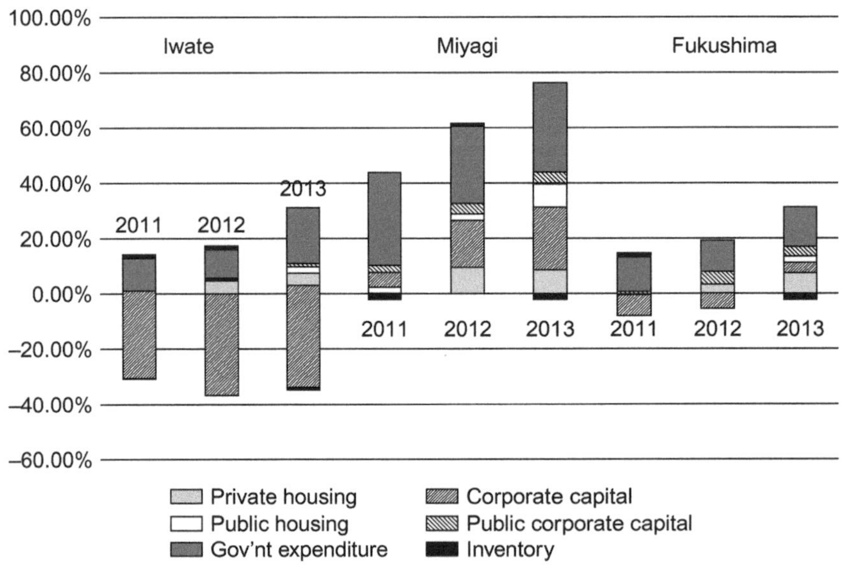

■ **FIG. 2.8** Contributing factors for the fiscal year 2010 prefectural total capital gain. *(Data from Annual Reports on Prefectural Accounts by the Cabinet Office of the Government of Japan.)*

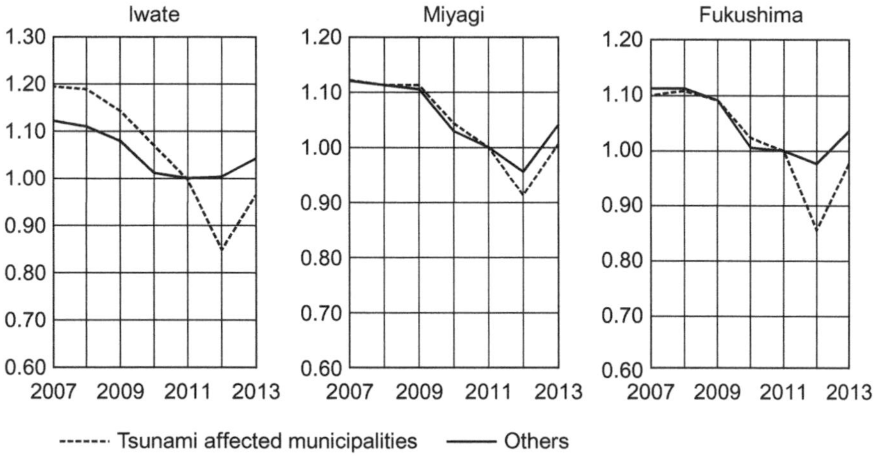

■ **FIG. 2.9** Change of taxable income in the three disaster-hit prefectures (2011 = 1.00). *(Data from City Tax Data by the Cabinet Office of the Government of Japan.)*

2.3.4 **Earnings**

Fig. 2.9 shows the change in taxable income in the three disaster-hit prefectures normalized to the 2011 value of 1.00. The taxable income for 2011 here is, in fact, the income during the previous year 2010. A great number of tax exemptions were given to the disaster-afflicted residents, but the taxable income was before these deductions so the numbers reflect the actual economical state of the victims.

First, let's compare the change in taxable income with the three afflicted prefectures with others. All three had large drops in the tsunami-hit and nuclear-afflicted areas. Miyagi, however, showed large drops in other areas as well. That means, for Iwate and Fukushima, the earthquake effect on income was limited to the tsunami- and nuclear-afflicted areas, whereas Miyagi faced drops over the whole prefecture.

Recovery of income is thought to have positive effects on economic recovery in general, but we have to be a bit more careful in the case of those hit by the Great East Japan Earthquake. Let's look at the change in taxable income for each community. Fig. 2.10 shows the 2013 taxable income ratio over the year immediately before the earthquake, 2011. The community average for the afflicted three prefectures was 1.035, and most communities in the tsunami and nuclear afflicted area had lower numbers.

The large values 1.55 for Katsurao and 1.37 for Iitate immediately attract our eyes. These increases need some explanation:

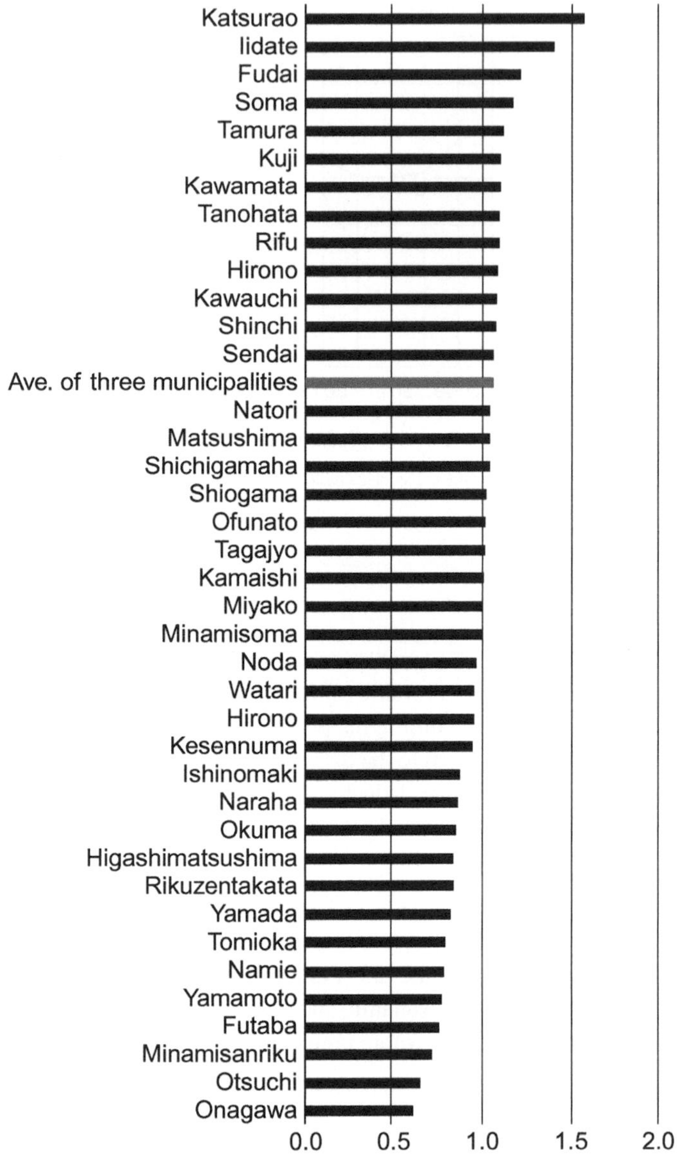

■ **FIG. 2.10** 2013 community taxable income (2011 = 1.00). *(Data from City Tax Data by the Cabinet Office of the Government of Japan.)*

The primary industry of the two villages was farming. Most farmers were self-supplying and had small cash income. The income per capita for the two villages were JPY1,645,000 and JPY1,697,000, surprisingly low when compared against the prefectural average of JPY2,743,000.

These number do not mean poverty in these villages, but rather ample crops from the land meaning they did not need cash for living.
Then residents of the two villages were forced into long-term evacuation and they could no longer produce even crops for themselves due to nuclear contamination. These people were forced into needing cash earning, and that caused the increase of taxable income in the two villages.
In particular, those in the nuclear-afflicted areas were probably under similar situations. Then we see that recovery of taxable income in tsunami- and nuclear-afflicted areas did not necessarily mean that recovery was on a good track. In the farmland, the situation was the opposite.

2.3.5 Savings and Loans

The analyses so far had limitations on available data and economic recovery in recent years was not clear. As preceding indicators of industrial recovery, let's look at activities of banks.

With almost all disasters in the past, when one hit, individual savings tended to go up. The savings went up immediately following the Great East Japan Earthquake as with other past disasters. For Fukushima Prefecture, the savings balances have been gradually rising, showing, most likely, continued damage compensation by TEPCO.

An interesting fact is that all prefectures had an increase in public savings in 2014 and after. A huge recovery budget was passed, but delays in its execution have caused the budget to be carried over to the following year. The recovery businesses, therefore, have not produced demand to the extent expected by their budget sizes and have not made much progress.

When we look at loans, all three prefectures do not show much increase since immediately after the earthquake. Capital demand in the afflicted areas not showing much growth is an indication of weakness in recovery demand in the private sector and the situation has not turned around yet (Fig. 2.11).

2.4 CONCLUSION—NEED FOR STRONGER RECOVERY SUPPORT

In this chapter, we reviewed the state of recovery in the disaster-afflicted areas, through statistical data, 5 years after the Great East Japan Earthquake.

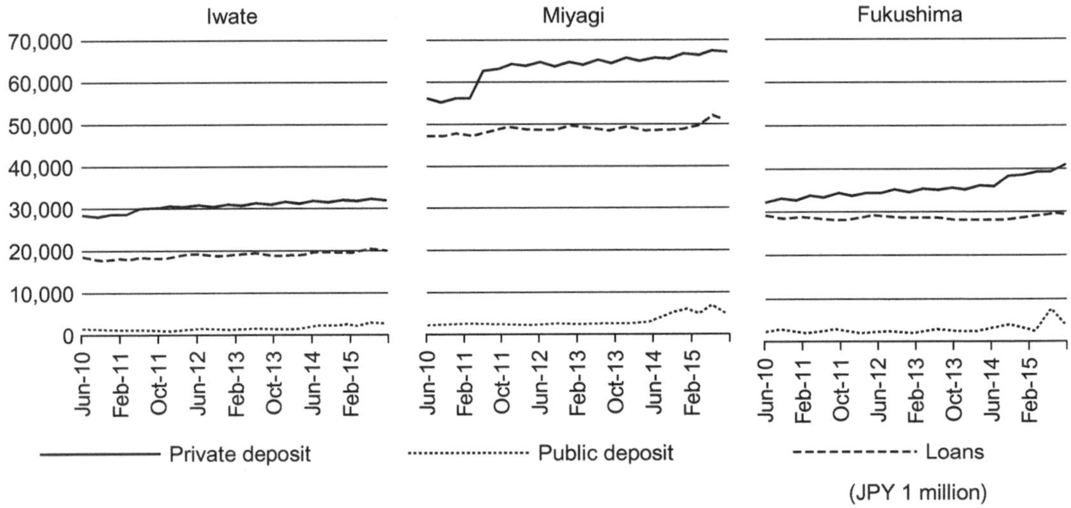

■ **FIG. 2.11** Change in individual savings, public savings and loan balance in the three afflicted prefectures. *(Data from Bank of Japan.)*

Prime Minister Abe's administration planned to spend about JPY25 trillion over 5 years for the recovery of disaster-struck regions. The whole budget has not been used up, and we shall hold back from making any judgment on the policies of recovery. This chapter, however, revealed that recovery from the Great East Japan Earthquake has had its own problems different from those with the Great Hanshin-Awaji Earthquake.

First of all, the economy in the disaster-struck areas, after at least 2 years from the outbreak, had hardly received any benefit from recovery demand. Reasons include the fact that demand shifted to the outside, but a bigger cause was no big growth in capital investment in the private sector. The stumbling investment, different from past disasters like the Great Hanshin-Awaji Earthquake, was due to the need for conditioning over a very wide area and extended evacuation orders caused by the nuclear accident.

Secondly, gaps of recovery status are widening among different communities in the afflicted regions. Within the afflicted prefectures, communities hit by the tsunami and those that were not show great differences in their state of recovery. In particular, communities hit badly by the tsunami are seeing drops in population and industry. Manufacturing, however, is expecting some increase in labor productivity in the disaster-struck areas.

Thirdly, loans by banks in the three afflicted prefectures indicate that even today, 5 years after the disaster, there is hardly any significant increase in capital demand in the private sector. Thus, the aforementioned situations cannot be expected to have improved recently.

From this analysis, I would like to state my personal view that difficulties in recovery with afflicted areas will continue. The situation might even get worse. When the infrastructures for wide-area reconstruction are complete and decontamination is finished, conditions for recovering the disaster-struck areas might be met. By that time, however, population and economic activities will be largely lost from the areas, and negative recoveries will be seen, compared to situations before the earthquake. In addition, public funds that have been sustaining the economy of the regions will go up.

Recovery action from this point will, no doubt, be centered around software instead of organizing the hardware. For recovery under such harsh conditions, however, simple local recovery plans are insufficient and we need something new. I currently lack the knowledge to point out what this might be. But at least I can say there is a need for gathering young people to create some new opportunities. For that purpose, support for social activities, equaling the amount of energy our nation has poured into recovering the hardware, should continually be given to these regions.

ACKNOWLEDGEMENTS

This chapter includes results from MEXT supported research "Sub-project 3 of Project: Reducing damage from disasters caused by city vulnerability: Study on policies for improving countermeasures against city disasters," head researcher Haruo Hayashi, and Scientific Research Fund, Grant Basic Research (S) "The worst scenario of damages to cause national suffering and measures for disaster reduction," head researcher Yoshiaki Kawata.

Chapter 3

Current Status of Rebuilding Houses and Reconstructing Local Regions After the Great East Japan Earthquake

CHAPTER OUTLINE

3.1 INTRODUCTION

About 5 years have passed since the Great East Japan Earthquake. However, traces of the tsunami destruction make it difficult to see the whole picture of reconstructing local regions. The most devastated areas (e.g., Fig. 3.1) are still in the midst of building levees, raising land, and preparing the construction infrastructure of land for buildings and roads. The situation still does not show any sign of people's new lives after recovery.

Inland areas, on the other hand, that did not suffer tsunami damage are now going through building public housing for victims and building new homes. Inland urban areas have now extended their plans for "private rental

The Fukushima and Tohoku Disaster. https://doi.org/10.1016/B978-0-12-814078-9.00003-0

■ **FIG. 3.1** Current scenes of areas hit hard by the tsunami (March/2015, Rikuzentakata, Iwate).

housing-turned-temporary-housing units" and plans for rebuilding victims' lives are made through efforts for each household instead of the entire region.

Parts of Fukushima Prefecture that suffered evacuation and restrictions on residential areas, as a result of the nuclear disaster, have further complications of relocation and decontamination in rebuilding local communities. These regions are facing further delays, compared to the areas hard hit by the tsunami, in laying out the picture for reconstructing communities and even laying down the road map for doing so. Victims that cannot return to their hometowns are still suffering and reports have been made that the reality of the situation has started to make big differences in their attitudes [1,2].

These great numbers of factors are burdening and complicating the reconstruction process. This chapter focuses on the reconstruction process of housing in the affected regions and points out the problems people are facing there.

3.2 DAMAGE TO HOUSING RECONSTRUCTION WITH THE GREAT EAST JAPAN EARTHQUAKE

We point out the following four aspects of the Great East Japan Earthquake in reconstructing housing in the regional recovery process:

- Effects of devastation over a wide surface area
  Damage from tsunamis and nuclear accidents stand out in their effects in that they uniformly destroy a wide area and make it unusable. The

damage is destructive and special reconstruction takes a long time. The damaged regional society has to recover altogether, so recovery of a single function of housing, commerce, or industry alone does not work, because these functions affect one another in a complex manner. Reconstructing local cities and regions has a close relation to rebuilding housing.

- Effects of spreading over a wide area
  A factor that affects the speed of recovery of a local community is its relationship with unaffected areas. Systems of disaster countermeasures in Japan are based on complementary assistance from neighboring communities, so in addition to an initiative by the primary region, backup support from a wide surrounding area and from the entire nation has a strong effect. This system developed in response to frequent small-size natural disasters in Japan. This time, however, the tsunami damage spread over the east Japan coastline of about 500 km, and took away the complementing effects from the neighboring communities. Modern Japan has never before experienced developing support for such a wide disaster area and it is still in process. The disaster also caused victims to relocate to distant locations. Resource transportation for emergency response, recovery, and reconstruction had to deal with extended distances, causing added time and cost in them. Even a small residential community faced difficulty in recovery and reconstruction from its damage.
- Effects of large quantity
  Another factor causing difficulty in the reconstruction process is the large volume of the disaster damage. Housing reconstruction is facing a number of never-before-experienced problems from the vast amount of damage. Housing damage over a huge area led not just to problems with quantity, but also with the uniformity of housing reconstruction for speed and fairness. The variety of the victims and uniformity of support are in conflict. To quickly support a large number of victims, the practice often is to break the victims into classes, separate the classes, and lead each class to a uniform support aimed at that specific class. The large quantity in this case had its own variety problems and the uniform support processes had their own concerns in leaving individual problems behind.
- Effects of multilayered damage
  Damage this time was not only caused by the tsunami, but also by the earthquake and landslides as well. In Fukushima Prefecture, damage from the nuclear accident has had great effects on the local residency. In addition to these complex damages from the disaster, preexisting problems of depopulation and an aging population in the regions

along the coast of Tohoku area added to the complexity. It is important not only to look at the damage caused, but also to give consideration to the state of the regions before the disaster. It can be said that mid- to small-size communities along the coast were already going through a slow suffering from these problems. Planning reconstruction takes a careful evaluation of these complex states of damage.

Reconstruction from the Great Hanshin-Awaji Earthquake was centered on a large amount of damage to housing in the urban area. For the 2007 Chuetsu offshore earthquake, it was on the reconstruction of the mountain regions. And for the 1993 southwest off-Hokkaido earthquake, it was about building tsunami preparation for the fishing communities. Solving the problem of reconstruction this time was not as simple as extending these efforts to the number of suffering regions and combining them. If we look at the problem from the area covered and the number of affected business units, the reconstruction looks more like that for the 1923 Great Kanto Earthquake or post-World War II. However, its nature differs even from them, so although based on past lessons of recovering from disasters, new solutions to meet this task are clearly needed.

3.3 SITUATION OF HOUSING RECOVERY

3.3.1 Population and Number of Households in the Affected Regions

We now review the transition, after the earthquake, of population and number of households in the regions devastated by the disaster. Table 3.1 shows the estimate of change in population and the number of broadcast subscriptions in the communities in Iwate, Miyagi, and Fukushima prefectures that were hard hit by the tsunami and nuclear accident [3–5]. The population estimate is based on the residential registry of each municipal government, but some have moved out without relocating their registries. Thus, limitations exist in estimating the real number of residents in the area. Residents in the nuclear accident–affected communities, especially, tend to have moved out with the registries still in the area, so we also show the number of broadcast subscriptions for our analysis.

The table shows that, for almost all communities in Iwate, both the population and number of households have not recovered; however, they were monotonically decreasing from before the earthquake and, in comparison to other prefectures, we evaluate that their recovery is not particularly slow.

Table 3.1 Recovery Progress of Population and Number of Households

	Tsunami Flooded Area (km^2)	# of Totally Collapsed Houses	Nuclear Accident Evacuation Zone (# of Households in 10, 2013)	Estimated Population				Recovery Rate	Household Count With NHK Subscription				
				2011-03	2012-01		2016-01		2011-03	2012-03		2016-03	
				pers.	pers.	%	pers.	%	House-Holds	House-Holds	%	House-Holds	%
Iwate Prefecture													
Miyako	10	2767		59,229	57,898	98%	55,870	94%	22,691	21,817	96%	22,044	97%
Ofunato	8	2789		40,579	39,068	96%	37,547	93%	13,562	12,420	92%	12,973	96%
Kuji	4	65		36,789	36,499	99%	35,147	96%	13,143	13,225	101%	13,297	101%
Rikuzentakata	13	3805		23,221	20,080	86%	19,473	84%	7673	6349	83%	6982	91%
Kamaishi	7	2957		39,399	37,200	94%	36,363	92%	16,210	15,090	93%	15,025	93%
Otsuchi	4	3092		15,222	12,510	82%	11,642	76%	5590	4018	72%	4238	76%
Yamada	5	2762		18,506	16,792	91%	15,578	84%	6540	5589	85%	5845	89%
Iwaizumi	1	177		10,708	10,534	98%	9643	90%	4407	4354	99%	4165	95%
Tanohata	1	225		3838	3737	97%	3424	89%	1408	1245	88%	1338	95%
Fudai	1	0		3065	3014	98%	2753	90%	1064	1050	99%	1009	95%
Noda	2	311		4606	4450	97%	4102	89%	1531	1443	94%	1484	97%
Hirono	1	10		17,775	17,504	98%	16,370	92%	6026	5973	99%	5880	98%
Total				272,937	259,286	95%	247,912	91%	99,845	92,573	93%	94,280	94%
Miyagi Pref.													
Sendai	52	30,034		1,046,737	1,052,476	101%	1,084,674	104%	342,520	351,236	103%	384,394	112%
Ishinomaki	73	20,036		160,394	150,177	94%	146,172	91%	53,008	45,662	86%	50,541	95%
Shiogama	6	672		56,221	55,674	99%	53,793	96%	17,873	17,783	99%	18,251	102%
Kesennuma	18	8483		73,154	68,875	94%	64,114	88%	23,627	20,913	89%	22,061	93%
Natori	27	2801		73,603	72,140	98%	77,322	105%	20,470	20,073	98%	22,406	109%
Tagajyo	6	1746		62,990	61,576	98%	62,127	99%	19,862	19,205	97%	20,145	101%
Iwanuma	29	736		44,160	43,738	99%	44,791	101%	12,980	13,059	101%	13,806	106%
Higashimatsushima	37	5515		42,840	40,365	94%	39,601	92%	12,985	11,823	91%	13,299	102%
Watari	35	2389		34,795	33,482	96%	33,547	96%	9831	9120	93%	9824	100%
Yamamoto	24	2217		16,608	14,358	86%	12,178	73%	4815	3745	78%	3812	79%
Matsushima	2	221		15,014	14,964	100%	14,240	95%	5455	5380	99%	5732	105%
Shichigahama	5	674		20,353	19,766	97%	18,572	91%	5803	5477	94%	5592	96%

Continued

Table 3.1 Recovery Progress of Population and Number of Households *Continued*

	Tsunami Flooded Area (km^2)	# of Totally Collapsed Houses	Nuclear Accident Evacuation Zone (# of Households in 10, 2013)	Estimated Population				Recovery Rate	Household Count With NHK Subscription				
				2011-03	2012-01		2016-01		2011-03	2012-03		2016-03	
				pers.	pers.	%	pers.	%	House-Holds	House-Holds	%	House-Holds	%
Rifu	0.5	56		34,279	34,768	101%	35,758	104%	9605	9727	101%	10,352	108%
Onagawa	3	2924		9932	8196	83%	6183	62%	3510	2066	59%	2217	63%
Minamisanriku	10	3143		17,378	15,141	87%	12,052	69%	5319	3897	73%	3987	75%
Total				1,708,458	1,685,696	99%	1,705,124	100%	547,663	539,166	98%	586,419	105%
Fukushima Pref.													
Iwaki	15	4644		341,463	333,336	98%	348,454	102%	113,391	115,072	101%	123,070	109%
Soma	29	1004		37,721	36,465	97%	38,448	102%	12,118	11,909	98%	12,557	104%
Tamura		19	117	40,234	39,490	98%	37,889	94%	11,115	11,068	100%	11,164	100%
Minamisoma	39	2323	3868	70,752	66,173	94%	56,979	81%	21,781	17,499	80%	22,393	103%
Kawamata		28	355	15,505	15,125	98%	14,162	91%	4923	4834	98%	5024	102%
Hirono	2	160		5386	5168	96%	4255	79%	1826	1201	66%	2309	126%
Naraha	3	147	2729	7676	7361	96%	0	0%	2571	1306	51%	802	31%
Tomioka	1	338	5691	15,959	14,809	93%	0	0%	5206	2052	39%	1197	23%
Kawauchi		8	188	2819	2701	96%	2016	72%	894	546	61%	813	91%
Okuma	2	61	3988	11,570	11,020	95%	0	0%	3662	1446	39%	781	21%
Futaba	3	103	2425	6891	6398	93%	0	0%	2146	774	36%	391	18%
Namie	6	759	7191	20,854	19,360	93%	0	0%	6131	2353	38%	1369	22%
Katsurao			462	1524	1483	97%	0	0%	437	247	57%	105	24%
Shinchi	11	439		8178	7875	96%	8250	101%	2348	2208	94%	2482	106%
Iitate			1902	6132	5977	97%	0	0%	1653	621	38%	342	21%
Total				592,664	572,741	97%	510,453	86%	190,202	173,136	91%	184,799	97%

Based on Geospatial Information Authority of Japan, Area (approximate) flooded by tsunami, 2011; Fire and Disaster Management Agency, About the 2011 off the Pacific coast of Tohoku Earthquake, 152nd report, 2015; Statistics Japan, Portal Site of Official Statistics of Japan. https://www.e-stat.go.jp/SG1/estat/eStatTopPortal.do (Accessed October 2015).

Miyagi, on the other hand, is divided into two groups in the recovery process: the large cities of Sendai, Natori, Iwanuma and the surrounding urban functional areas show good recovery; in contrast, the towns of Yamamoto, Onagawa, Minamisanriku, and similar communities that had their downtown areas totally destroyed and are not close to other towns show a slow recovery in population and number of households.

Fukushima was doubly hit by the tsunami and the nuclear accident, and communities hit only by the tsunami show marginally high recovery. On the other hand, communities designated as the difficult-to-return zone and evacuation-order zone have extremely low recovery rates when compared even with communities hard-hit by the tsunami in other prefectures.

As we have seen, each prefecture has its own transition in population and number of households. There were differences in the amount of damage suffered, but we can also see that plans for rebuilding housing, uncertainty in regional reconstruction, and differences in local situations are factors in reconstruction.

3.3.2 **Status of Reconstructing Public Infrastructure in Disaster-Hit Communities**

Next we look at the status of reconstructing ground-surface infrastructure in disaster-hit communities. The Reconstruction Agency showed the progress of reconstruction of 18 items, seen in Table 3.2 [6]. The basic infrastructure of lifeline, waste management, schools, hospitals, and fishing ports is almost complete, but coastline, disaster prevention forests along the coast, recovery housing, and recovery towns show hardly any progress in their recovery. Communities hit by the nuclear accident do not even have reconstruction plans yet and some are even omitted from the count.

Coastal disaster prevention planning and rebuilding of the towns are proceeding particularly slowly. Planning in this case means much more than recovering the original functions; this would come after acceptable tsunami risks are set, agreements are reached on town construction plans, land is secured for building new housing, relocation from disaster-prone areas is decided on, and decisions on how to use the land are made. This is not just "going back to where we were" but, in order to construct hardware for the new local environment, obviously involves complex agreements between the administration and the residents. Furthermore, the severity of this disaster has forced installment of countermeasures for much stronger and larger tsunamis and unprecedented large-scale construction is taking place over a wide area, extending the required timeframe for completion. The regional policies on planning community formation cover a wide range and the fact

Table 3.2 Recovery Progress of Public Infrastructures in June 2016

As of June 2016	Total	Started	Complete	Comp' Rate	Notes
Coastline	501	442	153	**31%**	Number of coastlines to start main reconstruction
Disaster prevention forest	164	155	47	**29%**	Length of disaster prevention forrest to start main reconstruction (km)
River banks	2115		2115	100%	Number of river banks to start main reconstruction
Water supply	184		178	97%	Number of water supply projects to start main reconstruction
Sewage	73		73	100%	Number of sewage plants that needs recover regular process
Debris management	1843		1831	99%	Amount of disaster debris estimate and disposed
Public housing for disaster victims	29,999	29,196	18,641	**63%**	Number of houses that require recovery
Group relocation upon disaster	333	330	275	83%	Number of sections with planned relocation
Land readjustment project	50	50	9	**18%**	Number of sections with planned readjustment
Fishing community disaster prevention	36	35	25	**69%**	Number of sections with plans
Medical facilities	182		174	96%	Number of hospitals that need to recover admission facilities
School facilities	2309		2266	98%	Number of schools that need reconstruction
Recovery and recovery support roads	570	560	241	**42%**	Extent of recovery roads and recovery support roads
Railway	2330		2226	96%	Extent of railroads that require track reconstruction
Ports	131	131	128	98%	Number of port facilities that require reconstruction
Farmland	21,480		16,770	83%	Farmland area that requires recovery for commercial farming
Fishing ports	319	313	251	79%	Number of fishing ports that require recovery
Acquafarming facilities	68,945		68,893	99%	Number of acquafarming facilities that require recovery

Bold numbers indicate raw rates relatively which appear delay progress.

Data from Reconstruction Agency, Current status and issues in reconstruction, 2016. http://www.reconstruction.go.jp/topics/main-cat1/sub-cat1-1/20161109_hukkounogenjou.pdf.

that this planning will take long years is clearly the biggest theme of discussion in rebuilding housing in these regions.

3.3.3 Status of Temporary Housing

The slow recovery of the population and the number of households in the affected communities, added to the delays in rebuilding basic infrastructure, means that a number of households are still forced to live in shelters, i.e., temporary housing. The Recovery Agency shows that, as of June 2016, the number of people evacuated and living in temporary housing was over 150,000 [7]. At 1 year after the disaster in March 2012, the number of people evacuated, including those living in shelters, was about 350,000. So, the number has gone down by about 56% in 5 years, and we can hardly say that this is rapid progress. A fair evaluation is to say that as many as 150,000 people are still left in a state of evacuation.

To review the current status of temporary housing, Table 3.3 shows the change in the number of evacuees sorted by the type of living in March 2014, when the population in shelters reached 0, and after [7]. The table also shows the number of people who fled to out-of-prefecture locations. Temporary housing means, not shelters such as schools, but housing built for temporary purposes, rented homes, houses of friends, relatives, or hospitals. The numbers are those recognized by public offices. Table 3.4 shows the transition in the number of people in temporary housing based on the current location of living [7]. We look at these tables to find the factors leading to the current situation.

First, the victims are not only spread through the affected local communities but are living over the 47 prefectures nationwide, and the administration can collect such information. In the past disasters before Fukushima, the administration had difficulties in tracing the evacuees who moved across the boundaries of local communities and they were blamed for not following those who should have been supported. Although not perfect, this time their tracking of where the evacuees are across the nation is noteworthy.

Reviewing the state of evacuation in the three main disaster-hit prefectures, 31% of the evacuees were evacuated in out-of-prefecture areas in June 2016. This ratio was an increase over the percentage in 2014. The severity of tsunami destruction over a wide area, the long lead time to recover infrastructure, and the effect of the nuclear accident have caused tens of thousands of people to be evacuated out of their prefecture. Although people were tracked, the legal system for supporting victims lacked support for out-of-prefecture evacuees. Support has been given after the status of evacuees was traced. The handling will probably not change in the near future and this means nothing has been decided on yet.

Table 3.3 Change in the Number of Evacuees

	2014-03	2014-06	2014-09	2014-12	2015-03	2015-06	2015-09	2015-12	2016-03	2016-06	Ratio Over 2014-03
Relative or friend's house	14,133	16,371	16,141	16,245	18,553	18,746	18,382	18,334	18,213	18,055	128%
Include house and hospital	249,825	235,048	226,899	217,267	206,624	188,386	176,411	163,666	152,628	136,727	55%
Total	263,958	251,419	243,040	233,512	225,177	207,132	194,793	182,000	170,841	154,782	59%
Evacuated out-of-prefecture											
Iwate	1477	1441	1451	1453	1566	1548	1509	1496	1426	1406	95%
Miyagi	7012	6813	6925	6810	7146	6944	6797	6533	6333	6072	87%
Fukushima	47,683	45,279	46,645	45,934	46,902	45,395	44,387	43,497	42,801	41,375	87%
Total	56,172	53,533	55,021	54,197	55,614	53,887	52,693	51,526	50,560	48,853	87%

Based on Reconstruction Agency, Number of evacuees over the nation. http://www.reconstruction.go.jp/topics/main-cat2/sub-cat2-1/hinanshasuu.html.

Table 3.4 Change in Evacuees Sorted by Household

Evacuation Destination	2014-03	2014-06	2014-09	2014-12	2015-03	2015-06	2015-09	2015-12	2016-03	2016-06	Ratio Over 2014-03
Iwate	34,494	33,221	31,714	30,289	28,950	26,673	25,230	23,525	21,687	19,515	57%
Miyagi	88,575	81,923	77,836	73,796	69,561	61,816	55,920	50,206	45,672	38,441	43%
Fukushima	84,221	82,657	78,577	75,440	71,399	65,485	61,800	57,775	53,983	49,140	58%
Other prefectures in Tohoku	12,555	11,287	10,641	10,233	9992	9148	8805	8600	8418	7830	62%
Other than Tohoku	44,113	42,331	44,272	43,754	45,275	44,010	43,038	41,894	41,081	39,856	90%
Total	263,958	251,419	243,040	233,512	225,177	207,132	194,793	182,000	170,841	154,782	59%

Based on Reconstruction Agency, Number of evacuees over the nation. http://www.reconstruction.go.jp/topics/main-cat2/sub-cat2-1/hinanshasuu.html.

Next we look at the change in the number of evacuees. Within a little over 2 years from 2014, the number gradually decreased by 41% but hardly anyone moved from their old addresses to out-of-prefecture. In the case of Fukushima, the effect of the difficult-to-return zone will be significant; however, comparing it to the change in out-of-prefecture evacuees from Miyagi and Iwate, many of the evacuees from Fukushima are still residing outside the prefecture. We can foresee that they will soon want to change their temporary locations into permanent ones. This trend of "turning the temporary location into a permanent one" was also pointed out after the Great Hanshin-Awaji Earthquake, and it is a factor that the administration has to take into account when planning housing reconstruction.

On the other hand, when we turn our eyes to the transition in Iwate and Miyagi, the rates of the drop in the number of evacuees are not fast at all. Six years has already passed since this quake, whereas in the case of the Great Hanshin-Awaji Earthquake, temporary housing was gone within 5 years after the earthquake. This fact now warns us that we will see the first-ever situation, in the history of disasters in Japan, of "extremely extended temporary housing for evacuees" and we will have to face this with a sense of crisis in our minds.

The nationwide development of temporary housing was largely due to "private rental housing-turned-temporary housing units" that allowed evacuees to relocate into private rental housing with the same conditions as temporary housing. The conventional framework of emergency evacuation viewed relocation into rental housing as self-recovery, but this time, the national system, for the first time, secured such relocation as an intermediate stage for evacuees in their self-recovery process. This plan can be praised for reducing the number of prefabricated temporary housing units and giving the evacuees a larger selection for their temporary housing. It will probably be made a standard policy for future large-scale disasters. Detailed studies have been made and their problems have been pointed out elsewhere [8,9], so we will not discuss this topic here.

Similarly, the conventional way of providing temporary housing with prefabricated houses underwent challenges based on past lessons: e.g., building houses with lumber from local mountains, comfortable housing built by house-building manufacturers, and three-story-type complexes [10]. These all made their noteworthy contributions, but the effects of the entire housing reconstruction process have been limited. In the overall picture, this time again, due to the harsh conditions after the earthquake, we are repeating the problem that temporary housing turns into a constraint for the evacuees in rebuilding their lives.

3.3.4 **Status of Building New Houses**

On the other hand, households that could afford to rebuild their houses quickly purchased and built new houses. We will now review the status of the self-supplied rebuilding group with data from new houses that started construction in the three prefectures of Iwate, Miyagi, and Fukushima. The data are prefecture based and a detailed analysis of regional trends relied on other research results.

The 4-year average of new house construction that started in 2008–11 was about 5500 in Iwate, 13,000 in Miyagi, and 10,000 in Fukushima. Table 3.5 shows the numbers after the earthquake with the breakdown of owner houses and rented [5].

Each prefecture shows greater numbers over the 4-year average; 1.5 times in 2012, and an even greater count in 2013. These numbers show the trend after the earthquake damage. The differences in the 4-year average over the 4 years sum to about 12,800 for Iwate, 43,800 for Miyagi, and 19,400 for Fukushima. In total, an additional 76,000 houses, roughly over the previous 4-year average, started construction in the 4 years after the quake, and they are assumed self-supplied. Among these newly constructed houses, there are more rented houses, about 56% over 32% owner houses. A closer examination reveals a large increase in owned single-family houses, rented condominiums and rented apartment houses. Miyagi showed a rapid increase in the number of single-family owner houses.

The counts are not uniform across each prefecture and results of local construction trends within each prefecture after the earthquake are added. Kondo and Karatani [11], for example, pointed out that "infill" type rebuilding with new houses after relocation was prominent. In other words, in the limited regions where new housing construction took place, the ratio of postearthquake housing will largely change and many of those that started construction probably changed their form of ownership and form of construction. Moreover, the added current special demand for rented housing for inflow of outside people and short- to mid-term residents caused a major change in the local housing environment in the areas that surround the tsunami-affected regions, including downtown Sendai.

3.3.5 **Summary**

Currently, the disaster-struck areas have reached their limits in "temporary" housing. The city and country areas need more time for reconstruction to gain sufficient safety against the devastation experienced in 2011. The uncertainty in the roadmap of spatial reconstruction for recovery of people's

Table 3.5 Change in New Housing Construction Sorted by Prefecture

							Index When Ave. is 100				
	JFY						JFY				
Total	**2008–11 Yearly Ave.**	**2012**	**2013**	**2014**	**2015**	**Increase Within 4 Years [a]**	**2008–11 Yearly Ave.**	**2012**	**2013**	**2014**	**2015**
Iwate	5651	8121	9870	9006	8422	12,815	100	144	175	159	149
Miyagi	12,867	21,177	25,746	24,476	23,921	43,852	100	165	200	190	186
Fukushima	9950	12,421	15,954	14,221	16,609	19,405	100	125	160	143	167
Total	28,468	41,719	51,570	47,703	48,952	76,072	100	147	181	168	172
Own house											
Iwate	3306	4522	5176	4433	4463	5370	100	137	157	134	135
Miyagi	5828	9575	9673	7438	7780	11,154	100	164	166	128	134
Fukushima	5396	6917	8269	7199	7247	8048	100	128	153	133	134
Total	14,530	21,014	23,118	19,070	19,490	24,572	100	145	159	131	134
Rented											
Iwate	1781	2978	3926	3847	3520	7147	100	167	220	216	198
Miyagi	4745	7527	12,516	12,356	11,593	25,012	100	159	264	260	244
Fukushima	3333	4529	6752	5713	7241	10,903	100	136	203	171	217
Total	9859	15,034	23,194	21,916	22,354	43,062	100	152	235	222	227

[a] *Increase within 4 years = Sum(2012 – 2015) – (AVE. before disaster × 4).*

Based on "Construction initiation statistics" in Statistics Japan, Portal Site of Official Statistics of Japan. https://www.e-stat.go.jp/SG1/estat/eStatTopPortal.do (Accessed October 2015).

lives and regions is causing misalignment between the "story" of recovering daily lives that the victims see and the actual recovery process of space and local societies. Past cases pointed to such concerns immediately after the disaster, but, to our regret, fundamental plans to work around the issues were never made and the results are just as expected.

Some of the victims who have reached their limit have started heading into recovering their own daily lives individually, without waiting for the regional spatial reconstruction. This movement makes it difficult to grasp the overall picture of victim recovery and affects the process of spatial reconstruction in the devastated regions. There surely will be a problem involving shrinking of the spatial reconstruction zone of the affected area. The regional infrastructure reconstruction is underway and almost complete in the fifth year after the disaster. The regions are now facing the problems of adjusting the final form of regional reconstruction.

3.4 SKETCHING THE OVERALL PICTURE OF THE HOUSING RECONSTRUCTION PROCESS

Rebuilding housing is the process for disaster victims of recovering their place of living. In the case of large-scale disasters, public support is needed for this process. If we review past disasters in and out of the country, many disaster recovery plans were about rebuilding housing for the people. We will now review the overall process of rebuilding housing in countering the Great East Japan Earthquake. Our discussion is based on recovery plans by the local communities, research results, and public announcements made.

The process of recovering housing normally proceeds in three steps: damage, temporary housing, and permanent housing. Table 3.6 shows, for the disaster-struck communities, the number of damaged houses, the supply of temporary housing, and the scheduled supply of public housing (postdisaster public housing and land). For evacuation due to the nuclear accident, the supportive supply was not from the base municipal government and the housing reconstruction plans are undecided, so Fukushima Prefecture is excluded from this analysis.

The Iwate Prefecture basic policy in housing reconstruction [12] planned to provide about 14,000 units of emergency temporary housing for victims, about 4500 units of temporary housing equivalent, and about 5000 units of permanent public housing. The policy also planned to have about 13,000 units of rebuilt owner houses and rebuilt rental houses. Among these self-supplied houses, about 8200 units are within the areal reconstruction land. From these numbers, the percentages of publicly supplied and

Table 3.6 Damage and Public Housing Provisions in Local Communities

	# of Household Estimate Before Quake	Flooding	Damage to Buildings			Number of Publicly Supplied Houses				Scheduled Business Supplied Houses			
		Household Count in Flooded Zone	Totally Wrecked	Half Wrecked	Rate of Total Wreckage	# of Temporary Houses	Regarded Temporary Houses	Total	Temporary House Rate	Public Houses	Private Houses	Total	Public Supply Rate
Iwate Pref.													
Miyako	24,332	7209	2767	1331	11%	2010	684	2694	97%	793	706	1499	54%
Ofunato	14,729	6957	2789	1148	19%	1811	718	2529	91%	801	631	1432	51%
Kuji	15,208	2553	65	213	0%	15	61	76	117%	11	15	26	40%
Rikuzentakata	8196	5592	3805	236	46%	2168	175	2343	62%	1000	2194	3194	84%
Kamaishi	17,561	5235	2957	698	17%	3164	693	3857	130%	1314	1475	2789	94%
Otsuchi	6348	4614	3092	625	49%	2146	135	2281	74%	980	1666	2646	86%
Yamada	7182	4175	2762	405	38%	1990	312	2302	83%	777	1244	2021	73%
Iwaizumi	4710	431	177	23	4%	143	17	160	90%	51	59	110	62%
Tanohata	1452	526	225	45	15%	186	30	216	96%	63	53	116	52%
Fudai	1121	380	0	0	0%	0	0	0	0%	0	0	0	0%
Noda	1674	1069	311	168	19%	213	91	304	98%	100	177	277	89%
Hirono	6838	932	10	16	0%	5	0	5	50%	4	17	21	210%
others	396,697	0	147	1698		133	1545	1678		27	0	27	
Prefectural total	506,048	39,673	19,107	6606	4%	13,984	4461	18,445	97%	5921	8237	14,158	74%
Miyagi Pref.													
Sendai	455,958	10,385	30,034	109,609	7%	1523	8580	10,103	34%	3179	717	3896	13%
Ishinomaki	60,928	42,157	20,036	13,045	33%	7297	6568	13,865	69%	4500	3297	7797	39%
Shiogama	22,165	6973	672	3278	3%	206	399	605	90%	420	104	524	78%
Kesennuma	26,601	13,974	8483	2571	32%	3504	1678	5182	61%	2155	2360	4515	53%
Natori	26,433	3974	2801	1129	11%	910	1283	2193	78%	716	301	1017	36%
Tagajyo	24,733	6648	1746	3730	7%	373	1407	1780	102%	532	75	607	35%
Iwanuma	16,003	2337	736	1606	5%	384	452	836	114%	210	171	381	52%
Higashimatsushima	15,080	11,251	5515	5559	37%	1753	1299	3052	55%	1010	717	1727	31%
Watari	11,442	4196	2389	1150	21%	1126	697	1823	76%	477	200	677	28%
Yamamoto	5561	2913	2217	1085	40%	1030	760	1790	81%	484	207	691	31%
Matsushima	5492	1477	221	1785	4%		78	78	35%	52	8	60	27%
Shichigahama	6568	2751	674	649	10%	421	224	645	96%	212	591	803	119%
Rifu	11,536	192	56	901	0%		127	127	227%	25	0	25	45%
Onagawa	3852	3155	2924	349	76%	1294	451	1745	60%	918	855	1773	61%
Minamisanriku	5362	4375	3143	178	59%	2195	326	2521	80%	738	863	1601	51%
others	217,479	0	1346	8502		79	1721	1800		360	0	360	
Prefectural total	915,193	116,758	82,993	155,126	9%	22,095	26,050	48,145	58%	15,988	10,466	26,454	32%

Based on Geospatial Information Authority of Japan, Area (approximate) flooded by tsunami, 2011; Fire and Disaster Management Agency, About the 2011 off the Pacific coast of Tohoku Earthquake, 152nd report, 2015; Statistics Japan, Portal Site of Official Statistics of Japan. https://www.e-stat.go.jp/SG1/estat/eStatTopPortal.do (Accessed October 2015); Iwate Prefecture Prefectural land development, Fundamental policy for Iwate housing recovery plan, 2011; Miyagi Prefecture Civil Engineering Department, Miyagi Prefecture Housing Recovery Plan, 2011; K. Koshiyama, A study on the geographical removing process of suffers whose houses demolished through the public housing system of contemporary and permanent for recovery, J. Soc. Saf. Sci. 9 (2007) 21–28.

reconstructed are as high as 97% for temporary housing and 74% for permanent housing. The overall percentage of postdisaster public housing among the total number of scheduled rebuildings is about 30%, which is about the same percentage with these as after the Great Hanshin-Awaji Earthquake.

Miyagi Prefecture, in its housing recovery plan [13], set the target number of housing units to about 72,000 with about 15,500 supplied publicly. During the stage of temporary housing, 22,095 units were built as emergency temporary houses, and 25,050 households are settled in temporary housing equivalents. For permanent housing, from the delivery of public funding for life reconstruction support, the prefecture estimated about 89,000 households in need of housing reconstruction. Among them, 12,114 households received their funds by August of 2012, and of the remainder, about 15,000 are in public housing and about 39,000 are estimated to be self-supplied. The areal reconstruction project has scheduled to reconstruct civil residential land for about 11,000 units. From these numbers, the percentage of public housing is about 58% for temporary housing and about 32% for permanent housing, so the percentage of postdisaster public housing is about 20%. In total, the rate of public supply is lower than for Iwate Prefecture.

The difference between the two comes from the housing supply in city areas neighboring the tsunami-hit regions, civil housing markets in the urban areas, and the existence of rental housing. Iwate Prefecture with severely damaged communities, including the central downtown areas, and without strong coordination with neighboring cities had to rely on public support, whereas Miyagi Prefecture with inland cities that escaped severe damage started early on to accept self-supplied people and as a result lowered the reliance on public support.

Community-based percentages, however, show a large variation, and some will face hardships. Japan's housing reconstruction support policy isolates each community with victims. Each community basically has to prepare its own sufficient numbers of temporary houses. After the Great Hanshin-Awaji Earthquake, which destroyed about the same number of houses, many communities quickly prepared the necessary counts of temporary housing. The principle of land use is based on each municipal territory and each community having its own temporary housing supply. When the supply is insufficient, temporary housing spreads into the neighboring communities and at times to neighboring prefectures. The area of temporary housing centered on the disaster-struck area spreads in this manner [14]. A spread-out temporary housing area like this makes it difficult to return as time passes, and this generates the possibility of ruining the scenario of supplying public housing and plans for regional reconstruction. It is difficult to build municipal plans that assume relocation to other communities, as well as to draw regional reconstruction plans with such assumptions. Even if we accept such moves, the prefecture level does not have systems or ways of handling such reconstruction. These problems in housing rebuilding and regional reconstruction are the most difficult to solve, and probably we will have to face them again with the next great earthquake.

3.5 STATUS OF RECONSTRUCTING TOWNS IN THE TSUNAMI-HIT AREAS

Reconstructing tsunami-hit areas requires rebuilding the basic land and providing a means of disaster prevention. After this disaster, and throughout the Japanese history of disasters, large-scale projects for reconstructing wide urban areas are developed in sizes that follow postwar projects [15]. These projects have a numbers of problems. The legal system surrounding methods of reconstruction, in particular, is causing a chain reaction of problems.

The wide area and severity of the damage reveal a large variety of types and forms of damage. Recovery and reconstruction plans should be specialized to each small municipal unit. Neither the legal system nor personal resources, however, are ready to handle the variety. Postdisaster spatial reconstruction projects are very costly, are extremely rare, require specific measures to be taken quickly, have hardly any preparation time, and take time in arranging people to carry them out. These facts make the projects executable only under the existing laws of the country concerning urban construction and disaster prevention. As a result, building plans and executing the plans for the needs of each community are difficult and the spatial reconstruction ends up the same across all communities [16,17]. The tsunami-struck areas, this time, include a variety of communities centered on fishing, agriculture, and urban areas, so this obstacle will surely cause a large impact in reconstructing the communities.

The primary actions for spatial reconstruction of postearthquake disaster regions include: (1) Specify disaster-prone areas and promote relocation of whole communities to safer land, (2) Enhance disaster prevention functions for fishing communities for better safety, (3) Apply general urban area reconstruction plans for primary roads, and zoning for parks and infrastructure, (4) Provide a central location for posttsunami reconstruction projects that functions as a center for housing planning and a facility for primary city functions. All these actions relate to land ownership, and obtaining agreement from the residents, developing the land, and reconstructing the base land takes time. These are the results shown in Table 3.2 and they are the main factors in the delay of reconstructing the regions.

These projects are currently underway in many communities; however, each community is facing its own problems. Here we discuss the primary three problems, among the large variety of them, in reconstructing communities:

- Gaps between safety and reconstruction
  The ultimate goal in reconstructing a posttsunami region is "tsunami protection measures." The size of the tsunami this time is said to occur once in a thousand years; however, since we have experienced it, spatial reconstruction must protect against it. The question in reconstruction is whether we prepare against the same level tsunami, or do we take the rate of occurrence into account. The central government took the lead in setting the guidelines for applying the concept of L1 and L2, used for setting the earthquake preparation standards, in coastal reconstruction of, e.g., levees. Each municipal government set its levee height against simulations of tsunami expectations based on this guideline. The two-level standard, however, affected not only the heights and strengths of structures, but also regional reconstruction behind the levees. It was the first time for planning spatial reconstruction to assume two levels of risks in the residential zone. There is no such areal planning to protect against low-frequency disasters of one in 100 years but be vulnerable to extremely low-frequency disasters of once in 1000 years. The preservation of local life came before disaster countermeasure, so that there were negotiating an agreement about acceptable risk of disasters at first. After that, those who living there set physical constructions for disaster prevention such as levees. This time, the tsunami expectation existed first to set the levee height, and it turned into a constraint in regional planning; in other words, it was a "reverse planning." Preparing against L1 level tsunamis was a premise; thus, the condition of levees higher than previously was the starting point for regional planning in discussions with the residents. It took a long time to recognize that this plan greatly changes the living environment from what it was before the disaster and this blocked the road to reconstruction. After this lead time, some communities ran into conflicts between the administration and residents. In some other communities a number of residents lost hope in the changes after reconstruction and requested relocation, giving up reconstruction in their residential areas.
  We can see a fundamental question in this issue. That is the discussion point of "Who decides the risk of disaster in the community?" It is the responsibility of the administration to build a safe environment,

especially after a disaster; it will execute reconstruction so that it will never see the same level of damage. The victimized residents most likely have the same idea, but when "rebuilding" and having "continuity" of the community come into the picture, the story turns complicated. When the time to reconstruct the community differed too much from the time to rebuild a living for the residents, the long lead time pushed people to move out. The result was a hardship in actually realizing the community image projected at the time of planning. Also, there are the questions of whether the same standards of living can be reached after sufficient hardware is in place and, concerning local continuity, of whether the local specialties can be restored. Maybe the disaster-struck residents now face the question of whether it is possible to carry on the old lifestyle of having the ocean within reach, with modern tsunami protection measures in place.

- Difficulty in reaching an agreement

  Tsunami damage is determined, regardless of existing class, by the depth of inundation, determined by the distance away from the sea and the elevation. Naturally, the ability of postdisaster reconstruction depends on the living conditions and household of each victim. The ages are different and so are the senses of values. In discussing regional reconstruction, these points tend to cause conflicts of interest and agreements are hard to reach. A great number of things must be done in rebuilding a community, and getting everyone to look in the same direction takes a great deal of knowledge, discussions, and time, in order to share the same ideas.

  These discussions were not in place before the disaster and the administration and people both lacked experience with them. The starting point was different with community reconstruction after the Great Hanshin-Awaji Earthquake. The administration and residents of local communities had to take on building plans for the first time, exchange ideas, negotiate interests to form agreed-upon plans. The task was difficult for small communities, who were all acquainted with one another, to complete in a short time. In this case, victims had to become used to quick support and slow planning.

- Limits to downtown reconstruction

  There are legal grounds for spatial reconstruction and support funds for it are collected from the country. Thus, municipal governments cannot freely execute their own plans that meet regional needs. There are specific rules, spatial requirements, specifications for progress, and so on. Most of these projects are for constructing urban areas with spatial reconstruction and development in mind. Postdisaster group relocation projects, which are said to be different in nature, also relocate people into new building complexes that meet the Building Standards Act; thus

unless some efforts are made, local features and variety will be lost. In order for the new area to have a sense of the region, the central players have to make some effort.
If one questions the application of this method of building urban areas to this damaged area, the answer is, it is executable but not suitable. We can see that rebuilding the damaged area in this way will result in a city-like network of straight roads and flat residential areas; however, it is hard to imagine the land being filled and further development needed. We can view this problem as such: in Japan, we do not have plans to execute areal reconstruction for nonurban communities in the hills, of farmers and fishermen. That is why they say, "Urban area construction methods determine the finished spatial image." We would like to avoid seeing a continuation of urban-like square scenery along the coasts of Tohoku, but that is exactly what is going on.

3.6 PROBLEMS THAT REMAIN

We discussed the current status and problems, mainly related to housing reconstruction, in the recovery process over the 5 years after the Great East Japan Earthquake. Here we list the problems that remain.

- Moving out from temporary housing equivalents
  For disaster recovery this time, more "private rental housing turned temporary housing" units were offered, compared to prefabricated temporary housing. It is a new support policy, but we can see the difficulties in moving the residents in them to the next step. Moving out from a temporary housing equivalent is the same as from temporary housing. The housing was originally a private house for rent and forcing the evacuees out for the reason of rental expiration has been known to cause problems in past disaster cases. Residents in temporary housing equivalents have a different level of making adjustments to their lives in the new neighborhoods and rebuilding their lives. The level of adaptation is hard to observe and courteous handling is needed. After all, it will be necessary to provide stepwise supports suitable for each evacuee's situation. One way is to shift the current residence into a "housing turned public housing" from "temporary housing equivalent." The plans should proceed under a thorough overall plan and with preparation for detailed support.
- Actual recovery of damaged areas
  As time passes, the number of people returning to their original homes decreases. The regions hit this time especially need to establish new industries for new employment and construct an environment for building new lives for new residents. Urban areas were able to recover naturally

by building new housing, but that is not the case with the Great East Japan Earthquake, as many of the local administrations are now aware. Each community needs, not just recovery, but an environment for a new society. However, leaders, planning, knowledge, and resources are lacking. For regional planning, there is a need for mid- to long-term support of resources in people, material, and ideas.

– Wide-area planning for wide-area disaster
  The local municipal governments want to take the lead in building and executing plans for reviving themselves, but they will need large support in resources for the execution. Major intellectual support is needed, especially in executing the plans. This type of wide-area disaster needs variety in local planning. Readying human resources to apply skills in spatial reconstruction into disaster recovery, and building the framework for doing so, is needed.
  On the other hand, in planning recovery from a wide-area disaster, including planning for regional and spatial continuity is important. A plan that covers a wider area but that sets the conditions for smaller communities is in need. In societies after disasters, victims move across boundaries of municipal territories as well as those of logistics and information. Perhaps giving authority to the prefectures to build and execute plans would do. In the case of severe disasters over a wide area, adjusting and supporting small communities alone will not be sufficient to produce creativity for the varying wide area.
– Recovery from nuclear disaster [18]
  This report did not carry out an analysis of the nuclear disaster–struck area in Fukushima. The problem was not just in physical damage, but many residents chose to relocate for emotional security and a number of problems surfaced that were not covered in past plans for supporting housing reconstruction, e.g., difficulty in evacuated lives and not being able to imagine the roadway to return home. Also in local planning, they had problems of double administration systems in the relocated areas, entire community relocation plans, and the recovery plans that followed. These totally unexpected situations came up, and in fact are still around. We believe it is necessary to continue to plan and execute solutions to handle the situation as not just a postnuclear accident case, but also as wide-area evacuation, systematic evacuation and acceptance for local communities, and building new residential areas for evacuees from other communities.

REFERENCES

[1] The City Planning Institute of Japan, Report on the Great East Japan Earthquake Disaster, City planning volume, 2015.
[2] Japan Housing Council, Housing and life recovery whitepaper 2011–2013, 2013.

[3] Geospatial Information Authority of Japan, Area (approximate) flooded by tsunami, 2011.

[4] Fire and Disaster Management Agency, About the 2011 off the Pacific coast of Tohoku Earthquake, 152nd report, 2015.

[5] Statistics Japan, Portal site of Official Statistics of Japan. https://www.e-stat.go.jp/SG1/estat/eStatTopPortal.do (Accessed October 2015).

[6] Reconstruction Agency, Current status and issues in reconstruction, 2016. http://www.reconstruction.go.jp/topics/main-cat1/sub-cat1-1/20161109_hukkounogenjou.pdf.

[7] Reconstruction Agency, Number of evacuees over the nation. http://www.reconstruction.go.jp/topics/main-cat2/sub-cat2-1/hinanshasuu.html.

[8] F. Meno, Problems on the housing provision for suffers after the disaster 2011, Document of the lecture meeting of the Building Research Institute 2012, 2012.

[9] F. Meno, Moving processes and conditions of households under the housing lease program for disaster victims in Iwate Prefecture: an analysis based on a questionnaire survey of households in the entire prefecture, Urban Hous. Sci. 87 (2014) 133–138.

[10] Ministry of Land, Infrastructure, Transport and Tourism, Report of the construction of temporary housing at the Great East Japan Earthquake Disaster, 2011.

[11] T. Kondo, Y. Karatani, The challenges of post-disaster recovery after the great east japan earthquake through the case of individual self-help housing recovery with relocation, Research paper of Housing Construction Research Institute No. 41, No. 1307, 2014.

[12] Iwate Prefecture Prefectural Land Development, Fundamental policy for Iwate housing recovery plan, 2011.

[13] Miyagi Prefecture Civil Engineering Department, Miyagi Prefecture housing recovery plan, 2011.

[14] K. Koshiyama, A study on the geographical removing process of suffers whose houses demolished through the public housing system of contemporary and permanent for recovery, J. Soc. Saf. Sci. 9 (2007) 21–28.

[15] K. Koshiyama, Points at issue of recent disaster reconstruction and disaster prevention town planning, in: Symposium of Safety Engineering, 2011. L-4.

[16] K. Koshiyama, Urban Reconstruction Plan after disaster over development and preventing disaster, Report on the Great East Japan Earthquake Disaster, City planning volume, 2015.

[17] K. Koshiyama, A study of the temporary housing provision after huge disaster, Proc. Soc. Saf. Sci. 37 2015 91–94.

[18] Fukushima Prefecture Headquarters for Disaster Control, Progress of temporary housing, leased public housing and public housing, 2015.

Chapter 4

Problems With Lifeline Recovery and Public Transportation

CHAPTER OUTLINE

4.1 THE GREAT EAST JAPAN EARTHQUAKE AND RECOVERY OF LIFELINE

4.1.1 Modern Society and Lifeline

In modern society, in order for citizens to lead healthy and civilized lives, stable supplies of wealth and services are indispensable. They are also required for economic activities of corporations and continuation of all other activities by social organizations. The necessary wealth and services to support civil lives, economy, administration, and cultural activities are supplied

The Fukushima and Tohoku Disaster. https://doi.org/10.1016/B978-0-12-814078-9.00004-2

through networks of infrastructures. The systems of supplying wealth and services necessary for the modern society are our lifelines.

Lifelines are categorized into four types. The first category is supply systems, including electricity, gas, and water. The second is transportation, which includes infrastructures and surrounding services of roads, railways, harbors, and airports. The third is communication via telephones, broadcast, and postal service. And fourth, we have waste management, such as sewage treatment. Public and private organizations build and manage these systems. Some are directly managed by the central and local governments, e.g., national roads and highways, municipal roads, airports (except privately owned ones), and harbors.

As we already discussed, lifelines form systems of networks. Some of these networks of supply, like electricity supply and ground-line phone networks, are visible above ground, and others such as gas piping and waterlines are buried underground. Facilities to produce them are necessary as well as networks for their proper supply.

The primary external forces that destroy lifeline facilities are natural disasters such as earthquakes or typhoons. When a strong earthquake hits, the damage to a lifeline can be enormous. Lifelines suffer damage when earthquakes of strength 5-strong hit, and when the intensity reaches 7, most of the lifeline systems are destroyed. When the Great East Japan Earthquake hit, accompanied by huge tsunami waves, lifelines in the devastated areas were completely taken down and civil lives, the economy, and social activities suffered serious damage.

4.1.2 The Great East Japan Earthquake and Recovery of Lifelines

The magnitude 9.0 great megathrust earthquake, its aftershocks and tsunami on March 11, 2011 caused massive damage over a wide area around the Pacific coastal area of Tohoku (northeast Japan). Damage to human lives, still not finalized, as announced on September 9, 2015 by the Tohoku Area Pacific Offshore Earthquake Emergency Response Headquarters, reached 15,893 dead and 2573 missing. In the modern history of Japan since the Meiji Era (starting in 1868), this disaster marked the second largest, following the 1923 Great Kanto Earthquake that took over 100,000 lives. The three prefectures of Iwate, Miyagi, and Fukushima suffered the most damage to the people and structures. Damage in Miyagi Prefecture was especially serious, with the death count there accounting for about 60% of the total fatalities of the disaster and the missing persons in Miyagi consisting of about 50% of the total.

The damage to lifelines was also huge, estimated and announced by the Cabinet Office of Japan three months after the quake to total about JPY3.5 trillion (about US$33 billion). The breakdown was about JPY1.3 trillion in lifelines (water, gas, electricity, broadcast, and communication), and about JPY2.2 trillion in infrastructure (rivers, roads, harbors, sewage system, airports, and so on). The definition of "lifeline" by the Cabinet Office, with some lifeline facilities regarded as infrastructure, does not necessarily match the general concept; thus we added the two for total damage to lifeline.

Table 4.1 shows lifeline damage announced by Miyagi Prefecture, the most damaged prefecture. As the table shows, the total lifeline damage was about JPY1.1363 trillion. This figure, however, does not count damage to Japan Railway (JR) East and Sendai Airport, which would make the sum of damages much larger. Damage that exceeded JPY100 billion was as follows: roads of about JPY394 billion, sewage system of JPY371.7 billion and harbors of about JPY108.8 billion.

At the time of this writing, almost 5 years have passed since the earthquake and we will review the states of recovery of the lifelines.

First, electricity. The sole electricity company in Tohoku is Tohoku Electric Power Company (Tohoku EPCO). Within its coverage, on March 11, 4.86 million units had a power outage. By March 18, within a week, most of these

Table 4.1 Lifeline Damage in Miyagi Prefecture

Lifeline Systems	Type	Damage (JPY1,000)	Notes
Supply	Water	83,481,403	Excludes industrial water
	Electricity	70,800,000	
	Gas	27,550,000	
Transportation	Railway	8,595,043	JR East have not announced damage by the prefecture; thus its damage is not included
	Bus	1,318,000	
	Ferry to remote islands	410,161	
	Highway	12,420,000	
	Other roads	394,044,000	Administered by central government and those with bridges
	Harbor	108,797,000	
Communication	Communication and broadcast	57,177,400	
Waste treatment	Sewage	371,690,000	

Based on the data of Miyagi Prefecture, "Damages by the Great East Japan Earthquake (December 12, 2016)."

households recovered their electricity supply. Excluding the households that had their houses taken away by the tsunami and the evacuation zones caused by the Fukushima Daiichi (Fukushima-1) nuclear accident, all electricity supply in Tohoku EPCO territory was recovered by June 18, i.e., 3 months later. Similarly, the affected area under Tokyo Electric Power Company (TEPCO) had a power outage affecting about 4.05 million households, but most of them recovered their electricity supply by the next day and all of them by March 19. Electricity was recovered quickly. Only TEPCO's Fukushima-1 nuclear power plant had a severe accident from losing all electricity supply due to the tsunami attack, and it is still suffering from long years of postaccident problems.

Next, gas. Damage to the production facilities and piping stopped gas supply in 19 communities within 8 prefectures. The number of households that needed recovery amounted to about 400,000, mainly in the three prefectures of Iwate, Miyagi, and Fukushima. The 1995 Great Hanshin-Awaji Earthquake saw 845,000 without gas supply and in the case of the Great East Japan Earthquake, the count was about half that. The difference was mainly due to the fact that consumers in the coastal regions in Tohoku, slow in catching up with urban-style gas supply through piping, were using liquefied petroleum (LP) gas in bottles.

Water supply was lost in about 1.6 million households in 187 communities in 12 prefectures. The damage was severe in the three prefectures of Iwate, Miyagi, and Fukushima, and also Ibaragi and Chiba prefectures suffered wide areas of land liquefaction. Water supply was restored within 2 weeks, except in these five prefectures. Water supply companies from the whole nation helped in the restoration and by March 28, the number of households without water went down to about 400,000, and by July 29, 134 days from the earthquake, all households had their water supply back, excluding the houses washed away by the tsunami.

The megathrust earthquake and tsunami destroyed or washed away cellular phone base stations, cut entrance lines, and caused power outages to stop 129 base stations delivering telephone communication services. They were mostly recovered by the end of April and cellular service coverage recovered to about the same level as that before the earthquake. Ground-line telephone communication suffered from damage to facilities, power poles and transmission lines. About 1.9 million lines were damaged but, similarly to cellular service, they were recovered to the preearthquake level by the end of April, except for the restricted zone around Fukushima-1 nuclear station.

Japan Postal Service suffered damage to 106 of the 1932 regional offices in the six prefectures of Tohoku, and 26 of the 648 community post offices. On the third day after the quake, March 14, 584 of them were closed. Reconstruction sped up after that point and as of August 12, 2011, the closed office count went down to 25 of 308 in Iwate, 37 of 363 in Miyagi, and 29 of 432 in Fukushima, and postal service was restored, including package delivery service, to the same level as preearthquake days. As of March 2, 2016, however, there were a total of 46 regional offices, 8 in Iwate, 17 in Miyagi, and 21 in Fukushima, and a total of 14 community offices, 3 in Iwate, 5 in Miyagi, and 6 in Fukushima, closed [1].

Finally, sewage management. Twelve prefectures, including Tokyo, had 48 sewage treatment facilities and 78 pump stations shut down. The total length of damaged sewage piping was about 946 km. The damage was especially bad in the four prefectures of Iwate, Miyagi, Fukushima, and Ibaragi, with 89 sewage treatment facilities and pump stations shut down and 846 km of piping damage. Reconstruction efforts lowered the number of nonoperating treatment facilities to 18, as of June 6. The status of Kitaizumi treatment center and nine others in Fukushima prefecture is still unconfirmed due to the nuclear accident. Damage to the sewage system totaled JPY370 billion (about US$3.5 billion) in Miyagi prefecture alone. The reason for such major damage to the sewage system was that many of the treatment facilities were being built on lowland near the coast and suffered direct tsunami attacks [2].

In the next section, we will review the recovery and state of reconstruction of transportation lifelines.

4.2 RECOVERY AND RECONSTRUCTION OF TRANSPORTATION INFRASTRUCTURES AND REMAINING PROBLEMS

4.2.1 Overall Picture of Recovery of Transportation Infrastructure

Table 4.2 shows the damage immediately after the disaster and current states of recovery of transportation infrastructures. Primary routes of transportation were damaged over a wide area; however, damage caused by the quake was small and recovery was quick. The tsunami damage along the coast, however, was devastating and roads, railroads, and harbors were still unrecovered as of September 1, 2015. Areas affected by the nuclear accident have factors different from normal recovery from an earthquake and the problem there is complicated.

We will discuss the Shinkansen (Japanese bullet train) and airports in separate sections.

Table 4.2 Damages and Current State of Recovery of Transportation Infrastructure

	Earthquake and Tsunami Damages	Status as of September 1, 2015
Roads	Closed: 15 routes of highways, 69 sections of national roads, 102 sections of auxiliary national roads, 540 sections of prefecture roads	Closed: 1 section of auxiliary national road, 13 sections of prefecture and ordinance city roads (excluding nuclear evacuation zone)
Railroads	Immediately after the quake, 6 lines of Shinkansen (Tohoku, Akita, Yamagata, Jyoetsu, Nagano, and Tokaido) stopped, as well as 177 lines run by 42 railway companies	Four lines owned by one railway company stopped
Buses	In 3 Tohoku main prefectures, 219 vehicular damages (62 areal transport and 157 charter). 115 buildings affected (30 demolished and 85 partially)	Partially closed with four companies
Airports	Sendai airport shut down by tsunami	Sendai and 12 other airports in the area all operational
Harbors	131 harbor facilities including 14 international hubs and important ports unusable	Restoration complete with 124 facilities and 7 underway
Ferries to remote islands	Damage like ships washed up on land and piers wrecked with four lines	All four companies recovered and restarted limited operation

Based on Cabinet Office of Japan "Damages and recovery status of infrastructure (Primarily in Iwater, Miyagi, and Fukushima) as of July 14, 2011" http://www.cao.jp.jp/shien/2-shien/1-infra.html.
Reconstruction Agency "Status of reconstruction, March 10, 2015" http://www.reconstruction.go.jp/topics/main-cat1/sub-cat-1/20150310_genjyo.pdf.
Ministry of Economy, Trade and Industry (METI) "The Great East Japan Earthquake (report 129) Summary, September 1, 2015" http://www.mlit.go.jp/common/001102754.pdf.

4.2.2 Status of Recovery and Problems in Reconstructing Roads

Within a month from the disaster, road closures were lifted except for one route of highway, 14 sections of national roads, 29 sections of auxiliary national roads, and 221 sections of prefectural roads. In July of 2012, those still closed were only 1 section of auxiliary national road and 47 sections of prefectural roads. The roads were recovered quickly because, learning from the Great Hanshin-Awaji Earthquake, bridges had gone through reinforcement to prevent them from falling. Nine bridges fell or had their upper sections washed away on national road 45, and 12 bridges on other roads, and from agreements that were in place for action in case of disasters, local construction companies quickly moved in and contracts were quickly made for this emergency case, without having a formal bidding [3]. Local construction companies before the earthquake had difficulties with the drop in the number of public projects; nevertheless, they played a big role at this time of disaster.

These statistics, however, exclude the nuclear warning area, and it took long years to lift the closures on national road 6 and Jyoban highway. National road 6 was fully reopened on September 15, 2014, and Jyoban highway on March 1, 2015. Parking is not allowed on sections of national road 6 and regular drivers need special permits to enter the warning zone. Also, due to the high dosage, pedestrians, bikes and motorcycles are still not allowed to enter the zone (as of September 23, 2015).

Although we said roads have been mostly recovered, there are still unrecovered community roads that are closed in the disaster area. On the other hand, there are a number of newly built roads due to Reconstruction Grant funded projects. These projects allow comprehensive support for regional areas that suffered damage so great that they would have had difficulty in reconstructing themselves just using disaster recovery actions. A total of nearly JPY 4 trillion (about US$38 billion) was prepared for the projects and funds greater than what were applied for have been distributed. One of the projects is "Road construction integrated with hometown recovery." Table 4.3 shows some road reconstruction projects and their funds for some

Table 4.3 Reconstruction Grants and Road Constructions (Unit: JPY100 Million)

	Rikuzen Takata	Kamaishi	Yamada	Otsuchi	Miyako	Ofunato	Ishinomaki
Reconstruction grant	1858	1545	1182	1029	943	886	4196
Road construction	131	95	112	48	108	130	683
Ratio	7%	6%	9%	5%	11%	15%	16%

	Kesennuma	Sendai	Higasi Matsushima	Onagawa	Minami Sanriku	Iwanuma	Yamamoto
Reconstruction grant	2984	1676	1676	1285	1195	866	846
Road construction	67	430	340	158	147	159	159
Ratio	3%	20%	20%	12%	12%	20%	19%

	Watari	Iwaki	Soma	Minami Soma	Shinchi	Total
Reconstruction grant	782	1653	699	590	492	26,903
Road construction	75	164	74	46	114	3192
Ratio	10%	10%	11%	8%	23%	12%

Road construction is abbreviation of "Road construction integrated with hometown recovery" from [4], "Reconstruction grant cases so far for primary communities" pp. 74–76.

major communities. The overall percentage of the Reconstruction Grant spent on road recovery is about 12%, and for some communities the percentage is as high as 23%. The total funds for the road recovery project are JPY319.2 billion (about US$3 billion) [4]. Some communities are busy building new roads with hometown recovery plans pushed aside.

Highways received recognition of their significance after the disaster. They barely suffered damage from the shaking or the tsunami, and some served as the second levee and others as routes for transporting supplies. Some of the Tohoku roads that were seriously damaged recovered as quickly as within 1 week, surprising the world. These factors led to "Recovery roads and recovery support roads" in the third supplemental budget of 2011. The budget aimed at quick reconstruction of damaged areas by building roads to connect missing routes, to form a network of high-grade roads. The total length is 570 km. To shorten the construction time, the contractors are now actively helping the administration in their preconstruction administrative work, including project progress management, process management, and presentation to the residents. The result is much faster project completion, with 95% already started and 39% complete.

Recovery and reconstruction of roads made progress rapidly and are still underway. Problems exist, however, in their policies, in that how to best make use of them is not well planned. We often hear "Once North-South Sanriku Highway is complete, we can connect our local communities with metropolitan areas using buses." However, no discussions are underway with the bus companies. The plans seem to be to first complete the roads and then beg for businesses to start. The reality is that local communities are busy with multiple projects and they do not have the manpower for planning these projects ahead of time.

4.2.3 Status of Recovery and Reconstruction of Railways

Unrecovered railways as of September 2015 are 223 km of four lines: Yamada, Ofunato, Kesennuma, and Jyoban lines, owned by East Japan Railway Company (JR East). We will discuss these four lines in the following subsections and, additionally, we will discuss the Senseki line and Ishinomaki line of JR East, and the Sanriku Railroad owned by a semiprivate company, Sanriku Railway. We added these additional lines because they were shut down for an extended time and recovered in unique ways.

Sanriku Railway quickly announced its intention to recover its service along the existing route. JR East also confirmed its plans for recovery; however, it did not take immediate action to rebuild its existing lines. This was not necessarily because JR East was "lazy." At the time, relocation of the central downtown area and administrative buildings was on the discussion table and JR East thought, "It is necessary to cooperate with the hometown rebuilding efforts." Some, however, disparaged this move, saying, "Recovery to where things were should come first." Once the railroads are reconstructed, "people will start to move in as well as logistics and information" [5]. This claim has its own justification; however, if it is decided later that the area should be abandoned, having built railroads there would be meaningless and wasteful. If rebuilt, people will gather around the stations and the town will start to form; however, that would be a town development without residents deciding where they wanted to live. The Ministry of Land, Infrastructure, Transport and Tourism, therefore hosted a reconstruction coordination meeting for each section of the railroad to discuss the railroad recovery with hometown development in the communities. This led, however, to a variety of reconstruction plans. The following sections discuss the problem.

4.2.3.1 Sanriku Railway

Sanriku Railway had many of its station buildings, bridges and tracks washed away. For a company in the red, the recovery was not an easy task. Six days after the disaster, however, all the people and communities in the area agreed and requested a total recovery of Sanriku Railway. The third supplementary budget of 2011 passed to cover almost all the recovery cost of semiprivate railway companies and it was only then that plans for reconstruction turned realistic. As a result, the South Rias line restarted on April 5, 2014 and North Rias line 6 days later.

The total number of passengers in Japan Fiscal Year 2014 (JFY, April 1, 2014 to March 31, 2015) reached about 690,000, a 42% increase over the previous fiscal year JFY 2013, and the red figure of JPY51.17 million in the previous fiscal year turned into a big black figure of JPY86.83. The ordinary loss, however, was JPY102.54 million, a continuous 21-year red number since 1994. The local communities are supporting the company with a number of subsidies, such as an operations subsidy. The number of passengers in 2009 was about 900,000; thus the company has not yet fully recovered to the predisaster state [6]. The restart of the service was welcomed; however, with the problem of having to acquire the JR Yamada line, the future forecast is not so bright.

4.2.3.2 JR Yamada Line

The 55.4 km section between Miyako and Kamaishi of the JR Yamada line lies north-south between the North and South Rias lines of Sanriku Railroads. Its reconstruction started in March of 2015, with a target completion of within JFY2018; however, there were a number of twists and turns in order to reach the plan.

JR East proposed a temporary recovery of the Yamada line by providing Bus Rapid Transit (BRT)[1] like the case of the Kesennuma line that we will discuss later. The local communities opposed the idea and requested recovery of the railroad. During the 6th Yamada line recovery discussion meeting in September 2013, JR East repeated its BRT idea, which was denied by the residents and JR East was forced to withdraw the BRT plan. The railroad recovery was estimated to cost JPY21 billion and JR East agreed to contribute JYP14 billion to recover the preearthquake state; however, it asked the central government and local support for the additional JPY7 billion for coordinating the recovery with that of the hometown. The central government did not change its position of "no support for profitable companies" and reconstructing the Yamada line was in a deadlock.

JR East, in January of the following year, 2014, in the 7th meeting, proposed to reconstruct the closed section of the Yamada line but then to transfer the operation to Sanriku Railway saying, "If Sanriku Railway connects all the way from Kuji to Sakari, there will be added convenience (e.g., in planning the time table to meet the local needs)" [7]. From then on, conditions were discussed between Sanriku Railway and local communities along the route. At the end, JR East committed to change over 90% of the tracks to high-grade rails, the railroad ties to concrete ties as with the Sanriku Railroad, transfer the cars at no cost to the communities along the route, and to provide workforce support to Sanriku Railway in maintaining the facilities. JR East also promised a one-time monetary support of JPY3 billion to the 12 communities along the route for operational support. The additional recovery cost of JPY7 billion was agreed to be paid by the local communities and the central government agreed to provide support [8].

And so, the reconstruction of the Yamada line was settled, however not as a JR East operation, but as part of a semiprivate (and half public) railway. As a result, the communities along the route will no longer receive fixed property tax but in turn will be obligated to support the operation cost. The future is not so bright. Increase in the passenger count is expected with the integrated

[1]BRT, in general, is an advanced bus system for rapid transit in urban areas that runs on designated bus lanes and roads, and in addition, does not receive the fares on the bus for faster boarding. The Curitiba bus in Brazil is well known and the system, often called Surface Subway, is quickly spreading around the world. MLIT defines it "an advanced bus system with articulated cars for Public Transportation Priority System (PTPS) that uses designated roads or lanes for rapid, on-time, and quantity transit." JR East uses the same definition; however, the plan only partly uses designated lanes and seems unfit for the term BRT.

operation by Sanriku Railway; however, the closed section only had 700 daily passengers before the quake, i.e., about 40% of when JR East started the operation. The city of Miyako has plans to build its new city hall at Miyako station, but the locals have a big issue to resolve as to how to continue the operation of Sanriku Railway in the future.

4.2.3.3 JR East Ofunato Line and Kesennuma Line

In contract to the communities along the Yamada line, locals along the Ofunato line (43.7 km along Sakari and Kesennuma) and Kesennuma line (55.3 km, Kesennuma to Yanaizu) agreed to temporary recovery with BRT. The Kesennuma line started its operation in August of 2012, and the Ofunato line in March of 2013.

The biggest factor for settling for a temporary recovery with BRT was their early recovery. Naturally there were voices that wanted railroad reconstruction; however, railroad reconstruction forecasted a long lead time to return to the preearthquake state, while BRT promised quicker recovery by using roads, but only temporary ones. JR East listed other advantages of (i) possible escape and evacuation on its own to a safe location upon an earthquake and tsunami; (ii) flexible arrangement in correspondence with the hometown rebuilding efforts; (iii) speedy and on-schedule operation by running the buses on the railroad track routes; and (iv) convenience by raising the frequency of rides [9].

Later, discussion on full recovery was repeated; however, as with the Yamada line, the cost in coordinating it with the hometown reconstruction was the problem. In the case of the Kesennuma line, returning to the preearthquake state would cost JPY30 billion; however, shifting the track inland would add another JPY40 billion. JR requested this cost to be covered by the central government. The government, however, turned the request down during the first "Ofunato line community mayor meeting" and "Kesennuma line community mayor meeting" during June, 2015. JR East, then, in July during the second mayor meeting, proposed permanent recovery with BRT. The town of Minamisanriku accepted the proposal on the spot; however, the other communities did not make clear replies.

Minamisanriku accepted the proposal because during the 4 years of postdisaster community reconstruction, the land had already been raised and the preearthquake track locations on lower land were no longer feasible. The existing bridges and station buildings that survived the disaster were then obstacles for hometown rebuilding. As long as the railroad was in the state of "temporary recovery" the town could not even dispose of the bridges and old station buildings.

The Kesennuma line has a history of "long awaited railroad." It was one of the last local railroad tracks built when JR East was still part of the nationally operated National Railway. Some residents, therefore, had strong feelings towards reconstructing the railroad; however, the passenger count before the disaster was low and the line was not very convenient. BRT, on the other hand, matches the needs at the time with a departure frequency of 1.5–3 times that before the disaster. Many praised this convenience. The situation was similar to other cities who were not "totally against" BRT. They are probably waiting for the right timing to make requests to JR East about continuity, convenience, and future development of BRT [10].

4.2.3.4 JR East Senseki Line and Ishinomaki Line

Reconstruction of the Senseki line (10.5 km, Takagi to Rikuzen Ono) and Ishinomaki line (2.3 km, Urashuku to Onagawa) was different from the previous line reconstructions and they had clear concepts of hometown reconstruction.

The Senseki line was an urban-type line connecting the City of Sendai, with a population of 1 million, to Ishinomaki, the second largest city in Miyagi prefecture. It ran nine trains an hour during the peak and was the nearest line to the scenic spot of Matsushima. The average number of passengers in 2009 was 22,000 a day. Even the postdisaster count was 17,000 daily (2013) on the 2-hour bus ride from Sendai to Ishinomaki and quick recovery was long awaited. JR East shifted part of the track inland by 500–600 m, the stations of Tomei and Nobiru to higher land and restarted the service over the entire stretch on May 30, 2015. The cost in reconstruction was not announced, but it was estimated to be over JPY10 billion [11]. Also, a new line, the Senseki-Tohoku line started operation using part of the Tohoku Main Line (Sendai to Matsushima) to connect Sendai and Ishinomaki. The 52-minute ride is 10 minutes faster than before the disaster, and also faster than the 1 hour 20 minute ride on the express bus. There is a plan to build a new station "Ishinomaki Ayumino" between Rikuzen Akai and Hebita (Ishinomaki City) and the City of Ishinomaki is developing the surrounding area to serve as an evacuation camp in case of disaster.

The last stop of the Ishinomaki line, the Shin-Onagawa station, will be shifted by about 200 m from the old station to 7–9 m higher land to concentrate commercial and public facilities as well as elementary to middle schools. This plan is along the Compact City Project by the national government. The 2014 Onagawa Hometown Reconstruction Council had plans to attract people from outside to recover from its population drop to about 7000, which was about a 30% drop from predisaster times. The community promotes "Ocean is visible from anywhere," "Scenery that makes you want to live, visit and to be proud of," "Safe, secure, and good living," and "Livings all centered in the middle." The new station is playing a symbolic role in this project.

4.2.3.5 JR East Jyoban Line

The Jyoban line had a totally different reconstruction from those we have discussed so far. The 5.0 km section from Watari to Hamayoshida restarted on March 16, 2013, and the 8.5 km section from Tatsuta to Hirono on June 1, 2014. The 22.6 km section from Hamayoshida to Souma, and the 46.0 km section from Haranomachi (City of Minami-Souma) to Tatsuta (Town of Naraha) are still shut down. The Hamayoshida-Souma section has plans to restart service in the spring of 2017 after shifting part of its tracks inland; however, the Haranomachi-Tatsuta section suffers a stronger influence from the Fukushima-1 nuclear accident and its recovery faces many difficulties.

This Haranomachi-Tatsuta section, since January of 2015, has substitute buses making two round trips daily to service the section. The 1 hour and 10–25-minute ride makes no stops at the eight station locations in between. Prime Minister Abe made a promise of full recovery, but without stating the time frame. JR East started serious research and design for reconstructing the section in JFY2015. The target is to restart Haranomachi-Odaka by the spring of 2016, Odaka-Namie within 2 years at the latest, and Tomioka-Tatsuta within 3 years. The Namie-Tomioka section with the worst radioactivity damage does not have a clear target date, using the words "will restart after decontamination and passenger safety secured at times of emergency." JR East is conducting test decontamination at six locations in the section between Yorunomori and Futaba with high dosage. The test involves decontamination by taking away the tracks and ballast and removing about 10 cm of soil, then replacing new tracks in place. The plan is to verify the effects of this decontamination and come up with a schedule for reconstruction [12].

The reconstruction is delayed; however, with the wide evacuation guidelines of restrictions and hard-to-return-to areas, hastening the recovery of railroads is meaningless. We need a firm scope of reconstruction for the whole area.

4.2.4 Current Status and Problems in Reconstructing Local Transportation

As we first discussed, transportation in the disaster-struck areas is mostly recovered. Nevertheless, railroads were stopped, and automobiles and houses were washed away. After the disaster, securing means of transportation for the people in temporary housing and building transportation in coordination with the changing hometown reconstruction plans were new problems. In the disaster areas, transportation by buses to meet such needs is increasing its share. Comparing the number of bus passengers during 2010 and 2013 in the six prefectures of Tohoku revealed that Aomori, Akita, and Yamagata respectively had a 4% increase, no change, and 10% decrease, whereas the three prefectures of Iwate, Miyagi, and Fukushima all had increases of 14%, 9%, and 5% respectively [13]. The number of bus passengers, however, is expected to drop with the reduced population and stabilizing lifestyle.

MLIT started the Project of Improving Security and Maintenance of Regional Public Transportation in 2011. A project for Supporting Disaster Struck Areas has its own position in the plan. The project involves "Project for Securing and Maintaining Inter-Regional Main Transportation for Disaster Areas" and "Research of Public Transportation in Special Disaster Areas (Research Project)." The former means operations support for the main means of transportation that connects disaster areas. The conditions of "scheduled transportation of at least 3 times daily" and "expected number of passengers of 15 to 150 daily" were relaxed for the three disaster prefectures and, in practice, half of the red figures are supported unconditionally. Since a great number of vehicles were washed away, support is given to depreciation and purchases (including second-hand vehicles). In JFY2014, the support in the three disaster prefectures amounted to about JPY870 million in operations support and JPY78 million in vehicle support. These figures consisted of 73% and 69% shares of the support for the six Tohoku prefectures. In JPY2010 before the disaster, the support amounts were JPY340 million (52%) and JPY21 million (41%); thus the support for the three prefectures is increasing in both amount and ratio [14].

The latter Research Project is mainly studying transportation means for daily life purposes within each area, from shelters, temporary housing, and remaining houses to hospitals, markets, and public offices. There are 39 communities in the Special Disaster Areas within the three disaster prefectures, and studies were made for 220 sections in 32 communities during JFY2014. A small number of communities stated that existing public transportation is sufficient because the damage was small.

This Research Project supports paid transportation with upper bounds depending on the number of stops, one at each temporary housing group: JPY35 million for less than 30 stops, 45 million for less than 60, and 60 million for 60 or more. The support has a single bound at JPY25 million for free transportation. The support can be applied to cost of transportation, purchase of vehicles, advertisement, and planning. The form of transportation is free to meet the needs of the groups, like buses, microbuses, or taxis with or without time schedules. The operators can freely revise the fee as well as routes and timetables. Application for support does not have to go through organizations and anyone in the transportation business can apply for and receive the support directly. It is a very flexible support project that never existed before.

Local communities are, nevertheless, not satisfied. Some complain that "Upper bounds of support should be different with severity of disaster and progress in recovery," or "It only fills in the transportation fee," and others about the calculation method and framework, asking for "household count based support instead of group based" or "set a separate category for purchasing buses." The biggest dissatisfaction, however, was with the time limit. The original plan was for the 3 years of 2011–13. It was extended by 2 years to 2015; however, the people cannot immediately stop using public transportation service just because the support stopped. A survey revealed that 14 communities (35%) wanted "transfer to existing support projects" followed by 11 (27%) saying "continue operation with funds within the community" [14].

Temporary houses are still filled to about 70% of the peak time. Disaster public housing being built in various locations and destinations for group relocation plans will produce new needs for means of transport. They are needed for reconstructing lives; however, the Research Project was not stable or continuous. Both Tohoku District Transport Bureau and Communities said that it is important to switch to regular means of transportation with transportation projects that coordinate with reconstruction plans. Easily said; however, "how to do so" is difficult. For now, the strategy is to continue the current project or work in ways that lessen the level of change, and later build policies for the entire local transportation to secure stable local transportation means.

4.3 SHINKANSEN AND AIRPORT

4.3.1 Shinkansen

For railways that run on tracks, natural disasters such as earthquakes, tsunamis, heavy rain, strong wind, and snow are major threats. The Tohoku Area Pacific Offshore Earthquake left devastating damage to the railroad facilities along the coast, as we discussed. Earthquakes destroy buildings and cause major damage to trains by derailing them and turning them over. If the fast Shinkansen is directly hit by an earthquake, risks for accidents and damage are great.

The Great Hanshin-Awaji Earthquake in January 1995 destroyed eight locations along the Sanyo Shinkansen track, including bridges, and stopped the bullet train service for 81 days in the section between Shinosaka and Himeji. Luckily the quake hit at about 5:46 in the morning before the first train departure. If the attack were during operation, a huge disaster could have taken place with derailed cars falling off elevated bridges or hitting tunnel walls.

The Chuetsu earthquake in October 2004 derailed a 10-car Shinkansen that was running. It was the first derailment accident of a Shinkansen train in service throughout its quarter century history of operation. Eight of the 10 cars derailed and luckily the derailed cars stayed close to the track and their wheels slid in the snow-collecting ditches to escape turning over or falling from an elevated track. This accident did not turn into a major catastrophe because the early earthquake detection and warning system triggered signals to slow down the trains and the tracks had snow-collecting ditches next to them as measures for the heavy snow where Jyoetsu Shinkansen served.

Among the natural disaster measures of Shinkansen, the most elaborate one is against earthquakes. There are three main preparations:

> The first is aseismic reinforcement of elevated tracks and bridges. Structures that support Shinkansen are mostly elevated trackways, bridges, and tracks on ground. Protecting these structures is extremely important in securing the Shinkansen systems. Since its start with Tokaido Shinkansen, National Railway (now deregulated into a number of JR companies) has implemented aseismic measures for structures following design engineering standards. These standards were reviewed after the 1978 Miyagi offshore earthquake for further enhanced earthquake resistance.
>
> When National Railway Company was deregulated in 1987, management and operations of Shinkansen were transferred to JR companies. Later when the Great Hanshin-Awaji Earthquake and Niigata Chuetsu Earthquake caused damage to Shinkansen structures, the Ministry of Transportation (now MLIT) announced guidelines for aseismic reinforcement and JR companies have been following them to enhance aseismic strength of the structures.
>
> The second is an early earthquake detection system.
>
> When an earthquake strikes, quickly slowing down and stopping the trains are essential in avoiding derailment and turning over. An early earthquake detection system is one such system that triggers emergency stoppage of the trains. Seismographs on beaches and along Shinkansen

tracks detect pressure waves (P-waves) and stop electricity supply to the trains so they can stop or at least start slowing down before the arrival of secondary waves (S-waves).

This system works well; however, it still has a problem. The system is effective when the epicenter is remote, but when an inland epicentral earthquake, like the Great Hanshin-Awaji Earthquake hits, S-waves arrive in no time after the P-waves and the system is hardly effective. In this case, the Shinkansen trains might be running at full speed when the electricity supply is stopped and there will be a big chance of derailment and turnover.

The third is antiderailment measures.

With limitations of the early earthquake detection system, a recent development is the installment of "derailment prevention guards" and "deviation prevention guards." The former prevents derailment upon earthquakes, and the latter installment between the two track lines prevents the derailed cars from sliding too far away from the tracks.

JR East has placed L-shaped railcar guides on all Shinkansen bogies so that a derailed car is restricted in its lateral movement. JR Tokai, West, and Kyushu are also installing deviation prevention stoppers that work the same way.

As we know, Tohoku Shinkansen runs through the area damaged by the Great East Japan Earthquake. When the megathrust earthquake hit, 27 Shinkansen trains were in service and they all succeeded in emergency stoppage. One out-of-service train running near Sendai, however, derailed. In terms of damage to structures, some power poles tilted and power cables were cut, but no serious damage was done to the elevated trackways, and on the 49th day after the earthquake the tracks were back in full service. Compared to the case of Sanyo Shinkansen after the Great Hanshin-Awaji Earthquake, the recovery was very quick. One factor was the epicenter being offshore in the Pacific, but the quick recovery owed more to the stronger earthquake resistant reinforcements compared to the time of the Great Hanshin-Awaji Earthquake, to keep the damage to the facilities small. The quick recovery of the Shinkansen service largely contributed to the reconstruction efforts by allowing people to move back and forth between the disaster area and Tokyo.

Japan has concerns over a Nankai Trough Megathrust Earthquake and Tokyo Metropolitan Area Epicentral Earthquake occurring in the near future. Tokaido Shinkansen, except for small portions, runs inland and Sanyo Shikansen has tracks inland so they will not suffer major damage from tsunami. The Tokyo Metropolitan Area Epicentral Earthquake or Tokai Earthquake will, however, directly shake Tokaido and Tohoku

Shinkansen and the quakes may cause major damage to Shinkansen. Further disaster mitigation plans are needed in both software and hardware to lower the damage from these earthquakes.

4.3.2 **Airports**

A runway, the basic facility for an airport, is built sturdy enough to withstand landing and takeoff of large passenger aircrafts that weigh hundreds of tons. In the case of asphalt coating, at the bottom is the subbase layer under the base layer and on top of it are the 2–3 m thick asphalt base and surface layers [15]. This structure would hardly be damaged even in the case of a megathrust earthquake. The airport building, though not as strong as the runway, has aseismic design and construction in place. During the 1995 Great Hanshin-Awaji Earthquake, Itami and Kansai International Airports hardly suffered any damage, and the same for Niigata Airport during the 2004 Niigata Chuetsu earthquake. Furthermore, in and after 2007, a report by the Best Form of Earthquake-proof Airports Committee led to enhancing earthquake resistance of runways and taxiways and preparing for soil liquefaction.

Each of the three main disaster prefectures of Iwate, Miyagi, and Fukushima hosts a public airport of Hanamaki, Sendai, and Fukushima, respectively, all with scheduled flights. These three airports did not suffer any large-scale damage from the Tohoku Area Pacific Offshore Earthquake; however, Sendai Airport, close to the coast, without much damage to the airport building, suffered the following serious damage.

Tsunami waves attacked the airport, reaching an inundation depth of 3.4 m, taking down the fences that surrounded the airport, and left soil, debris, automobiles and other objects on the runways and their surroundings. The tsunami inundated the emergency electricity-generating facilities and the internal power supply was lost as well as the external. The access railway and roadway tunnel were submerged and access to the airport became difficult [16].

The Great East Japan Earthquake revealed the vulnerability of airports against tsunami. MLIT in October 2011 set the "Guideline for tsunami measures at Airports [17]." The ministry pointed out seven airports along the coast of the Pacific at risk of flooding: Tokyo International, Chubu Centrair International, Kansai International, Kochi, Oita, and Miyagi. "Guideline for tsunami measures at Airports" set the two main actions: (1) Construct an emergency evacuation system for protecting human lives, and (2) construct quick recovery plans to recover the airport function after a tsunami attack.

Given these guidelines, these seven airports have already established their plans of emergency evacuation against tsunami.

Furthermore, in November 2014, MLIT formed the Committee on the Best Form of Disaster Preparedness at Airport Facilities in Anticipation of Widespread Disaster, including the Nankai Trough Megathrust Earthquake. Five months later, in March 2015, the committee announced new guidelines for disaster preparation including those against tsunami. Sendai Airport, which suffered tsunami damage from the Great East Japan Earthquake, with efforts by the people recovered its function for landing and takeoff of emergency life-saving helicopters of the Self-Defense Force and US military, as well as cargo aircrafts for transporting emergency supplies. On April 13, special passenger flights to connect to Tokyo and Osaka started service that greatly relieved the efforts of saving and recovering the damaged areas [16]. Airports, in this manner, can serve the function of providing an access point for emergency life-saving helicopters and aerial transportation of emergency supplies and personnel. The Committee met its expectations of providing guidelines for disaster countermeasures and maintaining a network of aerial transportation, and also providing a point of emergency transportation.

4.4 PROBLEMS AND LESSONS FROM THE GREAT EAST JAPAN EARTHQUAKE

4.4.1 Background of Recovery and Reconstruction Policies and Remaining Problems

In this chapter, we reviewed the reconstruction of transportation infrastructures for recovering, for the damaged areas, roads, railroads, local traffic, Shinkansen, and airports. Roads recovered quickly and construction of new ones is underway. Railroads are left with the problems, as of September 2015, of permanent recovery with BRT and the Jyoban line. However, they are pointing towards recovery. Recovery of local traffic alone had only ad hoc solutions and it is hard to draw its picture into the future. That, however, is only one facet of the problem. What we have to think about is the living and transportation conditions in the affected areas.

If we evaluate the recovery and reconstruction of transportation infrastructure in relation to the demand for transportation, we would say that road construction was easy, railroads had problems in the players staying profitable, and local traffic lacked principles. These facts, however, only mirror the transport policies of our nation. From this point on, we will have to face the problem of transportation with reduction in our population. Although there have been some concerns about the issue, some have said we will no longer need railroads, but nobody voiced concerns about oversupply of roads. Maybe we should discuss the problem of future transport policies separately from the problems in recovery and reconstruction; however, it seems that we may be pushing some troubles into our future [18].

Bold policies were implemented for local transports, they were only for one time, for recovery and reconstruction. The real needs in the damaged areas were for transportation that was resilient, stable, and flexible, to assure transportation means in the future. The problem is the same for future local communities in Japan. The 2013 Basic Act on Transportation Policy showed the basic understanding that "It is important to appropriately meet the basic demand for transportation." What that "basic demand" is and how it should be guaranteed, however, are not clear. That is a problem that we will all have to face; as they sometimes say, "What we see in the disaster area in Tohoku is what we will see over Japan in the future."

4.4.2 Lessons From the Great East Japan Earthquake About Transport

The Great East Japan Earthquake forced us to see the strength of tsunami and that we had not been thoroughly prepared against it at all. There are many aspects to discuss, but for here, we will keep the topics to logistics of corporations, earthquake preparations of transportation businesses, and automobile evacuation upon disasters.

After the quake, companies recognized the problem of logistics during a time of emergency, and they are requesting business partners to set their Business Continuity Plans (BCPs). Many companies, however, have not yet quantified the risk (probability of disaster occurrence × damage) to evaluate the trade-off. They are still at the level of evaluating split centers for production and distribution, constructing alternate cooperative service for disasters, and evaluating hazards [19]. There have been reports that fewer businesses rent the ground floor of warehouses in harbors, saying that they are threats to the business [20].

Companies in the transportation business, like JR companies and Tokyo Metro, are starting their preparations against the Tokyo Metropolitan Epicentric Earthquake, Nankai Trough Megathrust Earthquake, and other flooding by making their own tsunami evacuation maps, or running drills against various situations. As we discussed previously, MLIT is strengthening preparations at airports from lessons learned at Sendai Airport.

There were a number of reports about evacuees in cars causing traffic jams that ended up being swallowed in the tsunami during the Great East Japan Earthquake. Some say "It was the first ever mega-tsunami disaster that hit the automobile society" and concerns have been raised about its future. Scenarios of residents fleeing to heights and landscapes are different between communities and trying to establish a single rule to ban the use of cars is unrealistic. While the basic rule of evacuation is on foot, the central and local governments are discussing alternate methods for regions that have to rely on cars, and are conducting drills, e.g., the Town of Yamamoto has half of its area in the lowland. The town has been exercising evacuation drills since 2013.

In any case, we cannot prevent a "once in a thousand years" disaster. We should brace ourselves against those with a bit higher occurrence frequency and prepare for disaster prevention as well as disaster reduction.

REFERENCES

[1] Post offices temporarily closed due to the Great East Japan Earthquake (as of March 2, 2015), Japan Post Co., Ltd.

[2] S. Abe, H. Nishimura, Damages to the lifelines and issues in their recovery, in: Kansai University, School of Societal Safety Sciences (Ed.), Verification, The Great East Japan Earthquake, Minerva, Kyoto, 2012, pp. 103–107.

[3] Tohoku Regional Bureau "Why roads reopen quickly", and "Why were roads recovered quickly". http://www.thr.mlit.go.jp/Bumon/B00097/k00360/taiheiyouokijishinn/newindex.htm (Accessed 29 August 2015).

[4] Reconstruction Agency, Reconstruction efforts and related systems, June 24, 2015, p. 70.

[5] T. Hara, Earthquake Disaster and Railroad, Asahi Newspaper Publishing, Tokyo, 2011.

[6] Sanriku Railway, Corporate information. http://www.sanrikutetsudou.com/?p=360.

[7] Nihon Keizai Newspaper, February 1, 2014.

[8] JR East, About the basic agreement, memorandum and groundbreaking ceremony for reconstructing Yamada line (between Miyako and Kamaishi). https://www.JReast.co.jp/press/2014/20150205.pdf (Accessed 31 August 2015).

[9] JR East, Kesennuma line and Ofunato line BRT. https://www.JReast.co.jp/railway/train/BRT/system.html (Accessed 31 August 2015).

[10] Hearing to the town of Minami-Sanriku and city of Kesennuma, August 26, 2015.

[11] Nihon Keizai Newspaper, May 25, 2015.

[12] Tohoku District Transport Bureau, MLIT, data and hearing, August 24, 2015.

[13] Tohoku District Transport Bureau, Change in passengers on community busses. http://wwwtb.mlit.go.jp/tohoku/zudemiru/zu-index.html#ryokaku (Accessed 10 September 2015).

[14] Tohoku District Transport Bureau, Study of securing sustainable local public transportation in the disaster struck areas, 2014. http://wwwtb.mlit.go.jp/tohoku/ks/new%20page/ks-tyousa25-2%20honbun.pdf (Accessed 1 September 2015).

[15] T. Uehama, N. Sanaka, Waterfront & Air Front—New Construction Technologies for Harbors and Airports, Sankaido, Tokyo, 1994.
[16] W. Sudo, The Great East Japan Earthquake and recovery of Sendai Airport, Yobojiho, vol. 249, The General Insurance Association of Japan, Tokyo, 2012, 19.
[17] MLIT homepage, Guideline for tsunami measures at Airports. http://www.mlit.go.jp/common/000170137.pdf (Accessed 10 September 2015).
[18] K. Shiozaki, Reconstruction Disaster, Iwanami, Tokyo, 2014.
[19] T. Osada, Y. Yano, Discussion on disaster preparation in logistics from questionnaires to companies, J. Jpn. Logist. Soc. 22 (2014), 259–266.
[20] Hearing during a visit to "Prologis Park Kasugai" organized by Japan Logistics Society on February 6, 2015.

FURTHER READING

[1] About the 2011 Tohoku Area Pacific Offshore Earthquake (The Great East Japan Earthquake), Cabinet Office of Japan, September 10, 2014, p. 2.

Chapter 5

Supporting Health of the Great East Japan Earthquake Evacuees

CHAPTER OUTLINE

The Fukushima and Tohoku Disaster. https://doi.org/10.1016/B978-0-12-814078-9.00005-4

Search, rescue, and emergency transportation in the case of disasters had already been in place. The Great Hanshin-Awaji Earthquake on January 17, 1995 triggered the establishment of preparedness for providing extremely fast medical service and health support activities for evacuees forced to live in shelters or quickly built temporary housing for mid- to long term [1]. We also recognized the problems of disaster-related deaths due to the evacuation lifestyle and worsened living environment. The reason behind having special health-support activities available at the time of large-scale disasters was the regional community administrations that had started to take responsibility in activities for supporting the health of residents during normal times and to increase the number of caretakers. With these preparations, at the time of the Great Hanshin-Awaji Earthquake, caretakers from communities all over the nation were systematically dispatched for health support for the disaster-struck areas. From then on, at times of large-scale disasters, caretakers from communities around the nation have been dispatched for supportive activities. The Great Hanshin-Awaji Earthquake also led to arranging the system for emergency life-saving activities upon wide-area disasters.

The Great East Japan Earthquake offered a chance to test the health and medical support activities established after the Great Hanshin-Awaji Earthquake. As a result, the disaster revealed problems and where the constructed systems were insufficient in providing health support.

The Great East Japan Earthquake was a wide-area disaster with combined earthquakes, tsunami, and nuclear disasters, and it forced a review of the support framework established after the Great Hanshin-Awaji Earthquake. The review concluded that the current system cannot handle Tonankai earthquakes and the Tokyo Metropolitan Epicentric Earthquake, which are predicted to take place in the future, and building a system for securing public health is now underway [2].

5.1 SUPPORTING HEALTH OF THE FUKUSHIMA AND TOHOKU DISASTER EVACUEES

5.1.1 Dispatching Community Caretakers From Over the Nation

The Great Hanshin-Awaji Earthquake broke out when Japan was reorganizing its local healthcare system. In July of 1994, the Public Health Center Act was amended and reshaped into the Community Health Act. The Community Health Act was going to be enforced from April of 1997. Currently, there is one community health center in the City of Kobe, while at the time of the Great Hanshin-Awaji Earthquake, the city had as many as nine centers, one for each district. The centers served as hubs of health and hygiene after the quake. The number of people affected by the Great Hanshin-Awaji Earthquake was so large that the communities, e.g., the City of Kobe, were unable to provide service to all.

On the 10th day from the disaster, January 26, 1995, the Health Guidance Officer of the Health Policy Planning Bureau, Ministry of Health, Labor and Welfare (MHLW) requested the Prefectural Public Nurse Administration Bureau to dispatch nurses to the affected areas. Registered public nurses from communities all over the nation went into the earthquake-struck areas 15 days after the quake. A total number of 9732 nurses was dispatched [3]. The health center in each district carried out health support activities of medical procedures and providing consultation while going around the areas with the help of medical and nurse teams from all of Japan.

Ever since the Great Hanshin-Awaji Earthquake, at the time of a disaster, nurses are dispatched from the whole country for support. When the damage to the communities makes it difficult to arrange health centers for the supporters, the prefectures take over the task of arranging the support force. The prefectures first request dispatch to the municipalities within their jurisdiction, but if the force is insufficient, they will go beyond the border to spread the call for help. During the Chuetsu earthquake in 2004, nurse dispatching was quicker than at the time of the Great Hanshin-Awaji Earthquake.

There is a manual that specifies what health support actions the nurses can take at the time of the disaster [3]. Also, in the case of large-scale disasters like the Great East Japan Earthquake, an activity manual provides guidance for going around to shelters and temporary housing for health consultation and management, and also for what to do for secondary health issues of economy class syndrome, norovirus, heat attack, disuse syndrome, lone deaths, and posttraumatic stress disorder (PTSD) [4].

5.1.2 Caring for the Minds of Disaster Victims at the Time of the Great Hanshin-Awaji Earthquake

It is only recently that we have recognized the need for mental healthcare for victims of disasters [5]. The need to provide care for victims with posttraumatic stress disorder (PTSD) was recognized after the 1991 Mount Unzen-Fugen eruption and the 1993 earthquake offshore southwest of Hokkaido. The need was widely recognized after the Great Hanshin-Awaji Earthquake [6]. The area between the cities of Kobe and Osaka, struck by the earthquake, had a large number of psychiatric clinics, and a number of psychiatric patients had their homes in the area. The clinics suffered major damage from the earthquake and psychiatric treatment was needed in shelters and for isolated psychiatric patients at home.

When the doctors in the area announced their special needs, a number of psychiatrists and nurses gathered from in and out of the disaster area to provide psychiatric treatment and support from the health care centers. The 10 health care centers in the disaster area set "psychiatric rescue centers." The centers played major roles in providing medical support to the psychiatric patients that suffered from the earthquake. Later when the victims moved from shelters to temporary housing, mental health problems involving PTSD caused by shock and confusion at the time of the earthquake surfaced, with the victims calling for special mental healthcare in the case of disasters. Existing healthcare centers and medical establishments, which provided mental treatment during regular times, lacked the proper manpower and special caretaking skills. In response to the need, Hyogo prefecture established the Healthcare Center for the Mind and hired about 40 staff members, including clinical psychotherapists, mental health welfare consultants, and nurses. Later, in 2004, the center was established as the Hyogo Institute for Traumatic Stress, the first center in the nation to have a number of healthcare functions for mental health [7].

5.1.3 Disaster Area Support for Lifesaving Medical Treatment

In January of 1975, emergency services were assigned to each municipality, but the duty was only emergency transportation without medical treatment for lifesaving in the case of large disasters. The Great Hanshin-Awaji Earthquake destroyed medical functions in the area, blocked roadways and caused a large number of injuries. If wide-area lifesaving and medical treatment had been in place, many lives would have been saved [8]. Immediately following this disaster, lifesaving medical treatment systems started to take shape.

5.1.3.1 Preparing a Central Hospital in a Disaster Area

After the Great Hanshin-Awaji Earthquake, in April 1995, people from medical offices in the disaster area, medical associations, construction, machinery, communication and pharmacists gathered to form the "Study group, triggered by the Great Hanshin-Awaji Earthquake, for Medical Arrangement upon a Disaster," supported by the MHLW. Their report stated the need for a "central hospital for medical support upon a disaster" to handle wide-area transportation, emergency supplies of equipment, and to accommodate dispatched medical aid teams.

The Health Policy Bureau issued its Notification Number 451, "Enhancing initial emergency medical service upon a disaster," dated May 10, 1996. The notification formed a system for preparing to accept the injured and dispatching medical aid teams in response to a prefectural governor's request in cases when municipalities face difficulties in providing proper medical services.

A central hospital in a disaster area is assigned by the prefecture. When a disaster occurs, it is assigned to provide the following services: (1) High-level treatment to save lives of critically injured patients, (2) Provide wide-area transportation functions to accept and send out the injured, (3) Dispatch self-sufficient medical aid teams, (4) Supply emergency medical equipment to local medical facilities.

The central hospital must have a heliport. Each prefecture has one "primary" central hospital and one "local" central hospital in the secondary medical service area. The primary central hospital has training facilities for medical emergency personnel training.

5.1.3.2 Establishment of Disaster Medical Assistance Team

Many medical facilities lost their functionality in the Great Hanshin-Awaji Earthquake. "Study report of Disaster Medical Assistance Team (DMAT) in Japan," created in Japan Fiscal Year 2001 (JFY2001), led to the establishment of the DMAT. The DMAT consists of doctors, nurses, and assistants (emergency lifesaving technicians, pharmacists, clinical engineers, clinical lab technicians, occupational therapists, radiologists, social welfare counselors, healthcare providers, administrators and so on). It assumes lifesaving activities by teams with a variety of specialists. The established teams were "Prefecture DMAT" and "Japan DMAT."

A "Prefecture DMAT" performs disaster medical activities within its prefecture. Tokyo DMAT was the first established in 2004. MHLW established the "Japan DMAT" in April 2005. Its administration office was set inside the

National Disaster Medical Center in the City of Tachikawa, Tokyo. The Japan DMAT is dispatched from locations throughout Japan for wide-area medical transportation, opening staging care units (SCUs), hospital supports, regional transportation, and medical services in disaster-struck areas.

5.1.3.3 Problems With DMAT and Medical Support Team Coordination

When the Great East Japan Earthquake hit Japan, a number of DMATs immediately gathered to the disaster area. There were not too many bodily injuries and the need for lifesaving treatment by DMAT was small. The large-scale disaster crippled the communication, and the headquarters for dispatching medical teams did not function well because information on patient locations did not reach the teams. DMATs were specialized teams for urgent medical treatment immediately after the disaster. Their scope ended up leaving patients with chronic problems in shelters alone, and there were even reports of deaths of chronic disease patients after the DMAT teams pulled out from the disaster area without notifying other medical teams about the situation.

It was a problem of continuity from DMAT, which carried out initial disaster medical activities, to medical assistance teams that took over the service. MHLW amended the Japan DMAT activity guidelines on September 4, 2013. When a disaster strikes, a specially trained medical team will immediately enter the disaster area to grasp the medical needs of the affected area. Then emergency medical services are built into the disaster area. The team then supports emergency procedures and hospitals in the area, and arranges transportation of those injured in the area to proper facilities outside.

Discussions are underway for DMAT to coordinate its disaster emergency medical activities with medical support teams that are also dispatched to the disaster area, like the Japan Medical Association Team (JMAT) and other disaster medical teams sent by university hospitals, Japan Red Cross, National Disaster Medical Center, Japan Hospital Association, All Japan Hospital Association, Japan Dental Association, Japan Pharmaceutical Association, or Japanese Nursing Association.

5.1.3.4 Promoting the Use of Helicopters for Emergency Transportation Upon Disasters

The Great Hanshin-Awaji Earthquake caused large numbers of deaths and injuries. Medical facilities in the disaster-struck area lost their functionality, roadways were blocked and ambulances could not perform transportation. On the day of the earthquake, one helicopter made a single transport, and the next day two helicopters. Each helicopter carried one patient per transport to a hospital outside the area. Helicopters from the Self-Defense Force started making trips on the evening of the third day [8]. The first 48 hours are the most critical for emergency medical treatment during a large disaster; thus, the emergency transportation system was not functioning. The problem led to arranging a system to transport patients in need to outside the disaster areas.

To generalize emergency patient transportation using helicopters, the Fire and Disaster Management Agency added helicopter transportation to its "Standards of Emergency Team Formation and Equipment" in March 1998. The agency formed a "Study group for promotion of emergency transportation by helicopters" to develop guidelines for when to dispatch helicopters and promote emergency dispatch of fire and disaster management helicopters. MHLW experimented with Dr. Helicopter in October of 1999 and wrote "Act on Special Measures on Securing Emergency Medical Service with Helicopters (Act on Doctor Heli)." Prefectural emergency medical policies are starting to arrange for Doctor Heli with a doctor, nurse, and an emergency lifesaving technician on-board. Doctor Heli has the duty to support DMAT as needed during disasters.

5.2 CURRENT STATE AND PROBLEMS OF LOCAL HEALTH PROMOTION AFTER THE FUKUSHIMA AND TOHOKU DISASTER

5.2.1 Regional Health Policies for Supporting Disaster Victims

In Japan, the Act on Regional Health states that municipalities are responsible for providing basic health-supporting activities for the residents. There used to be more public health nurses in prefectural health care centers, but since the 1980s there have been more with the municipalities. Enforcement of the 1983 Long-Term Care Insurance Act and the 1997 Community Health Act established the municipality-based healthcare system in Japan; 85% of the administrative public health nurses are registered with municipalities. The JFY2014 Public Health Nurse Activity Surveillance resulted in 32,896 full-time public health nurses, of which 4941 were registered with prefectures and 27,955 with municipalities, and among them, 19,513 were registered with municipalities without health centers.

Article 3 of the Community Health Act states that municipal governments, including special wards, should prepare necessary facilities, secure personnel and make efforts to improve the quality for smooth operation of regional health measures. Prefectures and the central government provide, per request, technical and financial support to the municipalities in performing their duties specified in this act.

5.2.2 Basic Guideline for Promoting Regional Heath Measures and Its Amendment After the Tohoku Earthquake

Act on Regional Health requires the Minister of Health, Labor, and Welfare to announce the basic guidelines for promoting regional health measures. The first basic guideline in 1994 had not listed public health activities at the time of disasters. On March 31, 2000, after the Great Hanshin-Awaji Earthquake and other disasters, the basic guidelines were partially amended.

The regional health administration, for the first time, included "health crisis management." Health centers were required to: (1) In preparation against a health crisis, as the administration body of regional healthcare, comprehend the regional medical supply status, make arrangements with related organizations and groups, and establish systems of providing medical service to the community. Especially, collect health and medical information to establish a system for overall arrangement of corresponding measures. (2) Upon outbreak of a health crisis, collect and broadcast information regarding threats to health victim's lives, arrange support and overall management for related organizations in making coordinated activities for securing health and medical service to health victims, and (3) After outbreak of a health crisis, evaluate the response to the health crisis based on scientific reasoning and announce the findings.

After the Great Hanshin-Awaji Earthquake, the basic guidelines were again partially amended. Prefectures and municipalities are required to expect large-scale disasters and prepare arrangements for supporting and providing overall management for collecting information and providing healthcare in coordination with other municipalities and the central government. The central government is required to construct systems for training personnel for wide-range disaster health activities and quickly dispatching and arranging public health nurses.

5.2.3 Problems With Earlier Merging of Municipalities

The Local Autonomy Act was amended in 1994 and the Decentralization Act was enforced in April 2000. The Community Health Act now has a strong effect in terms of overall service for residents in health, welfare, and caretaking. Since 1999, local municipalities have been merged to form larger ones. In 2004, the Municipal Merger Act was enforced and the central and prefectural governments started active involvement in merging local municipalities.

The number of municipalities, from 3232 in March 1999, dropped to about half, or 1718, by April of 2014. The City of Ishinomaki merged with the Towns of Monou, Kanan, Kahoku, Kitakami, and Ogatsu in Monou District, and the Town of Oshika in Oshika District on April 1, 2005. The City of Minamisoma, affected by the Fukushima-1 Nuclear Power Plant accident, was born on January 1, 2006, merging the Towns of Odaka and Kashima, and the City of Haramachi.

Murosaki reported [9] that the mergers of these municipalities caused difficulties in providing health support after the Great East Japan Earthquake, because it was timed so soon after the mergers. The Odaka ward of the City of Minamisoma is almost within the 20 km caution area from Fukushima-1 and the Haramachi ward of the City of Minamisoma is the 20–30 km range of the emergency evacuation preparation area (later, the planned evacuation area) and the residents were evacuated. Minamisoma city is born on January 1, 2006, one city and two cities of former Odaka cho, former Kashima cho, and former Haramachi city merged. Minamisoma city consists of 3 regional districts, which are Odaka, Haramachi, and Kashima. The Kashima district of the Minamisoma city lies outside of the 30 kilometers from Tokyo Electric Power Fukushima Daiichi Nuclear Power Station TEPCO Fukushima Daiichi Nuclear Power Plant. The Haramachi district existing the city hall is within the emergency evacuation preparation area and, without evacuating the city functions, it is still maintained there [10]. If there were no merger, evacuees in the former Odaka cho area would have had to flee to other municipalities, like evacuees from Futaba cho. The Kashima district of Minamisoma city, although it was badly damaged in the tsunami attack, accepted the nuclear accident evacuees from the Odaka district. Some criticized that the move delayed support for the earthquake and tsunami victims in the Kashima district.

5.2.4 Reducing Number of Health Centers in the Disaster Areas of the Great East Japan Earthquake

After the Great Hanshin-Awaji Earthquake, the number of health centers has been largely dropping in Japan. It will be difficult for the workforce to serve as the center for health crisis management during disasters. The number of health centers in the prefectures of Iwate, Miyagi, and Fukushima is decreasing, while the coverage of health centers is expanding. It is a national trend that the number of health center doctors is dropping. In the Tohoku area, it was difficult to find even one doctor for each health center.

There was a health center in the prefecture of Iwate for which the doctor was also serving as the health center manager. Health centers in the prefectures of

Iwate, Miyagi, and Fukushima had merged with the prefectural welfare offices. The number of health centers in the Iwate prefecture was 10 (which had dropped from 15 in 1997), in Miyagi 7 (from 9 in 1997), and in Fukushima 16 (from 18 in 1994). Further merging of health centers in Fukushima dropped the count to 7. In 1999, the City of Iwaki turned into one of the major cities in Japan and the number of health centers in Fukushima dropped to 6.

All 12 municipalities in the coastal area of Fukushima prefecture affected by the Fukushima-1 Nuclear Power Plant accident were covered by one health center in Sousou. The health center manager was the only doctor registered there. Nuclear evacuees from Futaba district of Fukushima prefecture concentrated to the main city of Kooriyama's evacuation center, and later moved to the Aizu area. Temporary housing for Futaba residents was built not only in the Aizu area but also in the City of Iwaki. Health support activities for victims living in the City of Iwaki, next to Futaba, were provided by a local branch office of the Sousou health center.

When a wide-area disaster breaks out, health support for victims is hard for just the municipality level to cover. However, the number of health centers within the prefectures is dropping, making the support even more difficult for prefectures. For intense wide-area disasters, prefectures and MHLW should review the support system and make changes if necessary. Joint prefectural efforts, as in the case of the Kansai Wide-Area Alliance, are needed to shape the system for support.

5.3 DIFFICULTIES IN COVERING ALL MEDICAL NEEDS WITHOUT OPERATIONAL PUBLIC HEALTH CENTER FUNCTIONS UPON THE FUKUSHIMA AND TOHOKU DISASTER

At the time of the Great East Japan Earthquake, staff dispatched to the disaster area were not only public health nurses but also a number of professionals gathered for support, including nutritionists, mental health welfare consultants, and dental hygienists. Public health support teams that are self-sufficient and can support disaster areas are needed, along with a system to dispatch them when local communities and health centers are not functional [11].

5.3.1 Establishing Disaster Health Emergency Assistance Team (DHEAT) to Avoid the Problem of Poor Health Assistance

At the time of the Great East Japan Earthquake, shelters and rescue stations were designed to be led by the municipalities; however, many of these local communities lost their functionality during the disaster. Health support personnel did not enter the nuclear accident regions and there were big gaps in health support activities for the victims, depending on the area.

After these problems were pointed out, discussions then took place to build a health support system for disasters. In July 2011, "Research on status of dispatch of public health nurses, doctors, and managerial nutritionists in relation to the Great East Japan Earthquake – Report on problems in disaster areas found through supporting them" was published. In March 2012, another report, "Report on fact-finding of healthcare, medical and welfare support to the Great East Japan Earthquake disaster areas from municipalities of the nation," was published. The reports pointed out that health support in the disaster areas of the Great East Japan Earthquake was neither effective nor efficient, victim support was not made on an extended scope, and the support activities were different among different communities.

In May of 2011, the Disaster Support Public Health Forum, headed by Shigeru Omi, was established requesting a public health movement to break the walls between different professionals to contribute to regional reconstruction in the future. In response to its accomplishments, MHLW Grant-in-aid for scientific research formed a "Research group on establishment of regional health foundations in case of large-scale earthquakes," and a "Research group on formation of support organization for public health activities in case of disaster." The research reports suggest the establishment of the Disaster Health Emergency Assistance Team (DHEAT). The National Group of General Managers of Health established the "Standardization Committee of Health and Medical Support at the Time of Disasters" to discuss the realization of these teams.

5.3.2 Prefectures Not Prepared With Public Health Teams Upon the Fukushima and Tohoku Disaster

Some prefectures have built their own public health support teams for disasters. The Prefecture of Oita, in April of 2012, edited an "Outline for Setting Oita Prefecture Public Health Measure Team at the Time of Disasters" to provide dispatching support to disaster areas within the prefecture. Hiroshima Prefecture made their "Manual for Public Health Activities at the Time of Disasters" in 2012, and set a disaster public health team in February of 2013.

The Oita prefecture public health support team consists of employees of the local government. However, the Hiroshima disaster public health team has professional group members in it. On August 21, 2014, Hiroshima suffered a landslide disaster caused by heavy rain. The prefectural public health team supported the victims at the time.

5.3.3 Poor Psychiatric Assistance Upon the Fukushima and Tohoku Disaster

After the Great Hanshin-Awaji Earthquake, it has now become general practice to provide psychiatric support for victims of large-scale disasters and accidents. At the time of the Great East Japan Earthquake, the three prefectures of Miyagi, Fukushima, and Iwate, in that order, set up their Care Center for the Mind starting in December 2011, completing the last one in March 2012. The centers, however, are said not to have provided sufficient psychiatric care for the disaster victims of the Great East Japan Earthquake.

So the Care Centers for the Mind are now being advanced into the Disaster Psychiatric Assistance Team (DPAT). MHLW send a notice in 2013 to the prefectures to establish their DPAT. The notice explained actual activities to start effective psychiatric care immediately following disasters and requested the prefectures to arrange their DPAT to match their local needs. The Prefecture of Hyogo has made its own "Hyogo DPAT" activity manual, organized DPAT, and published its organization, basic rules, and activities.

5.3.4 Japan Dietetic Association Disaster Support Team (JDA-DAT)

With the extended evacuation after the Great Hanshin-Awaji Earthquake, concerns were raised about dietary and nutritional matters for those that need nursing care, such as infants, elderly, and handicapped, as well as the diet of diabetics and those with kidney problems. The concerns led to the necessity of having nutritionists in addition to public health nurses involved in the disaster health care activities.

When the evacuation extends over a long period, diet problems (shortages of vitamins, minerals, and fiber, and high sodium, etc.) could be discussed with managerial nutritionists when they visit the evacuees in rounds. At the time of the Great East Japan Earthquake, the Japan Dietetics Association took the lead in soliciting nutritionists (managerial or not) from the nation and gathered a total of about 1600 nutritionists to dispatch to the disaster areas to support victims in their nutrition and diet. The nutritionists provided wide-ranging support, from dietary support in shelters and facilities to visiting those at home. Based on this experience, the Japan Dietetics Association established The Japan Dietetics Association—Disaster Assistance Team (JDA-DAT), with highly flexible nutritionists to move quickly to support disaster-affected areas.

5.3.5 Disaster Dispatch Nursing (Welfare) Team

At the time of the Great Hanshin-Awaji Earthquake, supporting those that could not live like healthy evacuees, e.g., handicapped, elderly, and expecting mothers, turned out to be a major issue. Many shelters had no toilets that accommodated wheelchairs, and some psychiatric patients could not rest in congested areas. These problems led to the concept in 1996 of welfare evacuation centers for those in need of support. Later, as we entered the superaged society, communities began to have more healthcare facilities and at-home residents in need of health support receiving nursing insurance payments.

During a disaster, the communities not only need to support those in need of nursing support, but also need to support the staff that provides nursing care at home and in nursing facilities. After the Great East Japan Earthquake, MHLW issued a notification "Organization for dispatching support for caretakers (disaster dispatch nursing team)" to request building disaster dispatch nursing (welfare) teams. A disaster dispatch nursing team consists of caretaking staff of social welfare counselors, nurses, and physical therapists to reach disaster-struck areas in buses to support health-care facilities and shelters. The staff will take turns with the needs of the area to provide mid- to long-term nursing care in the disaster-struck area.

On March 26, 2014, the Prefecture of Iwate organized a maximum of 27 DCATs to gather information on those in need of nursing care and provide the emergency service in shelters and houses in the case of large-scale disasters. The prefecture registered those that went through formal welfare training for social welfare counselor, certified caretaker, and psychiatric social worker, so it can form teams of 4–6 members in the case of large-scale disasters to dispatch as DCAT teams for supporting disaster-struck areas.

5.3.6 Evacuation System Not Ready for Those in Need of Support

Upon the Great East Japan Earthquake, among the fatalities in the disaster area, about 60% were those at 65 or older and the death rate of the handicapped was about twice that of the death rate of the whole region. Not only those in need but many caretakers of staff in the fire stations, local firefighting volunteers, and social welfare workers lost their lives [12].

During the Great Hanshin-Awaji Earthquake, supporting evacuation and nursing in shelters for those in need of nursing support, namely the elderly and handicapped, surfaced as concerns. "Guideline for supporting evacuation of those in need of support at the time of disasters" was set in 2006 and communities were urged to make lists of those in need of nursing care and plans for supporting their evacuation [13]. During the Great East Japan Earthquake, however, information about evacuation was not thoroughly passed on to those in need, and the administration did not smoothly confirm their safety.

In March 2013, the "Report by Study Group on Supporting Evacuation of Those in Need at the Time of Disaster" was filed to amend part of the Basic Act of Disaster Control Measures. The amendment in June of 2013 now requires communities to maintain lists of those in special needs, among elderly, handicapped, and infants, during disaster prevention. "Guideline for Supporting Evacuation Action by Those in Needs of Evacuation Actions" include actual processes for building this list. This guideline replaced the 2006 "Guideline for supporting evacuation of those in need at the time of disasters." Municipality communities are required to make use of this guideline to arrange a system of providing support for those in need.

5.3.7 Infection Surveillance in Disaster Area After the Poor Sanitation Cases

When a large-scale disaster strikes, the chances of spreading infections increase because the quality of living conditions go down with a large number of people living together in shelters and visits by a number of supporters. Maintaining good living conditions in shelters is the basis of preventing the spread of infections. Laws about infections changed somewhat after the Great Hanshin-Awaji Earthquake: the Act on Infectious Diseases Control was repealed and the Act on the Prevention of Infectious Diseases and Medical Care for Patients Suffering from Infectious Diseases (Act on Infectious Diseases) was enacted on October 2, 1998. After the Great East Japan Earthquake attack, the National Institute of Infectious Diseases and infectious diseases departments of municipal governments started surveillance of infections in shelters in some communities.

In the case of the Great East Japan Earthquake, one of the shelters, Big Palette Fukushima in the City of Kooriyama in Fukushima Prefecture, saw a spread of norovirus causing vomiting and diarrhea among a large number of people. From hearing and observations, the following problems were identified: During the early stage of the spread, each floor was overpopulated, sewage and contaminants were not properly treated, people did not follow instructions for keeping the hands clean, water from the public restroom tap was used, toilets were not kept clean at the beginning, the air was dry causing dust from carpets and blankets, and there was a shortage of ventilation facilities.

Also, during March 2011 to March 2012, after the Great East Japan Earthquake, the prefectures of Iwate and Miyagi reported 10 cases of tetanus infection [14]. In the case of disasters in large shelters, we should have organized efforts to prevent the spread of infections and quickly identify the spread if one breaks out; if one does break out, we need a support team to take measures.

5.3.8 Nuclear Disaster Medical Dispatch Team Not Ready for the Fukushima Disaster

Tokyo Electric Power Company (TEPCO) owned Fukushima-1 where a nuclear accident spread radioactive material over a wide area. The designated hospital also suffered contamination and planned medical treatment for exposure did not function at all. Based on the lessons from the Great East Japan Earthquake, the Nuclear Regulation Authority set up a Discussion Team on Emergency Medical Treatment for Exposure on November 15, 2012. The team has been the Discussion Team on Proper Medical Organization at the Time of Nuclear Disasters since April 24, 2015. The team made proposals to build an Advanced Exposure Treatment Support Center, a Medical and Total Support Center for Nuclear Disasters, and to designate one to three hospitals as "nuclear disaster central hospitals" to play central roles of exposure treatment in the communities.

The central hospitals will take those injured regardless of whether the patient is contaminated or not and, if exposed, will provide proper medical treatment. The team requested that a Nuclear Disaster Medical Dispatch Team be ready at the central hospitals to handle local medical procedures of the area. The Medical and Total Support Center for Nuclear Disasters will provide support for the central hospitals and build networks with related medical institutions. It also covers the role of dispatching nuclear disaster medical dispatch teams at the time of nuclear accidents. "Nuclear disaster medical dispatch teams" are affiliated with the central hospitals to carry out emergency medical procedures within the prefecture where the hospital is built. Upon a nuclear disaster, the Nuclear Disaster Medical Administrator is appointed to quickly direct where to transport patients based on the levels of contamination and exposure.

There was much confusion about exposure testing and decontamination while people were evacuating from the Fukushima-1 nuclear power plant accident affected zones. The confusion led to revising the conventional body surface contamination screening and decontamination procedures. Examinations of residents that arrive in automobiles or buses start by first examining the vehicle, and if it exceeds Operational Intervention Level 4 (OIL4), one of the passengers from the group that took the same evacuation route is examined. If this person exceeds OIL4, then examination is made for all the passengers. This way of gradually increasing the subject of examination was from the lesson that body surface contamination screening for evacuees from the danger zone during the Fukushima-1 nuclear power plant accident resulted in long evacuation routes and some patients even died in the process of making the evacuation.

5.3.9 Managing Crisis Management Teams for Public Health at the Time of Disasters

After the Great East Japan Earthquake, a number of teams for activities in various fields for disaster support have been built. Local health centers are expected to manage activities for health support and health crisis management in communities. Health centers in Japan, however, are administrative organizations; thus, for them to handle health crisis management, the regular line of command has to change to handle those at the time of health crisis. We need standardization of these emergency systems.

MHLW formed a "Research Group about Functional Evaluation and Quality Improvement of System for Regional Health, Safety, and Crisis Management," to evaluate the Japanese version of Standard Health Center ISC/IAP/AC. The target is to standardize the basic activity system for the initial activities and coordination with other organizations and groups. The system is modeled after the Incident Command System (ICS) of the US Federal Emergency Management Agency (FEMA). The Japanese version of ICS (natural disasters) followed the basic framework of the US ICS in command, operation, logistics, planning and administration/finance.

Health centers in Japan assume the headquarters will handle logistics, planning, and administration/finance, and thus the HQ needs to enhance its capabilities in command and operations functions. Furthermore, DHEAT by MHLW and the research group on crisis management will join their efforts to build an executive organization to operate at the time of disaster.

5.4 CURRENT STATE AND PROBLEMS WITH FUKUSHIMA NUCLEAR POWER PLANT ACCIDENT AFFECTED AREAS

5.4.1 Expected Health Support Upon Nuclear Accidents Before the Fukushima-1 Accident

Japan has experience in health support during a nuclear accident after the JCO Uranium processing plant criticality accident that broke out on September 30, 1999. Based on this accident, health centers with possible nuclear accidents in their covered area were prepared for arranging notifications and rescuing and consulting with those exposed, based on legal planning and disaster prevention guidelines from the Nuclear Safety Commission. Shelters were to be set by communities and medical service sites were to be arranged by the prefecture and health center. Radiation contamination of residents and screening were to be performed as needed and stable iodine tablets were to be administered as needed [14].

The reality was, however, that information about the nuclear power plant accident was not properly transferred and the severity reached an unexpected state of meltdown, and evacuees from the nuclear power plant accident did not receive proper handling.

5.4.2 The Nuclear Accident Obstructed Planned Health Support Activities

The offsite center that was supposed to serve as the local emergency response headquarters was contaminated with radiation and became unusable. The sudden threat of exposure and evacuation was over the capability of proper screening. Medical staff at the Fukushima prefectural hospital also needed to evacuate. Residents in the area near the Fukushima-1 nuclear power plant had hardly any medical services of public health nurse support and dispatch for several months.

Most local health and medical staff evacuated the area and the nuclear disaster area did not offer any health support services and, moreover, emergency medical service was restricted. Health and medical service providers who remained in the area had to perform their services without external help and support [15]. The situation was very different from sending aid and support to natural disasters. Evacuees from the nuclear accident, quite differently from those from natural disasters, have no schedule for returning home and without plans for rebuilding their lives; as time passes, many of them move away to restart their lives.

The situation makes it difficult to determine how long to provide health support to the victims. In the designated evacuation area of the nuclear disaster, only the local public health nurses were requested to provide services. A few of the public health nurses in the City of Minamisoma volunteered to provide service, but it was provided for only a short time [16]. So far, there have been no clear standards on health support systems for residents of a nuclear accident–affected area. How to manage health support personnel in case of a nuclear accident is a topic we must further discuss, especially after this Fukushima-1 Nuclear Power Plant accident.

5.4.3 Shortage of Health Workers in Minamisoma, Fukushima Prefecture

In the Kashima district in the City of Minamisoma, Fukushima Prefecture, a number of temporary housings have been built for those evacuated from the Alert Zone of the nuclear power plant accident. Many still live in the area after 5 years. The City of Minamisoma was split into three zones: the evacuation order lift preparation zone, the residence limitation zone, and the hard-to-return zone. Later, as decontamination took place, some areas have relaxed their restrictions. Some areas that were designated as an Alert zone at the time have scheduled to have that zoning lifted.

As people return home after decontamination, the area is in need of enhanced support for health consultation about the impact of the radiation. For this reason, in 2015, the city hall of the City of Minamisoma assigned a new department manager in charge of radiation health studies in the department of health building. An officer in charge of radiation health was stationed in the Kashima district, where a large number of evacuees live in the emergency temporary housing, and now there is another officer with the same title in the main city hall as well. Radiation consultants were newly hired to be stationed in the districts of Kashima, Haramachi, and Odaka. The plan was to hire personnel with expert knowledge. However, there were few applications and the city made a public announcement about hiring them from the general public, to be trained and then to be stationed.

Caretaking of mental health was contracted to the Fukushima Prefecture Psychological Health Welfare Association and stationed in the Fukushima Care Center for the Mind. The City of Minamisoma received service from a satellite office in Soumahoubu Center. Total mental health caretaking related to damage to the victims and concerns with their future lives is handled by guidance counselors in the emergency temporary housing area. If a case is over the head of the counselor, a psychiatrist takes over. Since the disaster, the staff at the City of Minamisoma are kept busy with health consultations and support for people's concerns. The city needs more help from specialized public health nurses.

5.5 CONCLUSION

Health support services in Japan for disaster victims started to take shape after the 1995 Great Hanshin-Awaji Earthquake. At the time, medical treatment alone was the subject. As time passed, however, needs were recognized for mid- to long-term services of health, medical procedures, welfare, and caretaking for the victims in temporary housing. The Great

Hanshin-Awaji Earthquake was 10 years after local municipalities had been assigned to take main roles in providing health services to the residents. The number of public health nurses had increased and that fact enabled proper service at the time of the disaster. After the Great Hanshin-Awaji Earthquake, the Act on Regional Health and the Long-Term Care Insurance Act were enforced and social systems to support the victims began to take shape. The Act on the Prevention of Infectious Diseases and Medical Care for Patients Suffering from Infectious Diseases (Act on Infectious Diseases) enacted in 1999 takes care of countering the spread of infections. The Medical Care Act underwent amendment in 1992, 1997, 2000, 2006, and 2014 to handle medical care services. At the time of the Great East Japan Earthquake, private organizations and groups offered many services in medical treatment, health, welfare, and caretaking.

Some local municipalities, which were to offer services to the disaster victims, lost their functionality at the time of the Great East Japan Earthquake. Some municipalities and organizations in the disaster-struck areas could not offer sufficient health support service to the victims. After the quake, thus, the administration was in search of an overall health support plan. As a result, public health activity support teams have been organized and put into the system. Plans for health support services during a nuclear accident are still in the maze.

Japan is now a superaged society. More people in need of caretaking increasing all over the country including countryside. Laws assuring community-based integrated care systems are being promoted. Preparing to support the evacuation of local elderly in need of nursing and to provide coordinated health, medical, and caretaking services during disasters is an issue. The June 2013 partial amendment of the Basic Act on Disaster Control Measures enforced municipalities to compose lists of those in special need of support in cases of disaster followed by evacuation among the elderly, handicapped, and infants. This amendment led to the composure and announcement of the Guideline for Evacuation Support for Those in Need of Support in August of the same year. The big burden on our shoulders now is to turn this guideline into an effective reality.

REFERENCES

[1] K. Tatara, T. Takatorige, Research report on manual making for regional health actions upon disasters, 1996.

[2] N. Sakamoto, Framework of wide-area (prefectural) support upon large-scale disasters—about the reality of health and medical support by local communities during the Great East Japan Earthquake and efficient and effective support to prepare

against megathrust earthquakes in the future, J. Natl. Inst. Public Health 62 (4) (2013) 390–404.
[3] Discussion on effective systems for health support at times of earthquake disasters, Japan Public Health Association, 2007.
[4] Japanese Association of Public Health Nurse, 2013.
[5] M. Noda, Forming a culture of disaster rescue, in: Okushiri and Shimabara, 1994. Iwanami Booklet #360, Tokyo.
[6] H. Nakai (Ed.), Psychiatrists During the Great Hanshin-Awaji Earthquake, Misuzu Books, Kobe, 1995.
[7] Research report on Hyogo Institute for Traumatic Stress, 2015.
[8] T. Sugimoto, Emergency Medical Treatment and Citizen's Lives—Learning From the Great Hanshin-Awaji Earthquake and Tokyo Subway Sarin Attack, Health Publishing, Tokyo, 1996.
[9] Y. Murosaki, M. Koda (Eds.), Loss of Strength in Disaster Prevention With Community Annexation—Negative Effect Surfaced With the Great East Japan Earthquake, Minerva Shobo, Kyoto, 2013.
[10] T. Takatorige, Health support activities in the City of Minamisouma, Saf. Sci. Rev. 2 (2012) 10–11.
[11] National Institute of Infectious Diseases, Ministry of Health, Labor and Welfare, The Great East Japan earthquake related tetanus, Infect. Wkly Rep. 14 (44) (2012) 12–16.
[12] Cabinet Office (in charge of disaster prevention), Report by the study group of supporting evacuation for those in need of support during the disaster, 2013.
[13] Cabinet Office, Ministry of Internal Affairs and Communication, MHWL, Study group of evacuation of those in need of support at the time of disasters, Guideline for Supporting Evacuation of Those in Need of Support at the Time of Disasters, 2006.
[14] T. Ogata, Role of health centers at times of nuclear disasters, J. Public Health Pract. 76 (12) (2012) 951–956.
[15] K. Sasahara, F. Kusano, T. Takatorige, Nuclear power plant accident and health center services—learning from the first nuclear power plant accident in Japan, J. Public Health Pract. 76 (12) (2012) 966–973.
[16] T. Takatorige, Actual activities and discussion of problems in dispatching public health nurses to the evacuation zone of a nuclear accident, Saf. Sci. Rev. 3 (2013) 85–98.

Chapter 6

Seismological Research and Earthquake Risk Assessment After the Great East Japan Earthquake

CHAPTER OUTLINE

6.1 IMPACT OF TOHOKU AREA PACIFIC OFFSHORE EARTHQUAKE ON SEISMOLOGY

The Tohoku Area Pacific Offshore Earthquake, which triggered the Great East Japan Earthquake, was a megathrust earthquake of magnitude 9 (M9), the biggest since the 1875 start of seismic observation in Japan. Globally rare M9 earthquakes have occurred five times since the 20th century. After the three huge earthquakes of Kamchatka (M9.0, 1952), Chilean

The Fukushima and Tohoku Disaster. https://doi.org/10.1016/B978-0-12-814078-9.00006-6

(M9.5, 1960), and Alaska (M9.2, 1964) broke out in the 1950s to 1960s, there were no such large ones until the Sumatra Andaman Earthquake (M9.1, 2004), a little over 40 years later. Then, in 2011, the Tohoku Area Pacific Offshore Earthquake hit Japan.

The 40 years of "no great earthquake" in the second half of the 20th century and the three M9 or larger quakes taking place in locations far from Japan had led most seismologists in Japan not to expect one near Japan, not until the Great East Japan Earthquake.

There were, however, opportunities to recognize the possibility of huge earthquakes near Japan even before the Great East Japan Earthquake. The Sumatra Andaman Earthquake in 2004 was an over M9 megathrust earthquake, the first in 40 years, and its location, the Indian Ocean, caught the eye. Up until then, M9 or larger earthquakes took place on the rim of the Pacific Ocean. The level smaller M8.5-class earthquakes have all occurred on the rim of the Pacific except for the 1950 M8.6 Assam-Tibet Earthquake. Uyeda and Kanamori were the first to point out that megathrust earthquakes take place in limited areas of the Pacific Rim such as Chile and Alaska [1]. Their hypothesis was based on the "plate tectonics" theory that had marked effects on earth science in the 1960s. The theory claimed that megathrust earthquakes take place in plate subduction zones and where they happen is strongly related to the oceanic plate age, thus whether they are likely or unlikely to happen depends on the area. This theory was well accepted by seismologists over the world and had made its way into educational textbooks.

Hypotheses in earth science take years to verify. Laboratory experiments can prove little and the verification requires firm evidence of past events and data when one such event takes place the next time. For the latter, especially, we have to wait for the phenomenon to occur. Even when we are lucky to see the event to collect observational data, the hypothesis may fail to explain the phenomena. Then we have to reconstruct another hypothesis or make major modifications.

It took a little over 25 years after the hypothesis to see the 2004 M9.1 megathrust earthquake in the Indian Ocean. This earthquake broke out on the plate subduction boundary; however, it was hard to explain with the hypothesis. After this Sumatra Andaman Earthquake, some researchers started to evaluate chances of earthquake occurrences at locations without records of historic megathrust earthquakes. This move was not intense enough to affect the disaster prevention administration in Japan.

The history of earthquake observation using recording machines is only about 150 years long. Ancient records of earthquake disasters in writing may go back some thousands of years. Earthquake history in Japan, with a number of historic records and earthquake activities, is only 1500 years long. This means that the earthquake records that mankind possesses are over a relatively short period with respect to the frequency of huge megathrust earthquakes. We can only rely on theoretical probabilities to forecast huge earthquakes and brace ourselves against disasters. We also need to collect observational data as much as possible with the next huge megathrust earthquake and pass these on to the next generation so they can prepare for them.

The Great East Japan Earthquake left unprecedented damage of over 19,000 dead and missing. Social factors that expanded the damage cannot be ignored, but the biggest factor was the huge magnitude of the earthquake itself. In this chapter, we will focus on the earthquake, its characteristics, and factors that made the damage grow. These findings were made through research that started immediately after the quake. Then, among the countermeasures in place after the quake, we will discuss the two points of change in the seismic origin model for disaster forecast and enhanced earthquake observation.

6.2 SEISMOLOGICAL CHARACTERISTICS OF THE GREAT EAST JAPAN EARTHQUAKE

6.2.1 Rupture Process of the Earthquake Source Fault

The earthquake source area of the Great East Japan Earthquake was as large as 350–500 km north to south and 150–200 km east to west. This earthquake source was offshore the Oshika peninsula and this was just where the destruction started and does not cover the whole earthquake source. The earthquake tremors and tsunami waves were generated from over a wide area rather than a point. The large tsunami waves along the coasts of Iwate and Fukushima prefectures started from their respective offshore areas where the faults slipped, and the strong earthquake shakings in Ibaraki and Tochigi prefectures were caused by strong earthquake movement at the southern end of the source. News often still describe the "earthquake source" as a point, perhaps because a quick calculation gives the point. In order to identify the wide source region and all the fault slips, we need to run a complex analysis with earthquake waveform data from a number of observation points.

The fault slips were found from earthquake waves recorded over the world and ground distortion (tectonic movement) reported at over 1000 Global Navigation Satellite System (GNSS, Global Positioning System (GPS)) recording points in Japan [2,3]. The observation data could only be explained by a 25–50 m fault slip at the Miyagi prefecture offshore location. Tohoku University and the Hydrographic and Oceanographic Department of the Japan Coast Guard had installed their ocean floor crustal movement observation machine before the earthquake and it also provided results of fault slip of several tens of meters, confirming the size of the slip [4,5].

It was the biggest fault slip ever recorded in the world and it probably was the biggest factor in the huge size of the tsunami waves induced by the Tohoku Area Pacific Offshore Earthquake. A number of hypotheses have been reported and the studies still continue. Many scientists point to "tsunami earthquakes" that cause huge tsunamis while the earthquake itself may not have been so significant, e.g., the 1896 Meiji Sanriku Tsunami, the 1946 Aleutian Islands Earthquake, and the 1992 Nicaragua Earthquake.

6.2.2 Strong Ground Motion

Records from the so-called strong-motion seismograph networks revealed that strong earthquake waves generated at five or so locations on the fault with large cycles that were sensible to humans and caused great effects on average height buildings [6]. Videos recorded in and around Miyagi Prefecture all showed two strong tremors that were about 1 minute apart. They were caused by two discharges of earthquake waves from two different locations. Videos recorded in Fukushima and Iwate prefectures did not show two distinct tremors. The two strong shakings in Miyagi Prefecture were released under the ground and decayed off before reaching the remote locations of Iwate and Fukushima.

Before the Great East Japan Earthquake, such locations that discharged strong tremor energy were called asperities. However, researchers used the word for different things and the definition was unclear. Since the Great East Japan Earthquake, these are called strong-motion generation areas (SMGAs).

The earthquake waves recorded in areas with major damage during the Great Hanshin Earthquake showed strong waves with a 1-second cycle that were later named "killer pulses." This cycle causes major damage to wooden buildings and it was the main factor for causing building damage during the Great Hanshin-Awaji Earthquake. The earthquake waves during the Great East Japan Earthquake, in contrast, did not record large waves with 1-second cycles but had large ones with smaller cycles. The Great East

Japan Earthquake and the Great Hanshin-Awaji Earthquake showed differences in terms of building damages; the former destroyed a smaller number of wooden buildings. The factor for this difference was the earthquake wave characteristics [7].

6.2.3 Movements of Multiple Faults

We started Section 6.2 with the description of the wide-area earthquake source with the Tohoku Area Pacific Offshore Earthquake. Hardly anyone expected an earthquake of this size. Parts of this fault region in the past had repeatedly caused earthquakes, collectively called "Miyagi prefecture offshore earthquakes." The Headquarters for Earthquake Research Promotion made long-term evaluations and estimated the probability of Miyagi prefecture offshore earthquakes taking place within the next 30 years at 99%; the highest within Japan. The estimated magnitude was M7.5 and even with multiple faults moving together, the estimate was M8. This estimate was about 1/30 of the actual strike of the Tohoku Area Pacific Offshore Earthquake at M9. Moreover, the source range went far beyond the expectation of moving faults in areas offshore Fukushima and Ibaraki prefectures as well. Furthermore, areas close to the trench usually do not generate strong earthquake waves; however, the crustal movement was a series of strong fault movements combined.

Combined fault movements were estimated to have taken place during the 1707 Hoei Earthquake or offshore Ecuador in the early 20th century. It is an important research topic in seismology. It was one of the National Research Projects, at the earlier time of The Earth Simulator, to run combined and uncombined simulations. Simulations, however, are largely affected by the initial assumption numbers; thus, in applying them to real case disaster prevention, the researchers have to cope with a large number of possibilities and varieties in the results.

6.2.4 Earthquake Prediction

Some interesting observations were made after the earthquake about its prediction. The earthquakes that occurred between the main earthquake and the M7.3 quake on March 9, i.e., 2 days before the main earthquake, were of relatively large scales. That means the parameter b in the Gutenberg-Richter law was small. Among past major earthquakes, many distinct preactivities had small b values. Those after March 9 turned the attention of some seismologists towards the seismic activities in the area [8].

GPS recorded an elevated count of electrons in the ionosphere above the earthquake source area during 60–40 minutes prior to the main earthquake

[9]. The mechanism that causes this phenomenon is unexplained; however, the same happened with many other M8.5 or larger quakes that caught the attention of seismologists.

Long-term crustal movement records with GNSS (GPS) in the Tohoku area showed accumulation of strain with strongly bonded underground plates. Some seismologists thought this bonding would peel off occasionally to slide slowly without causing earthquakes [10]. GNSS systems have accumulated records for only 20 years or so, which is a short period compared with the long intervals of earthquakes. Many explanations were possible about this data.

These earthquake predictions are the so-called postdictions and no warnings were issued before the earthquake. The Japanese earthquake prediction system has special measurements and observations made in Suruga Bay to issue warnings under the Act on Special Measures for Large-scale Earthquakes. Even if these preevent phenomena had been detected offshore of Tohoku, there were no routes for spreading warnings. Maybe someday in the future we'll be able to issue practical warnings before earthquakes, but as of now, there are postdictions with hardly any precursory symptoms. During the 1970s and 1980s, high expectations of observing signs even led to making a law, but later people lost interest in research on and detection of precursory symptoms. Some even say the study spreads optimistic ideas about earthquakes among the citizens such as "If predictable, there is no need to prepare against them," and the reputation of the study of earthquake prediction has lost ground after the Great East Japan Earthquake.

6.3 CHANGE IN EARTHQUAKE SOURCE MODEL FOR DAMAGE ESTIMATION

6.3.1 Standard Method of Damage Evaluation

The unprecedented damage from the Great East Japan Earthquake pushed administrations throughout the nation to review earthquake and tsunami damage estimations during these 5 years. Damage estimations lay the basics for quantitative disaster prevention measures; thus they were made for all levels of municipalities, prefectures, and the whole nation.

The following gives the standard procedure of damage estimation. First, an earthquake source fault model is assumed based on earth science knowledge. If an active fault forms an earthquake source fault for an inland epicentric earthquake, the length of the fault gives an estimate of the magnitude at which to set the earthquake source model. The process is the same for a trench-type earthquake. Parameters of the form of the subduction plate boundary, like estimated fault layer, magnitude and so on, set the fault model. The recent trend is then to hypothesize locations that generate large earthquake waves and make large slips for inhomogeneous earthquake source fault models.

Once the earthquake source model is set, physical formulae give the processes of earthquake wave propagation and tsunami to estimate local earthquake movements and tsunami heights. The calculated earthquake strengths and tsunami heights can predict damage to buildings and structures, fire breakout, damage to waterline, electricity, and roads. Death and injury counts can be estimated then from damage to buildings. The factors of the computation affect each other and damage simulation is made with appropriate equations based on past data, and so on.

The Great East Japan Earthquake, as we stated in the previous section, gave us new knowledge that a fault can slip over a length of several tens of meters along trenches. Damage estimations are in demand now that use the latest knowledge and are based on the maximum possible earthquake for our future preparations.

6.3.2 **Earthquake Level Estimate**

Announcements by many of the municipalities made after the Great East Japan Earthquake stated two levels of damage estimates, Level 1 and Level 2. Before then, most municipalities picked out one earthquake based on the possibility of occurrence and level of damage and announced damage estimates from that. Some announcements showed a "hazard map" of seismic intensity distribution and inundation depths. Some also estimated the number of deaths and amount of damage to the buildings in the municipalities.

The fact that the Great East Japan Earthquake was much larger than the expectations led municipalities to make two levels of damage estimations. It is now common to assume two earthquakes with different magnitudes for each expected earthquake.

During the 1995 Great Hanshin-Awaji Earthquake, structures such as bridges and buildings could not withstand the tremor and they collapsed to cause major damage. Engineers adopted the idea of setting two levels of earthquake expectations: Level 1 and Level 2. This two-level hazard evaluation for disaster prevention spread well in civil engineering and construction. A Level 1 earthquake is one that a structure is likely to encounter at least once during its life. The structure is required to stay almost unharmed against an earthquake of this level. In other words, a structure will not need a major repair and can still serve its purpose even after a Level 1 earthquake.

A Level 2 earthquake is a stronger one with maximum strength. A structure may collapse and turn unusable after a Level 2 earthquake; however, the destruction level shall stay at a degree not to take human lives. Structures that can stand earthquakes to a certain level, but with unpredictable behavior against those that go beyond that level, are not allowed. We may interpret the requirement means structures must sustain some level of endurance after they start to disintegrate until they are completely destroyed. People accepted this two-level evaluation well because it was easy to understand that "dying is not so easy" even with a bigger Level 2 earthquake.

Countermeasures against trough type megathrust earthquakes after the Great East Japan Earthquake also probed this idea of Level 1 and Level 2 earthquakes. However, its acceptance among the people was quite different. Level 1 and Level 2 events after the Great East Japan Earthquake were earthquake source models for countering tsunami attacks. Level 1, "tsunami with frequent occurrences," was modeled after large earthquakes with existing records within the past several hundreds of years, i.e., the earthquake that is "largest on record." Level 2 "maximum class tsunami," on the other hand, was modeled after a supermagnitude earthquake at the maximum level that can occur even without clear historic evidence of its occurrence, like the Tohoku Area Pacific Offshore Earthquake that caused the Great East Japan Earthquake.

The dynamics of earthquake occurrence, however, still remain unclear. Seismologists had started to point out the possibility of extremely large slips at shallow earthquake source faults along trenches that cause large tsunami waves at the time of the Sumatra Andaman Earthquake. This earthquake generated huge tsunami waves in the Indian Ocean around 2004. Such slips, however, were actually first recorded with the Great East Japan Earthquake using the ocean floor crustal movement observation system. The conditions and locations of how and where such large fault slips occur are unclear. At this time, there are only some hypotheses.

The Level 2 earthquake source fault model for the Nankai Trough Megathrust Earthquake was constructed with this new model, without much polishing up, and by applying an M9 class earthquake equivalent to the Great East Japan Earthquake source. This fault model gave results that were extremely high at some locations: e.g., the Cabinet Office calculated maximum tsunami heights of 34 m at Kuroshio in Kochi Prefecture and 33 m at Shimoda City in Shizuoka Prefecture.

Frequencies of tsunami attacks are orders of magnitudes smaller compared to those of earthquakes. Even the "frequent occurrences" with Level 1 attacks are once in over 100 years. Structures like buildings and bridges have life expectation of about 30–75 years; thus, even Level 1 tsunami attacks have cycles longer than those lives. Plate boundary earthquakes with magnitude 8 or larger only occur at cycles of 100 years or longer.

6.3.3 **How People Perceived "Level 2"**

Damage expectations made after the Great East Japan Earthquake resulted in high tsunami waves with even Level 1 modeling. Some Level 2 analyses led to unimaginable heights. The aforementioned 34 m for Kuroshio was one such example that often appeared in the news. The City of Shimoda in Shizuoka Prefecture relocated its city hall to higher land, but many people acknowledged the damage expectation results at "levels that we could no longer escape." Most communities had to face Level 1 results that far exceeded tsunami expectations made before the Great East Japan Earthquake. Their situations are quite different from post-Great Hanshin-Awaji Earthquake communities that reinforced bridges to prepare against Level 1 and Level 2 earthquake models.

In implementing earthquake and tsunami countermeasures, engineers in charge of new or reinforcing construction of buildings and bridges are given the conditions of earthquake strengths or tsunami heights to counter and come up with the best designs under those constraints. Level 1 and Level 2 models defined after the Great Hanshin-Awaji Earthquake were seismic movement, and structural design engineers would directly access the conditions. In determining seismic movement, the evaluation takes into account earth scientific conditions such as which earthquake source fault to use or how to evaluate inhomogeneity of the slip. However, these processes are for the small number of experts, and many engineers and citizens are not concerned about this part. Level 1 and Level 2 seismic movements are given conditions for the engineers to clear. Engineering outcomes are new buildings or reinforcements that withstand these conditions. The customers, administration or citizens, pay for these buildings and structures with a high level of aseismic strengths. Once in their hands, the customers have "aseismic capabilities" without having to take special action at the time of earthquakes.

Necessary external conditions for setting tsunami countermeasures are maximum wave heights and waveform transitions that are calculated from Level 1 and Level 2 earthquakes. This mechanism of evaluating conditions for the designers to counter were about the same for post-Great Hanshin-Awaji Earthquake and post-Great East Japan Earthquake. In the case of tsunami expectations, however, detailed information as to "how Level 1 and Level 2 tsunami were calculated with which earthquake source model" was made available at the same time. The Disaster Management Office of the Cabinet Office announced the calculation results of the tsunami fault model (equivalent to Level 2) for the Nankai Trough Megathrust Earthquake by expecting 11 different fault cases. Having 11 different calculation results for the single earthquake of Nankai Trough explains the complexity of the process. Each prefectural administration typically took three or so results that seemed to mostly affect that prefecture from the 11 results and found the maximum value for each point of interest along the coast. These processes were all made public and then the engineers and citizens became conscious of the uncertainties in the earthquake source models. People started to look at the variance in fault models, thinking, "Maybe other fault models may result in higher tsunami waves," or "Some of the many fault models may lead to tsunami heights that are no so high." People took Level 1 and Level 2 to mean different fault models and engineers, administration and citizens became overly concerned about fault models. The people ended up spending too much time looking in the wrong direction instead of planning tsunami countermeasures at each location for given tsunami heights and speed.

"Enhancing sustainability" to counter Level 2 tsunamis means building seawalls that do not collapse upon their attacks to gain time to evacuate and build the necessary number of evacuation facilities. These preparations require people to evacuate when an earthquake hits; thus, even after the preparation is complete, people have to be aware of tsunami threats. Citizens therefore have to take the active measures of fleeing upon an earthquake, even after the counter-tsunami construction is over. This leads some people to think that there is not much difference whether with or without the tsunami countermeasure.

Tsunami countermeasures of Level 1 and Level 2 are defined as a difference in earthquake sizes and this piece of information is shared with the citizens. This is a large step forward such that an extremely large earthquake of Level 2 is now an "expectation." Building countermeasures against damage from a Level 2 earthquake, however, requires a big budget, years of construction, and the understanding of life in a town surrounded by huge structures. Constructing these countermeasures against Level 2 earthquakes needs understanding and agreement by the citizens, and when we think about their frequency of occurrence we have to continue the efforts of these long-term projects.

6.4 ENHANCING OCEAN FLOOR EARTHQUAKE MONITORING AND ITS SIGNIFICANCE

6.4.1 Importance of Earthquake Monitoring in the Ocean

Enhancing earthquake and tsunami monitoring in the ocean is a policy being rapidly implemented.

Most large-scale earthquakes in Japan start in the ocean area, as was the case with the Great East Japan Earthquake. It is thus effective to expand the monitoring network over the oceanic area. The network will probably allow quicker emergency earthquake warnings.

Earthquake monitoring in the ocean means at the ocean bed, not at the surface. The ocean surface is sensitive to weather conditions and waves, and thus is not suited for earthquake monitoring. In addition, there are physical limitations of earthquake waves being slow in transferring through ocean water; especially S-waves do not travel through water. On top of that, there is an obstacle in that radio waves do not transmit through water and we cannot depend on underwater wireless data transmission. High-speed transfer of large data in real time takes routing ocean floor cables to the measurement points.

If we set "water pressure gages" in addition to ocean-bed seismographs, they will record the change in ocean surface levels to detect tsunami waves. The ocean-bed earthquake monitoring network for Japan has installed these measurement points as tsunami gages since the late 20th century. During the Great East Japan Earthquake, ocean-bed tsunami gages, offshore of Kamaishi, of the Earthquake Research Institute of the University of Tokyo started making records of the tsunami waves immediately after the earthquake and very unusual waveforms were recorded from 12 minutes after the quake [11].

Major development of the ocean-bed earthquake and tsunami measurement network began because the Great East Japan Earthquake taught us that our earthquake and tsunami measurements before then were insufficient and real-time measurement data, like that recorded offshore of Kamaishi, can clearly contribute to disaster prevention. Among the many ocean areas nominated for the measurement network development, installation started in 2011 in the area along the Japan Trench, where the Tohoku Area Pacific Offshore Earthquake started.

6.4.2 Dense Seafloor Observation Network for Earthquakes and Tsunamis Along Japan Trench

After a major earthquake attack, the next big one will probably take place in tens, hundreds, or thousands of years later. Placing a monitoring network on the same spot immediately after a major earthquake may look unreasonable, but there is a high probability of a major one in a nearby location and targeting such an earthquake is rational.

The Pacific Plate on the east side of the source area of the Tohoku Area Pacific Offshore Earthquake had its interplate strain relieved by the earthquake, and the pressure inside the Pacific Plate weakened. This is a condition for a "normal fault" earthquake that splits the plate itself to break out. This type of earthquake is called an "outer rise earthquake" known to follow major earthquakes in numbers. Predicting their occurrence is difficult because the timing of their start in past cases ranges from immediately to decades after the main earthquake. What we can tell is that the possibility of M8 class outer rise earthquakes on the Pacific side of the Tohoku area is higher now.

The Tohoku Area Pacific Offshore Earthquake raised the possibility of major earthquakes in other places. These places are plate boundaries adjacent to the earthquake source fault of the main earthquake. They are the Pacific coastline along Ibaraki and Chiba prefectures in the south and the Pacific coastline along the northern area of Iwate and Aomori prefectures in the north. These areas have higher possibilities of earthquakes than before the Tohoku Area Pacific Offshore Earthquake.

The Ministry of Education, Culture, Sports, Science and Technology (MEXT) started the "Seafloor Observation Network for Earthquakes and Tsunamis along the Japan Trench" project in 2011. The plan places 150 monitoring points in the ocean from offshore Hokkaido to offshore Boso. The observation network consists of six regional subsystems. Each subsystem has a single optical submarine cable laid out unicursally along the seabed. The cable has, on the average, 25 monitoring devices with seismographs and other sensors for every 20–30 km. The two ends of the cable climb to the land at different locations where they connect to network cables and power supplies. The total system length reaches as long as about 5700 km.

The landing stations are at the following six locations: (1) Minamiboso City in Chiba Prefecture, (2) Kashima City in Ibaraki Prefecture, (3) Watari-cho in Miyagi Prefecture, (4) Miyayo City in Iwate Prefecture, (5) Hachinohe City in Aomori Prefecture, and (6) Hamanaka-cho in Hokkaido. Each subsystem connects (1) and (2), (2) and (3), and so on. The subsystem that connects (1) and (6) is a straight line lying in the east of the Japan Trench.

Once in place, an undersea seismograph is difficult to repair if it fails. Its repair takes a special ship to pull the cable up to recover the seismograph, repair it or replace it with a new one and sink the device with the cable again. The cost for this repair operation will be significant. The Seafloor Observation Network for Earthquakes and Tsunamis along the Japan Trench has a number of smart features and it appears that it can provide stable long-term monitoring capabilities.

Each monitoring device receives power from the land stations at the two ends and transfers data towards the two stations. This arrangement lowers the possibility of data loss caused by cable breaks and power outages. The land stations are strategically located on highland in preparation against tsunami disasters. For physically protecting the optical submarine cables, divers placed protection tubes on them and buried them underneath seabed deposits where the ocean was 20 m or shallower. Remotely controlled machines routed the cables along deeper locations down to depths of 1500 m.

The observation network is expected to meet three tasks. One is highly accurate immediate tsunami prediction. Regular tsunami warning works by analyzing fast transfer earthquake wave data and issuing warnings against tsunami attacks that transfer slower than earthquakes. Once the observation network operation starts, it will directly detect tsunami swells on the spot and there will be no need for analyzing earthquake waves to make tsunami predictions; thus, the accuracy is higher. The direct information of "sensing tsunami swells" will also increase the number of people that evacuate after receiving the information. The second task is accurate and faster emergency earthquake alerts. Having seismographs near earthquake sources will detect earthquake waves quicker than before. The third task is scientific advancement in research about the process of occurrence of major earthquakes. It is a dense monitoring network placed over an earthquake source region within only a few years after a major earthquake attack. Such an arrangement was made for the first time in the world. Monitoring the aftershocks that are still taking place will reveal detailed mechanisms of earthquake wave propagation, i.e., physical properties of bedrocks, changes in the sticking between plates, and so on. Advancements are expected in research for clarifying the mechanism of earthquake occurrence.

The Nankai Trough is a location where major earthquakes are expected to break out. There, another type of ocean bed earthquake monitoring network is shaping up. The Japan Agency for Marine-Earth Science and Technology (JAMSTEC) started installing this monitoring network, Dense Oceanfloor Network system for Earthquakes and Tsunamis (DONET), in 2006 from the offshore Kii Peninsula and completed the installation in 2011 to start the monitoring. Now a similar monitoring network, DONET2, is underway off the shore of Cape Muroto in Shikoku.

6.4.3 Improving Earthquake Monitoring Networks After Huge Disasters in Japan

The "Seafloor Observation Network for Earthquakes and Tsunamis along the Japan Trench" is a giant project with an initial budget of JPY32.3 billion. Such a large project is unlikely for earthquake disaster prevention under normal circumstances, and it was pushed by the Great East Japan Earthquake. Monitoring networks have expanded following great disasters in Japan and we can say that the networks in Japan have shaped up after a number of earthquake disasters.

There are over 5000 earthquake monitoring points in constant operation in Japan. The density of earthquake monitoring points is one of the greatest in the world. If we look at the distribution map of earthquake monitoring points in Japan, the lowland areas are tightly packed with seismographs without gaps. The Seafloor Observation Network for Earthquakes and Tsunamis along the Japan Trench will cover the untouched mountain and ocean areas as earthquake monitoring in Japan has targeted. The optical cable ocean-bed seismograph systems started in the 1990s with a small one placed offshore to the east of the Izu Peninsula in Shizuoka Prefecture and gradually grew bigger and in a stable manner. The start of the big project was not an abrupt one.

The earthquake monitoring network developed as a national project boosted at times of big disasters and the number of earthquake monitoring points increased rapidly within a few years following each disaster. The "tsunami earthquake early detection network," a basic system for Japan Meteorological Agency (JMA), for example, started to take shape after the 1993 southwest offshore Hokkaido earthquake. From 1994 to 1996, about 180 monitoring points were set over the nation for constant connection with the JMA headquarters with the central data processing function to complete the modern earthquake monitoring network. Before then, earthquake monitoring occurred at meteorological stations and observatories in urban areas and suffered from noise from human activities. The tsunami earthquake detection network moved the earthquake monitoring spots from the city to remote mountains. The move knocked down the artificial noise and high-quality earthquake monitoring records have been gathered at central recording devices.

The number of earthquake monitoring spots grew into the thousands after the 1995 Great Hanshin-Awaji Earthquake. The extremely strong shaking with the Great Hanshin-Awaji Earthquake destroyed a large number of buildings; however, there were barely any records of the earthquake waveforms. Seismographs that can record strong tremors are called "strong motion seismographs" and there were only about 100 of them in the nation installed at meteorological stations and observatories. Some railway, electricity, and gas companies had some, however, at locations where they needed the monitoring. Given these conditions, the National Research Institute for Earth Science and Disaster Resilience (NIED) set about 1000 strong motion seismographs over the nation to form an unmanned network called K-NET.

Seismic intensity during an earthquake is an important piece of information that determines the initial action by administrative bodies; thus, the Fire and

Disaster Management Agency (FDMA) supported municipalities over the nation to install seismographs with strong motion–recording capabilities. At least one such seismograph was installed for each municipality and most of them had wave-recording functions. This move placed at least one seismograph in each municipality before the big trend of municipality mergers in the Heisei era. JMA also decided to add more strong motion seismographs to places other than meteorological stations and observatories and within a few years, starting from 1995, the number of strong motion seismographs in the nation reached over 4000.

High sensitivity seismographs that monitor small earthquakes that are insensible to human also started to spread. Monitoring extremely small earthquakes was a regional activity for university research centers up until then. NIED arranged a standardized monitoring network called Hi-net with about 800 monitoring points over the nation, except for Okinawa and remote islands. The data are not only for research but they are fed in real time to JMA's earthquake monitoring process for early emergency earthquake warning.

Measurement data from Hi-net and K-NET, arranged after the Great Hanshin-Awaji Earthquake, are now made available to the public via the Internet and this way of utilizing data is a new feature of the system. Seismographs with standardized performance record wave data in standardized format. The data are made available freely via the internet to researchers in the nation and over the world, after some simple user registration. These new data led to the discovery of "deep low-frequency earthquakes" that have a strong relation to the occurrence process of major earthquakes. The discovery quickly went around the globe and its research is now a hot topic all over the world.

Earthquakes originate inside the earth's crust where different rocks have uneven characteristics, and such differences in the structure play important factors in conditions for earthquake occurrence. Hi-net monitoring data are also used for research to unveil the complex underground structure of the Japan Islands.

New monitoring networks provide data that were unavailable before, and scientists have their scientific curiosity aroused to see new discoveries. The outcome, to produce any practical use at the citizen level, takes years and the number of researchers and engineers to bridge the gap is small. Hi-net has generated attention in the scientific research field, but it has not yet produced much to make visible changes to our disaster prevention. Seafloor Observation Network for Earthquakes and Tsunamis along the Japan Trench is expected to lead to not just scientific findings but also new results that have a practical impact on disaster prevention for the citizens and administration. With the revolutionary advancements in and overwhelming spread of IT technology, immediate and automatic information supply is most likely to have big effects. Alerts about disaster prevention are limited by the Meteorological Service Act and the same alerting system will probably be put in practice over the nation by JMA. To make full use of dense monitoring by the Seafloor Observation Network for Earthquakes and Tsunamis along the Japan Trench, high expectations are there for special regional services. Having hands-on raw monitoring data in regions with added service to urge evacuation will lead to high value for NIED with the best earthquake and tsunami monitoring network in Japan and one of the best in the world.

6.5 MATTERS THAT HAVE IMMEDIATE EFFECTS AND FUTURE EFFECTS

Disaster prevention is a matter of human life, so people expect immediate effects. This point is very much different for seismologists as compared to scientists from other scientific research fields, such as space science or quantum physics.

Earthquake disaster prevention has many results that can immediately and effectively reduce disasters. One such example is reinforcement or replacement of inadequate buildings or structures, such as wooden housing. Immediate spread of these counteractions will have a great effect in disaster prevention.

On the other hand, the science of earthquakes still has many unknowns. In the 1960s, with the rise of space exploration, Project Mohole was dreamed up to drill a hole through the earth to reach the mantle; however, it still has not yet reached the mantle. In the field of space exploration the interplanetary space probe New Horizon reached Pluto, but in our studies of the earth to discover where earthquakes originate, we have not made great advances.

In terms of the mechanisms of how earthquakes occur, the phenomena have not yet been clarified. There will always be events that go over our heads, as we take measures based on the stack of knowledge we have gained so far. When the next major earthquake hits, our efforts to collect as much continuous monitoring data before the breakout as possible and to pass the data to our next generation will surely prove useful for disaster prevention in the future.

The knowledge we have now about earthquakes and that we take for granted came about through detailed analyses of earthquake waves after the 1960s, which is very recent. As earthquake wave analysis advanced, seismologists restudied past earthquakes that took place decades ago and piled up evidence on which to base much of what we now know. For disaster prevention in the future, we shall keep "records of raw data" without the filter of our contemporary sense of value. We have to reach our conclusions with the best analysis available at the time, but at the same time pass the "records of raw data" to the future. That data may contain important signals that we cannot detect with our current technology and knowledge.

REFERENCES

[1] S. Uyeda, H. Kanamori, Back-arc opening and the mode of subduction, J. Geophys. Res. 84 (B3) (1979) 1049–12061.

[2] S. Ide, A. Baltay, G.C. Beroza, Shallow dynamic overshoot and energetic deep rupture in the 2011 Mw 9.0 Tohoku-Oki earthquake, Science 332 (2011) 1426–1429, https://doi.org/10.1126/science.1207020.

[3] S. Ozawa, T. Nishimura, H. Suito, T. Kobayashi, M. Tobita, T. Imakiire, Coseismic and postseismic slip of the 2011 magnitude-9 Tohoku-Oki earthquake, Nature 475 (2011) 373–376, https://doi.org/10.1038/nature10227.

[4] Y. Ito, T. Tsuji, Y. Osada, M. Kido, D. Inazu, Y. Hayashi, H. Tsushima, R. Hino, H. Fujimoto, Frontal wedge deformation near the source region of the 2011 Tohoku-Oki earthquake, Geophys. Res. Lett. 38 (2011), https://doi.org/10.1029/2011GL048355. L00G05.

[5] M. Sato, T. Ishikawa, N. Ujihara, S. Yoshida, M. Fujita, M. Mochizuki, A. Asada, Displacement above the hypocenter of the 2011 Tohoku-oki earthquake, Science 332 (2011) 1395, https://doi.org/10.1126/science.1207401.

[6] S. Kurahashi, K. Irikura, Source model for generating strong ground motions during the 2011 off the Pacific coast of Tohoku Earthquake, Earth Planets Space 63 (2011) 571–576.

[7] Report on the Great East Japan Earthquake, General volume, Editing Committee of the Great East Japan Earthquake, Maruzen, 2014, p. 233.

[8] K. Oike, Huge Earthquakes of the Japanese Islands, Iwanami, Tokyo, 2011. p. 128.

[9] K. Heki, Ionospheric electron enhancement preceding the 2011 Tohoku-Oki earthquake, Geophys. Res. Lett. 38 (2011), https://doi.org/10.1029/2011GL047908. L17312.

[10] I. Kawasaki, What a Slow Earthquake Is, NHK, Tokyo, 2006. p. 269.

[11] K. Satake, Mechanism of huge tsunami waves, in: N. Hirata, K. Satake, K. Meguro, Y. Hatamura (Eds.), Huge Earthquakes and Huge Tsunami-Review of the Great East Japan Earthquake, Asakura Publishing, Tokyo, 2011, p. 196 (Chapter 2).

FURTHER READING

[1] The Headquarters for Earthquake Research Promotion, Long-term evaluation of Miyagi prefecture offshore earthquake, Nov. 27, 2000. http://jishin.go.jp/main/chousa/kaikou_pdf/miyagi.pdf.

Chapter 7

Concerns and New Measures for Tsunami Disaster Mitigation in Preparation for the Next Major Earthquake

CHAPTER OUTLINE

7.1 GREAT EAST JAPAN EARTHQUAKE AND NANKAI TROUGH MEGATHRUST EARTHQUAKE DISASTER

The earth's surface is covered with more than 10 large tectonic plates. Where the plates meet, the earth's internal energy breaks out to form spectacular scenery, e.g., mountains like the Himalayas or arc-like island chains of the West Indies in the Caribbean Sea. The enormous energy, however, sometimes produces earthquakes or volcanic eruptions that cause damage to human civilization.

Japan is at a unique location where four of the large tectonic plates meet: they are the Pacific Plate, the Philippine Sea Plate, the Eurasian Plate, and the North American Plate. The collision of these plates formed the rich natural environment of the Japan Islands but, at the same time, is the source of the large numbers of earthquakes and volcanic eruptions that we suffer from.

The southern boundary where the Pacific and North American Plates meet shaped the Japan Trench, where magnitude 8 level earthquakes have been

The Fukushima and Tohoku Disaster. https://doi.org/10.1016/B978-0-12-814078-9.00007-8

breaking out in an approximately 100-year cycle. At 14:49 on March 11, 2011, the Off the Pacific coast of Tohoku Earthquake occurred. Its magnitude of 9.0 far exceeded expectations. The tsunami induced by this quake reached all the way from Hokkaido to Kyushu along the Pacific coasts, and in the Tohoku area the huge tsunami run-up height reached almost 40 m. The earthquake and its tsunami left over 18,000 dead or missing and the damage reached as high as JPY 16.9 trillion [1].

Other plate boundaries near Japan are the Chishima Trench where the Pacific and North American Plates meet, the Nankai Trough where the Philippine Sea and Eurasian Plates meet, and the eastern margin of the Sea of Japan where the Eurasian and North American Plates meet. People are concerned that earthquakes in these areas can cause tsunamis in the ocean area above them. In particular, at the Nankai Trough a magnitude 9 class megathrust earthquake is expected and the Central Disaster Management Council has estimated the potential damage at 230,000 deaths and JPY214 trillion at the maximum [2,3].

The prefectures of Iwate, Miyagi, and Fukushima suffered the most physical damage and the major cities of Tokyo, Nagoya, and Osaka had relatively small damage in the 2011 earthquake. The Nankai Trough Megathrust Earthquake, however, will damage these big cities, hampering the rescue activities and recovery actions that follow. The Great East Japan Earthquake, in addition, clearly showed how a major disaster over a wide area turns into a compound disaster. Fires broke out in the cities of Kesennuma and Natori. We already had repeated experiences of tsunamis compounded with fires on Okushiri Island with the 1993 southwest-off Hokkaido earthquake, in Niigata City with the 1964 Niigata earthquake, and in Kamaishi City with the 1933 Showa Sanriku earthquake. Furthermore, the Great East Japan Earthquake invoked a tsunami that inundated the Fukushima Daiichi Nuclear Power Plant (Fukushima-1 NPP), which lost power and had a core meltdown. It was a major accident with radioactivity release; a compound disaster of tsunami and a nuclear plant accident is an event that should never happen. The coasts of Japan have, in addition to nuclear power plants, a large number of important facilities such as oil storage stations and petrochemical complexes. If they experience an accident, it would spread the disaster over a wide area for an extended time. Many coasts in large cities have areas that are below sea level and these areas extend along the rivers. Strong tremors can cause liquefaction, which brings down or damages levees and if tsunami waves attack such areas, that would be a compound disaster of earthquake and tsunami. Preparations against different types of compound disasters are absolutely necessary in preparing for the Nankai Trough Megathrust Earthquake.

We, unfortunately, cannot bring the damage down to zero with huge earthquakes like the Nankai Trough; however, we can reduce the damage. The Central Disaster Management Council, which made the preceding estimates, also pointed out that proper disaster management measures can bring down the death toll to a fifth, or 46,000, and halve the damage to about JPY112 trillion. In order to mitigate this damage, we have to counter the problems we experienced with our past disaster mitigation plans, uncovered by the Great East Japan Earthquake, and implement the lessons learned in our disaster countermeasures.

About 90% of the human suffering during the Great East Japan Earthquake was caused by tsunami waves [4]. Tsunami waves induced by the Nankai Trough Megathrust Earthquake will attack a wide range in Japan, including the metropolitan areas of Tokyo, Nagoya, and Osaka. This means huge earthquakes at plate boundaries will cause major damage by tsunamis. In the following sections, we will review the tsunami damage caused by the Great East Japan Earthquake, and discuss problems with our past tsunami preparation. We will then introduce our new tsunami preparation, which has already started. Fig. 7.1 summarizes the problems and new solutions in tsunami preparation.

7.2 TSUNAMI DAMAGE FROM THE GREAT EAST JAPAN EARTHQUAKE

The Sanriku offshore area to offshore of Boso have histories of repeated tsunami damage caused by large earthquakes near the Japan Trough. From the past earthquake occurrence patterns, this ocean area has been divided into eight areas of "Sanriku offshore north" facing Aomori and the northern part of Iwate Prefecture, "Sanriku offshore central" facing the southern part of Iwate Prefecture, "Miyagi offshore" facing Miyagi Prefecture, the further offshore area of "Sanriku offshore south near the trench," "Fukushima offshore" and "Ibaraki offshore" facing Fukushima and Ibaraki prefectures, "Boso offshore" facing Chiba Prefecture, and the area along the trench "Sanriku north to Boso offshore near the trench." Among these areas, "Sanriku offshore central" and "Ibaraki offshore" were expected to possibly rupture together, but other areas were expected to have only single fault ruptures. The earthquake source of the 2011 Off the Pacific coast of Tohoku Earthquake, however, caused ruptures over a wide range of six ocean areas, including "Sanriku offshore central" to "Ibaraki offshore" in the north-south direction and reaching the Sea of Japan in the east-west direction. In other words, a huge earthquake with multiple ruptures far exceeding expectations broke out.

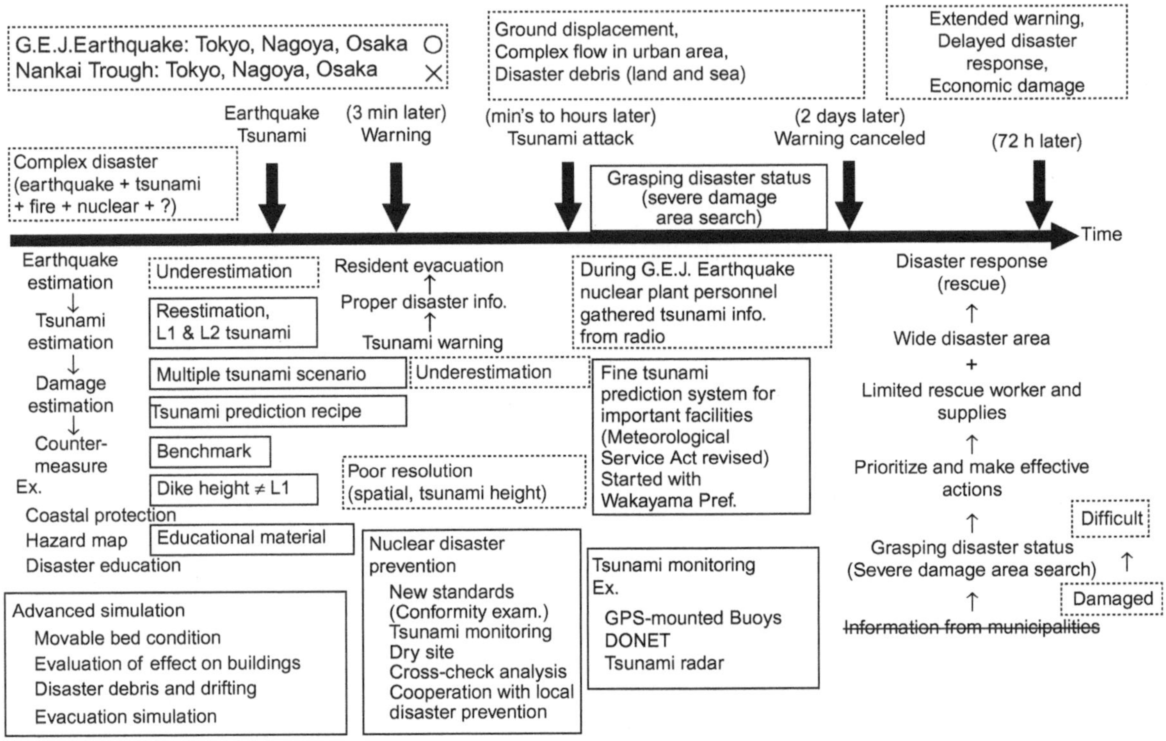

■ **FIG. 7.1** Problems in tsunami preparation identified with the Great East Japan Earthquake and our new solutions.

Fault motion under the ocean lifts or sinks the seabed. The 2011 Off the Pacific coast of Tohoku Earthquake had a maximum displacement of about 50 m. The fault motion lasted for several minutes; however, the slow-moving seawater shifted its surface over a wide range in response to the displacement at the bottom. This phenomenon is the tsunami wave source that causes tsunami waves. The initial condition of a tsunami is this fault motion and, when the earthquake is underestimated, so are the following tsunami waves.

Underestimation of earthquakes affects the tsunami alert for tsunami disaster mitigation. Tsunami warning systems are in effect over the world in many countries along oceans, including the United States, Indonesia, France, and Chile. The tsunami warning system in Japan has the highest level of technology. The Japanese system has a database of results of tsunami simulations. Once an earthquake breaks out, the system analyzes the seismic waves to find the hypocenter and magnitude of the earthquake and then searches for the closest earthquake in the database. This method allows a tsunami warning to be announced within 3 minutes from the start of the earthquake, and the system has been proven effective with magnitude 8 or less earthquakes, which are frequent in the ocean area near Japan. Unfortunately, the system carries the danger of underestimating some types of earthquakes. One such type is a huge earthquake like the 2011 Off the Pacific coast of Tohoku Earthquake. When a huge magnitude 9 class earthquake strikes, the immediate magnitude estimate from the system is too small and the searched-for tsunami is also underestimated.

The 2011 Off the Pacific coast of Tohoku Earthquake struck at 14:46, and the major tsunami warning was issued 3 minutes later. Such a quick response is only possible from the Japan Meteorological Agency (JMA). JMA, however, estimated the earthquake magnitude at 7.9. The tsunami height announced then was 6 m in Miyagi Prefecture, and 3 m in Iwate and Fukushima Prefectures. JMA quickly recognized the underestimation and raised the tsunami heights with its second and third announcements. However, many regions had power outages after the first announcement, which kept the information from spreading.

This event showed that precise tsunami source estimation at the time of occurrence is technically difficult and that this problem lowered accuracy of the disaster mitigation information and ended up magnifying the damage level. In preparation for the next huge disaster, we must solve this technical issue.

The preparatory estimation that was too small affected the hardware of the coastal levees. Coastlines of Japan have levees to protect against tsunamis, high tides and high waves. Rivers have levees to protect against inundation as well and they are effective against tsunamis that work their way up rivers. These levees have firm structures with low possibility of damage from the wave force. The wave breakers at the levee crown (upper end of levee) sometimes are destroyed but, even in that case, the levee itself remains solid. If, however, the tsunami waves go over the levee, a different mechanism of destruction takes place. The forward wave that goes over the crown may wash away the concrete or rocks that cover the back slope or scour the ground at the bottom of the back slope. Then the soil inside the levee is washed out and eventually the levee breaks. The same thing happens on the front side with the retracting wave. Damage to the levee thus depends on whether the tsunami goes over the crown or not.

Now, what determines the crown height? Normally authorities assume tsunami wave heights with a high probability of attack in the areas and build the levees based on these wave heights, using them to determine the crown heights. Iwate Prefecture determined the levee crown height based on the

1896 Meiji Sanriku tsunami, the 1933 Showa Sanriku tsunami and the 1960 Chilean Earthquake tsunami inundation depths. The northern part of Miyagi Prefecture had its levee height determined based on the 1960 Chilean Earthquake tsunami. They expected the same level of tsunami attack as the largest on record. The southern parts of Miyagi Prefecture and Fukushima Prefecture had higher storm surge attacks than tsunamis, so their crown heights were based on storm surges. Nevertheless, the tsunami waves from the 2011 Off the Pacific coast of Tohoku Earthquake were higher than any of the expected tsunamis or storm surges. The tsunami attacked and destroyed many of the coastal ocean levees. Of the approximate total length of 300 km of levees and seawalls in the prefectures of Iwate, Miyagi, and Fukushima, about 190 km were destroyed [5]. The tsunami waves that overtopped and destroyed the levees entered the city area to swallow the houses and the people.

Local communities in Tohoku along the coast had tsunami hazard maps and disaster management and evacuation drills had been conducted. The Sanriku area, with its frequent tsunami attacks, probably had more residents, compared to other regions, that had checked the tsunami hazard map to see expected inundation zones and shelters. These tsunami hazard maps, like the levee crown heights, were also based on the largest recorded tsunamis. The actual tsunami that attacked these regions inundated wider areas and even shelters on the hazard maps. For example, in the Shizugawa area of Sanriku-cho, Miyagi Prefecture, water reached twice as far inland and even a shelter building hospital was attacked by the tsunami and not just evacuees but even the inpatients were killed.

This discussion reveals the importance and difficulty of estimation. Basing estimates on the largest on record has its justification, because it is assuming a cyclic behavior of natural phenomena. This reasoning, however, has a necessary condition: i.e., that the largest on record is the largest ever. The earthquake that we believed was the largest ever was not. We have to seriously look back at our mistake of looking only at recent earthquakes and underestimating the earthquake risk in the area.

In addition to underestimating the earthquake and tsunami, the Great East Japan Earthquake brought the limitations of our rescue and support activities to the surface. Once the tsunami warning was canceled, rescue and support activities started in the inundated area. The damaged area with this earthquake was extremely widespread and the number of people to help and the relief supplies were limited. This means the people had to prioritize areas for rescue and support activities. While there were some areas that needed immediate help in order to avoid a large number of deaths, others had grave difficulties but could wait for the support activities. Prioritizing requires an understanding of the damage situations in the different areas, and this is normally accomplished by prefectures or the national administration collecting reports from municipalities. The Great East Japan Earthquake destroyed many administrative buildings and many workers in them were killed or injured, so the damage reports were not available immediately. This means, in general, that areas having the most severe damage cannot be determined quickly in times of large-scale disasters. The rate of survival rapidly drops 72 hours after the disaster. The rescue activities that take place after cancelation of the tsunami warning and sooner than 72 hours after the strike make a big difference in the count of human damage. To reduce the amount of human damage, we need to develop a way of finding the severely hit areas in large-scale disasters.

Currently cancelation of tsunami warnings tends to be on the conservative side, so warnings tend to last longer with the larger tsunami disasters. With the Great East Japan Earthquake, the delay in the earthquake size estimate and the damage to tide monitoring stations inhibited the estimation of the tsunami decay and the tsunami warning was in effect for as long as 2 days. Longer tsunami warnings do not just delay the disaster response but also expand economic loss. We need the technology to estimate the optimum warning cancelation period. From an economic point of view, quick recovery and reconstruction are also important. The Great East Japan Earthquake, however, left huge mounds of disaster debris that hindered recovery and reconstruction. The debris was not just piled up on land but also great quantities of debris were in the ocean, carried by retracting waves. This disaster debris deposited within the ship traveling routes delayed the ships traveling to the ports, and that in the pier areas obstructed the ships from docking at the piers, further delaying rescue activities. We need to prepare for this disaster debris and have methods to handle it in the local disaster management plans.

Lastly, concerning tsunami preparation at nuclear power plants, problems occurred in both prior estimations and status monitoring after the earthquake. The big problem with the former was that the estimations were made based on the largest on record. With the latter, there was too much emphasis on earthquake monitoring and the plant was not well prepared to monitor tsunami waves. The plant, therefore, did not have a clear understanding of the scale of the tsunami attack for action management. When we conducted a hearing involving a worker at the plant, to our question of how he received information on the tsunami, the answer was from the JMA tsunami announcements on the radio. The JMA tsunami announcements were targeted for a wide area, so the information was over a coarse grid.

An important facility like the plant, which can cause great secondary damage once an accident breaks out, only having information-gathering methods similar to those used by the general public indicated a major problem.

So far, we have discussed the tsunami damage during the Great East Japan Earthquake. In summary, we have the following issues in tsunami disaster management:

(1) Underestimation of prior earthquake and tsunami
(2) Underestimation of tsunami warnings
(3) Delay in understanding the damage caused
(4) Delay in canceling the tsunami warning
(5) Delay in rescue actions caused by disaster debris
(6) Poor tsunami information at important facilities

7.3 NEW TSUNAMI DISASTER PREPARATIONS FOR THE NEXT MAJOR EARTHQUAKE DISASTER

The following are needed to counter the problems we had with tsunami preparation listed in the former section.

(1) Prior earthquake and tsunami estimation from multiple scenarios
(2) Prevent underestimation of tsunami warnings and improve their reliability
(3) Find severely damaged areas within 24 hours
(4) Adequate tsunami warning cancelation and estimating the cancelation time
(5) Ways of finding disaster debris in the ports within 48 hours
(6) Dense grid tsunami forecast for important facilities

Takahashi discussed items (2) and (6) in Ref. [6]. A number of research organizations and researchers are onto (3), (4), and (5) and some research results are being reported. Thus in the following, we will introduce related developments concerning (1) and the new tsunami estimations that began after the Great East Japan Earthquake.

The earthquake estimation before the Great East Japan Earthquake was too small and based on this earthquake estimation the tsunami was also underestimated. These estimations directly affect the levee crown heights and hazard maps that play important roles in regional disaster management. They were factors in increasing human damage in the Great East Japan Earthquake.

Since the Great East Japan Earthquake, most tsunami estimations have been made in two levels: The level 1 tsunami, which occurs at the frequency of once every 100–150 years, and the level 2 tsunami, which is maximum scale with low frequency of occurrence but one that can physically break out. Tsunami level 1 is also called the "tsunami protection level" and is the level that the administration aims at building hardware, such as levees, to protect against loss of life and property of the citizens. Tsunami level 2, on the other hand, is called the "tsunami disaster mitigation level," which will inundate the land, but people's lives can be saved by total disaster mitigation plans with hazard maps and evacuation routes. Ever since the Great East Japan Earthquake, many municipalities have employed this bilevel estimation and set the former expected tsunamis at level 1. Setting a level 2 tsunami is difficult for the municipalities and they often reply on expectations announced by the Cabinet Office and other ministries.

The bilevel tsunami estimation, compared to the conventional single estimation, takes the uncertainty of natural phenomena into account and is an advancement in disaster management after the Great East Japan Earthquake. It adds complexity for the administration, however, within a range that it can handle. Several concerns have been pointed out. Examples include the discontinuity between levels 1 and 2, and how to handle uncertainty with large slip zones and super large slip zones. The former concern pointed out the major difference between a level 1 tsunami being several meters while a level 2 is tens of meters. Some residents when faced with the astonishing level 2 numbers quickly give up on disaster management. The disparity in large and super large slips comes from uneven tsunami occurrence such that a fault has local areas that will produce major tsunami waves and others that cause even greater ones. This is a conventional method of describing asperity or tsunami earthquakes that had advantage in practical tsunami estimation; however, the method still lacks a standard.

The method of multiple scenarios [7,8] is effective in improving these problems. We applied this method to the Nankai trough and came up with 324 types of tsunami faults. Fig. 7.2 shows the tsunami faults for the case of Mw9.1 and Mw8.8.

The multiple scenario model will add its importance to events other than tsunami disasters, and its application to real disaster management administration will require verification with record data. For tsunamis, research is underway on numerical models of the mechanism of occurrence and propagation since the 1960s and, in addition, experiments concerning the related phenomena of sand transport, wave force, floating debris and so on have been conducted. A large number of local studies are ongoing in and out

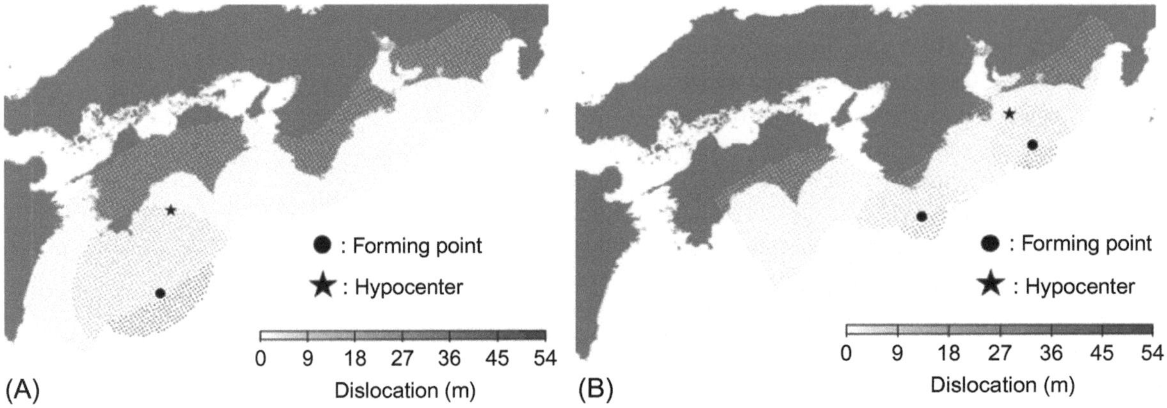

FIG. 7.2 Example of multiple scenario model applied to the Nankai trough. (A) Sample results with Mw9.1 and (B) sample results with Mw8.8.

of the nation; in particular, the 2011 Off the Pacific coast of Tohoku Earthquake has left us a vast amount of disaster data. These data do not just help us unveil the mechanism of the phenomena, but they will also play an extremely important role as base material for verifying numerical models for predicting huge tsunami attacks in the future.

Much of these data, however, are results of individual efforts by researchers and research organizations, and the data are scattered over a number of papers and reports. To make use of these valuable data in disaster mitigation research, we must collect them in a systematic way. Such efforts started after the Great East Japan Earthquake, and we will discuss one such example from the Coastal Engineering Committee of Japan Society of Civil Engineers. In July of 2015, we established a subcommittee for review and promotion of research on tsunami effects and discussed how to use existing data, as Fig. 7.3 shows.

We categorized the phenomena into four groups of behavior types of the tsunami itself, including generation of tsunami source, propagation in the ocean and run-up on land, wave force that damages buildings, sand transport that accompanies tsunami with scouring and accumulating, collapse and drifting of buildings and disaster debris. Reviews were made of published experimental data, field data, theoretical analyses and detailed computation results. Data formats and publication schemes are being discussed and the data will be assembled into a database. Some of the quantitative and highly reliable data will be used to create benchmarks. Benchmarks here refers to collections of calculation conditions and necessary data for numerical simulation, and also a dataset that summarizes output results with proper

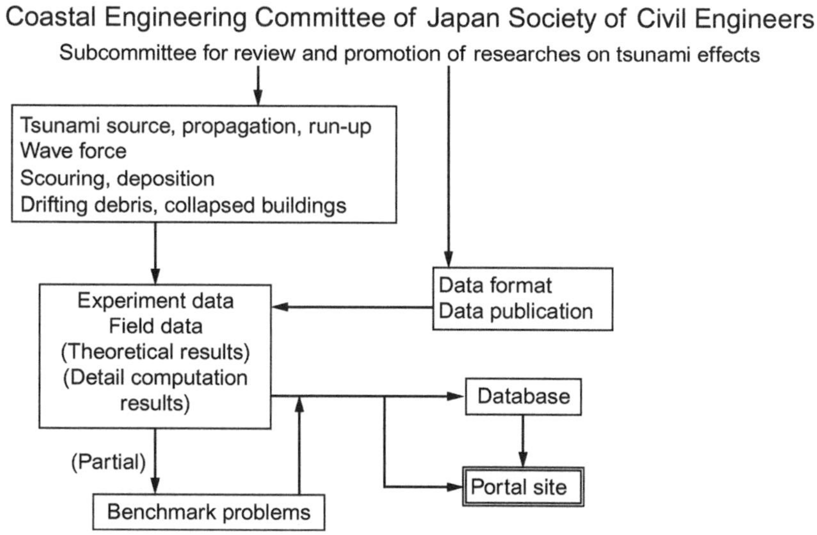

■ **FIG. 7.3** Making effective use of tsunami data records.

computation. A portal site will be opened to access the database and benchmarks.

The benchmarks have been in high demand from the field of disaster management. Ministries and local administrations estimate tsunami damages and produce tsunami hazard maps from the results. These actions are necessary for tsunami disaster management and they are based on numerical simulations of tsunamis. After the Great East Japan Earthquake, many tsunami simulations have been made, but there are doubts with regard to their reliability. For example, when different consulting firms run tsunami simulations under the same conditions, the results can vary. Tsunami simulation has a number of inputs, such as basic formulae, computation scheme, boundary conditions, initial conditions, topography models, external force models, and so on. Each firm codes its own algorithms to replicate the phenomena. Much of the input is reported openly, but the algorithms and codes are proprietary information that a third party cannot verify. The lack of benchmarks for overall evaluation of the accuracy of tsunami simulations has led to these differences.

Standardization of tsunami simulation techniques is also needed for higher reliability. Some technical references and guidebooks have been proposed [9,10] and efforts are being made to develop tsunami estimation recipes based on these publications. A recipe is a technical report that summarizes

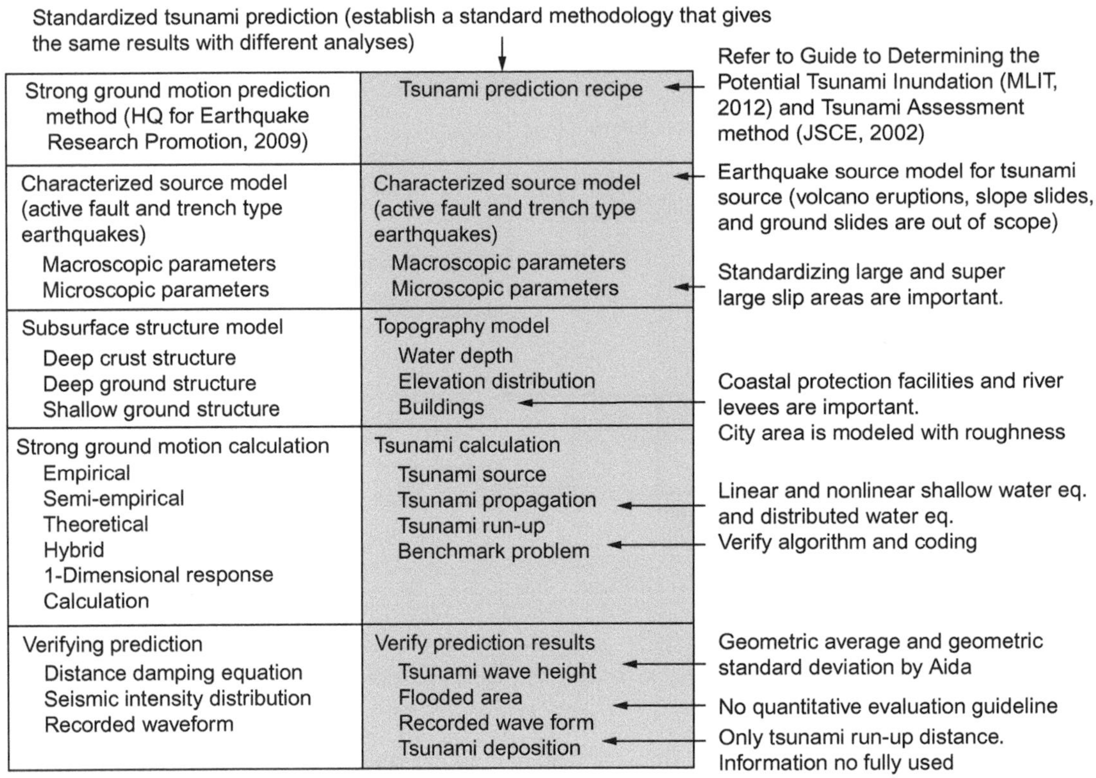

FIG. 7.4 Proposed tsunami estimation recipe.

the method for predicting a certain natural phenomenon and is intended to explain the method so that anyone who runs the simulation will come up with similar results.

For damage estimation upon strong ground motion, the Headquarters for Earthquake Research Promotion has created a recipe [11], available for standard estimation. Fig. 7.4 shows a plan for a tsunami estimation recipe that accompanies this strong ground motion recipe.

The strong ground motion recipe consists of four stages: setting the external force model, setting the field model of external force propagation, actual calculation methods, and verification of estimation results. A tsunami estimation recipe can have the same structure. Tsunamis can be caused by several types of triggers, such as volcano eruptions, landslides, or meteor strikes, but 90% of them are caused by earthquakes; thus modeling the oceanic active faults and trench-type earthquakes is important. This model has factors of both macroscopic source characteristics for the entire tsunami source and microscopic source characteristics for localized conditions. Setting of large slip zones and super large slip zones in the latter largely affects the final tsunami height, thus enough uncertainty should be taken into account, as discussed with multiple tsunami scenarios. Modeling a field takes both topographies of the ocean area where the tsunami travels and the land area where the tsunami runs up into account. In particular, coast line protection facilities such as levees and water gates determine the tsunami behavior regarding inundation of the city, so they require precise modeling. Using a basic formula with known results to establish a benchmark for verifying total reproduction is important. Verifying estimation results will requires comparison with tsunami wave height distributions and inundation area records.

The preceding standardizations of tsunami estimation recipes and benchmarks are important in increasing the reliability of damage estimation at the time of a tsunami disaster. The efforts are making solid progress. Tsunami simulation technology is needed for problems (3), (4), and (5) we pointed out earlier and we expect further advancement.

The results of tsunami simulation are complicated and may be difficult for nonexperts to quickly grasp. They, however, contain a good deal of useful disaster management information and the citizens need this information as well. Those in charge of disaster management in the administration need simulation results of tsunami risks in the city and these results are also complex. For example, if liquefaction from a strong motion earthquake lowers levees, how wide would the inundation area spread when the tsunami attacked? Or if disaster debris floats in the area, what percentage of the roads will become difficult to use for evacuation? The simulation can give the results of change in risk but the results have to be understandable in order to improve the strength of local disaster management. We, at the end, introduce a technology to directly visualize these complicated tsunami risk results.

Tsunami simulations are based on computer computations, so they work well with computer graphics. A recent computer graphics technology called augmented reality (AR) is receiving attention today. AR uses computer graphics to enhance the real environment. Kameda and Takahashi [12–14] applied AR to develop a system to easily attach or modify information about tsunamis, water surface, ground, roads, and buildings. Fig. 7.5 shows the outline of this system. Information about each entity is placed with a card called a marker and the user, with a head-mounted display, sees the graphics related to each marker. The user can switch the markers to those for other expected tsunamis or to a marker for after a building has been taken

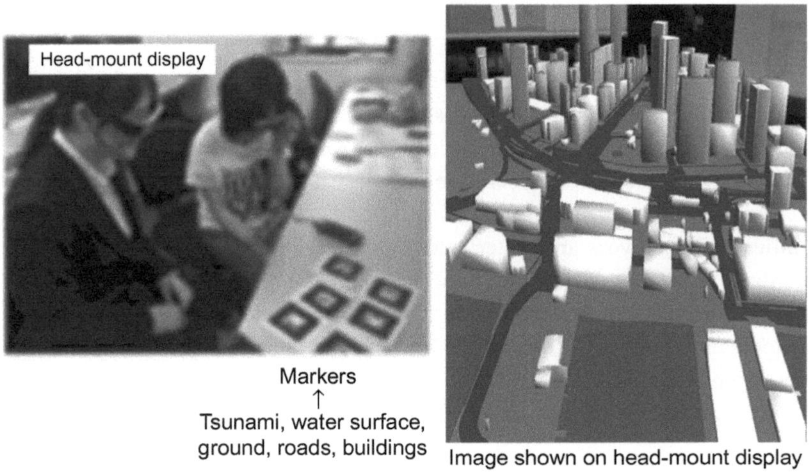

FIG. 7.5 Example of visualizing tsunami risks.

down by a strong motion earthquake, to visually see the change of risk. Using this system in training allows those in charge of disaster management to discuss tsunami risks under various conditions and the citizens can understand how the tsunami attack changes over time in their hometown.

The main players in tsunami disaster mitigation are the citizens, and those in-charge of disaster management with municipalities are there to help the citizens. Tsunami estimates and hazard maps will continue to increase in accuracy and new disaster management measures will be employed. The administration disaster management force will need to understand these new developments as they arrive. Thus, in the field of tsunami disaster management research, applying these visualization techniques to explain the research outcomes will gain more importance.

REFERENCES

[1] Cabinet Office, Government of Japan, White paper on disaster management, Tokyo, 2012. p. 135.

[2] Central Disaster Management Council, (Report 1) Damage estimate with Nankai Trough megathrust earthquake, Tokyo, 2012. p. 191.

[3] Central Disaster Management Council, (Report 2) Damage estimate with Nankai Trough megathrust earthquake, Tokyo, 2013. p. 21.

[4] Cabinet Office, Government of Japan, White paper on disaster management, Tokyo, 2011. p. 135.

[5] Headquarters for the Reconstruction from the Great East Japan Earthquake, Summary of construction plans schedules from the ministries, Tokyo, 2011. p. 36.

[6] T. Takahashi, Building a strong social foundation against large-scale tsunami, in: Social Safety Science for Disaster Prevention and Mitigation, Minerva Bookstore, Kyoto, 2014, p. 19.
[7] S. Seto, T. Takahashi, Recipe of predicting multi tsunami scenarios in the Nankai trough and its application, J. Jpn. Soc. Civ. Eng. Ser. B2 (Coast. Eng.) Tokyo 70 (2) (2014) 5.
[8] S. Seto, T. Takahashi, Model of multiple tsunami scenarios considering uncertainty of seismic magnitude and its application, J. Jpn. Soc. Civ. Eng. Ser. B3 (Ocean Eng.) Tokyo 71 (2) (2015) 5.
[9] Japan Society of Civil Engineers, Tsunami assessment method for nuclear power plants in Japan, Tokyo, 2002. p. 72.
[10] Ministry of Land, Infrastructure, Transport and Tourism, Guide to determining the potential tsunami inundation Ver.2.00, Tokyo, 2012. p. 85.
[11] The Headquarters for Earthquake Research Promotion, Strong ground motion prediction method ("Recipe") for earthquakes with specified source faults, Tokyo, 2009. p. 40.
[12] C. Kameda, T. Takahashi, Numerical modeling on tsunami inundation with detailed urban model and its visualization by using AR technology, in: Ocean Sciences Meeting, American Geophysical Union, Honolulu, 2014.
[13] C. Kameda, T. Takahashi, The development of Augmented Reality visualization technology for tsunami risks with a camera-embedded eyeglass, in: 26th IUGG General Assembly, International Union of Geodesy and Geophysics, Prague, 2015.
[14] C. Kameda, T. Takahashi, The development of visualization method for tsunami risks by using AR technology, Trans. Vis. Soc. Jpn. 36 (5) (2016) 8.

Chapter 8

Liquefaction With the Great East Japan Earthquake

CHAPTER OUTLINE

8.1 OVERVIEW OF LIQUEFACTION DAMAGE IN THE GREAT EAST JAPAN EARTHQUAKE

The Tohoku Area Pacific Offshore Earthquake (Mw9.0) struck on March 11, 2011 and liquefaction was observed in 190 municipalities in 13 prefectures from Tohoku to Kanto areas [1]. Liquefaction started mainly along the shores of large rivers along the Tokyo Bay coast, Tone River and Abukuma River. Housing, lifelines, roads, levees, and port facilities suffered major damage from the liquefaction. The housing damage, according to the Ministry of Land, Infrastructure, Transport and Tourism (MLIT), amounted to 26,914 cases, with 18,674 in Chiba, 6751 in Ibaraki, 1043 in Fukushima Prefectures, while the count was 140 in Miyagi Prefecture, which suffered the biggest tremor [1]. Kanto has a much larger total area of lowlands and landfills that are vulnerable to liquefaction compared to the hills and mountains dominant in the Tohoku area, and liquefaction damage was only minimal in Tohoku.

When we review the microtopography of Kanto and points where the liquefaction started, delta regions and lowlands along the coast (29%), landfills (15%), natural banks (17%), and backswamps (10%) made up over half of

The Fukushima and Tohoku Disaster. https://doi.org/10.1016/B978-0-12-814078-9.00008-X

them, with a clear contrast to the Tohoku area, where backswamps (21%), natural banks (17%), and delta and lowlands along the coast (14%) made up over half [1]. Liquefaction started where the seismic intensity was 5 (the Japanese seven-stage seismic scale) or more and the maximum surface acceleration was 150 Gal or more [1].

The next section discusses the mechanism and conditions for liquefaction, and highlights the differences of liquefaction during the Great East Japan Earthquake from other liquefaction phenomena in the past, explaining how it affected the damage in the area. Then, we discuss types of liquefaction countermeasures and their principles, and summarize current problems. Lastly, we review how we can prepare for the highly probable Tokyo Metropolitan Epicentric Earthquake and Nankai Trough Megathrust Earthquake, given the lessons from the Great East Japan Earthquake. Details of the liquefaction and change in ground conditions caused by the Tohoku Area Pacific Offshore Earthquake are found in references [1–3].

8.2 MECHANISM OF LIQUEFACTION

The word *liquefaction* was first used with the Niigata earthquake on June 16, 1994. During this earthquake, the ground, loosely packed with sandy soil and saturated with groundwater, was subjected to repeated shearing stress from the earthquake movement. This movement changed the soil phase from solid to liquid when effective stress was lost with the rise of excess porewater pressure. In the following, we describe the liquefaction mechanism in detail (Fig. 8.1).

When pores among soil grains are saturated with groundwater and the loosely packed sandy soil (with roughly uniform grain diameter) experiences repeated shear stress from an earthquake movement, the loosely packed sand, trying to pack denser, decreases in volume, i.e., negative dilatancy takes place. Dilatancy is a typical behavior of grain material like soil, meaning volumetric change upon shearing. Volume expansion with shearing is "positive dilatancy" and contraction is "negative dilatancy" (Fig. 8.1).

Sandy soil has relatively higher permeability and water in the pores is quickly discharged to pack the grains densely. If a sudden shearing force is applied within a short time, however, as in the case of an earthquake, the water movement is largely limited. The condition is due to the water not being able to be discharged and the pores become subject to excess porewater pressure, the partial pressure in excess of the static water pressure. In other words, when water in the pores cannot drain freely, a shearing force causes a contraction in loosely packed sandy soil and the trapped porewater

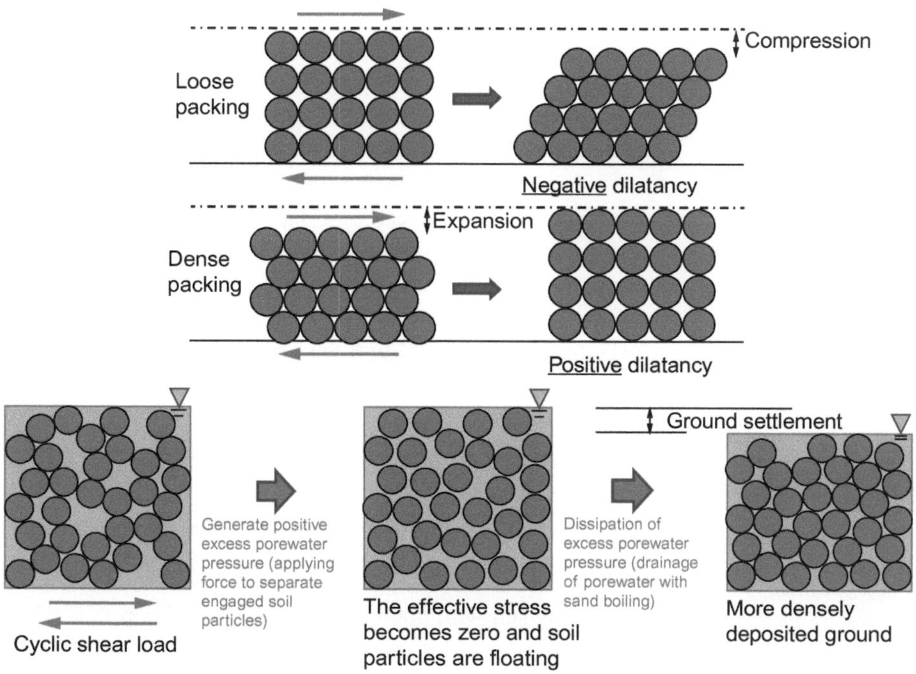

■ **FIG. 8.1** Mechanism of liquefaction.

raises its excess porewater pressure because the force tries to expel the water. This rise in the excess porewater pressure lowers the effective stress, which is the stress transferred via contact of the soil grains, among the soil grains, and this stress eventually reaches zero. That is, the positive excess porewater pressure that tries to pump out the porewater works to release the interference among the soil grains and eventually the interference is totally freed and the intergrain contact force goes to zero (no force transfer among grains). The soil grains then enter a state the same as floating in water. This is the mechanism of liquefaction.

When liquefaction occurs, the ground supporting force drops significantly and it can no longer support structures that stand on it. In contrast, when the grains are densely packed in sandy soil, the shearing force causes grains to go over other grains and the volume increases, i.e., positive dilatancy takes place, which tries to expand the gaps and suck water from outside. This phenomenon produces negative excess porewater pressure, or suction. Suction causes grains to pack even tighter to increase interference among them and the effective stress in the soil increases as well as the strength to support whatever stands on it. Liquefaction, therefore, can hardly take place with tightly packed sandy soil.

Liquefaction changes the soil phase to liquid and the mudlike soil acts as a heavy liquid (specific weight of soil grains is about 2.65, whereas that of muddy water is about 2.0). Underground structures, such as a maintenance hatch or water pipes buried in such soil, receive buoyant force and they come to the surface. Structures with a large specific weight, on the other hand, will sink into the soil. These phenomena were frequently observed with the Great East Japan Earthquake. When the earthquake stops, the high excess porewater pressure is dissipated when the groundwater bursts up over the ground surface and as the water takes soil grains with it, the phenomenon is called *jetted sand*. Soil that went through liquefaction is compacted and the ground level sinks when the pore-filling water is lost. The soil ground gains strength and stabilizes again as soil ground.

8.3 CONDITIONS FOR LIQUEFACTION

The general conditions for liquefaction to take place are as follows:

(1) Sandy soil with grain diameter of 0.1–1.0 mm
Sand grains have diameters in the range of 0.075–2 mm. Soil with grains in the range of 0.005–0.075 mm is called silt, 0.005 mm or less clay, and 2–75 mm gravel. Soil with more gravel allows water to pass through easily and it is difficult for it to reach the state of no water drainage, and therefore difficult for the excess porewater pressure to rise, making it hard to reach liquefaction. Silt and clay have a sticky adhesive force among the grains and this force keeps the interference among grains intact within soil with more silt and clay (called cohesive soil). Thus, even if the excess porewater pressure rises, the soil may lose some strength but sudden destruction does not take place, i.e., liquefaction rarely happens with cohesive soil. Soil composed of grains of various diameters has more contact points among the grains and the grains rarely rotate or shift place. This stable soil increases its density when compacted and the soil skeleton will transfer larger forces; this soil is called well-graded soil. In contrast, soil with grains of similar diameters do not have more contact points and packing it does not increase its density. Such soil does not become strong even when compacted and is called poorly-graded soil; it easily enters the state of liquefaction.

(2) The ground is loosely sedimented and the soil density is low
As we discussed earlier, loosely packed sandy soil sees negative dilatancy that causes positive excess porewater pressure to lose effective stress in the soil. In general, ground formed underwater isloose

and can easily enter liquefaction. When the *N*-value[1] for ground stiffness is less than 20–25, especially when it is below 10, it is easy for the ground to enter liquefaction.

(3) Groundwater level is shallow and soil is saturated
The ground being saturated is a precondition for liquefaction to take place and when the groundwater level is within 10 m of the surface, i.e., when the groundwater is close to the surface, liquefaction can easily start.

(4) Vibration (repeated shearing force) of a certain strength is present, such as an earthquake
In general, an earthquake that causes liquefaction has a surface acceleration (lateral vibration) of 100–150 Gal or more. A vertical vibration of 50 Gal or less can cause liquefaction as well. Also, other than an earthquake, a wave force can sometimes cause liquefaction in the beach sand.

8.4 SPECIFIC OUTBREAK OF LIQUEFACTION AND DAMAGE DURING THE GREAT EAST JAPAN EARTHQUAKE

The liquefaction and the damage it caused with the Tohoku Area Pacific Offshore Earthquake showed the following special behavior:

(1) Liquefaction occurred in areas far away from the epicenter
A wide area from Tohoku to Kanto liquefied. The furthest point from the epicenter was up to 440 km in the City of Hiratsuka in Kanagawa Prefecture [1].

(2) Liquefaction took place in a wide area along the coast of Tokyo Bay
The earthquake intensity along the coast of Tokyo Bay was not so strong, at "5-strong" (the Japanese seven-stage seismic scale), but liquefaction took place over a wide area from Koto-ku in Tokyo Prefecture to the City of Chiba. Large amounts of jetted sand were observed and the ground subsided 30–50 cm in places. Damage to houses and underground pipes was serious and the roads uplifted and bulged. Bridges, tall buildings, piers and seawalls with piles in the

---

[1]Standard penetration tests are made, in general, to find the strength of the ground soil. The test drops a 63.5 kg hammer from 75 cm above on a sample tube with outer diameter of 51 mm, inner diameter of 35 mm, and a length of 81 cm. The *N*-value is the number of blows necessary to drive the sample tube into the ground by 30 cm. The *N*-value is larger for harder ground and, regularly, an *N*-value of 50 is the guideline for supporting ground. Empirical formulae relate *N*-values to strength parameters of soil such as internal friction angle or uniaxial compression strength.

ground, however, suffered relatively little damage. The liquefaction scenerios were different from those seen during the Great Hanshin-Awaji Earthquake. This time, ground subsidence caused big wall-like steps and damage to the lifeline.

The acceleration amplitude recorded at the coast of Tokyo Bay and the maximum values from the main tremor were, from K-NET, e.g., N-S: 125.1 Gal and E-W: 157.3 Gal in the City of Urayasu in Chiba Prefecture (K-NET Urayasu [CHB008]), 178.7 Gal and E-W: 139.7 Gal in the City of Chiba (K-NET Chiba [CHB009]), and N-S: 219.0 Gal and E-W: 157.6 Gal in Tatsumi in Koto-ku, Tokyo (K-NET Tatsumi [TKY017]). The earthquake, however, continued for over 2 minutes and the ground was subject to a long cyclic shearing force. The strong motion of 50 Gal or above, said to have major effects on liquefaction, lasted for almost 1 minute and this contributed to the liquefaction as well. Some areas were liquefied by the Ibaraki Offshore aftershock (M7.6) that started 29 minutes after the main shaking. For example, the biggest (combined) acceleration in this aftershock, 82.3 Gal (174.3 Gal in the main earthquake), was recorded in the City of Urayasu in Chiba Prefecture, and the cities of Chyoshi and Tohgane in Chiba Prefecture had larger accelerations than in the main earthquake. (The main earthquake marked 184.0 Gal and 225.8 Gal in the cities of Chyoshi and Tohgane, respectively, and those of the aftershock were 315.9 Gal and 285.3 Gal).

(3) Fine-grained sand, not classified as susceptible to liquefaction in the past, liquefied

Fig. 8.2 shows the grain size accumulation curve[2] from sand collected from traces of jetted sand at each location liquefied in the coast of Tokyo Bay [4]. The figure also shows the range of grain diameter (0.1–1.0 mm) known to carry the danger of liquefaction. Most of the samples remain in the liquefaction danger zone, but some soil contained grains of smaller sizes. Most of the soil is of poorly-graded soils showing a quick rise of the grain size accumulation curve in the danger zone; however, a relatively well-graded soils can also be seen with a flat grain size accumulation curve.

This soil is called "silty sand" or "sandy silt" and had been categorized as difficult to liquefy as compared to sand. Soil with a fine grain has

[2]The quick and easy way to find the grain size of soil is to use a sieve. The practical smallest hole size of a sieve is 75 μm. Grain-size analysis of sand and gravels is made with sieves, while sedimentation analysis is used for silt and clay with grain size smaller than 75 μm. The results are shown in grain size cumulative percent curves with the cumulative mass percentile that passed the sieve in the vertical axis versus the sieve size in a logarithmic scale in the horizontal axis. These plots are called "grain size cumulative curves."

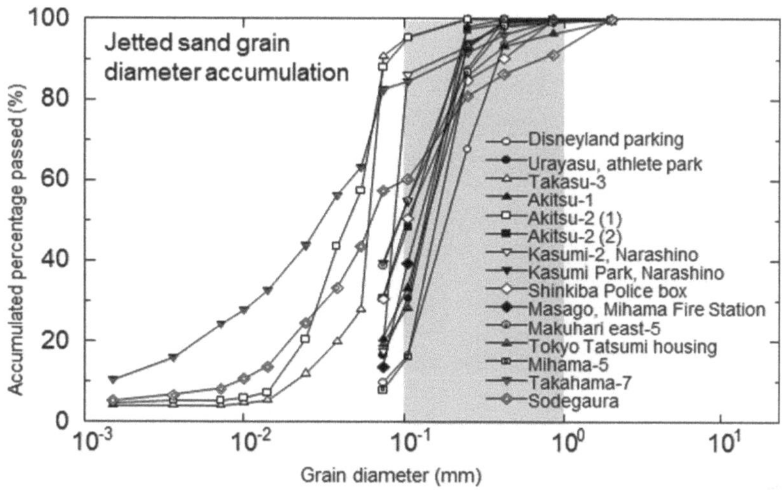

■ **FIG. 8.2** Grain size distribution of soil from traces of jetted sand in the Tokyo Bay area. *(Based on S. Yasuda, K. Harada, Liquefaction-induced damage in the reclaimed lands along Tokyo Bay, Geotech. Eng. Mag. Jpn. Geotech. Soc. 59 (7) (2011) 38–41 (in Japanese).)*

a small hydraulic conductivity and, once liquefied, discharging the water takes time and high excess porewater pressure causes water fountaining and jetted sand. The smaller grain size caused the sand to easily move upward to the surface with the fountain of water, which affects large amounts of jetted sand and large ground subductions. Cases of liquefaction with soil having fine or coarse grains, outside of the liquefaction danger zone, were also reported with the January 17, 1995 Southern Hyogo Earthquake [5]. The Southern Hyogo Earthquake marked a short 10–15-second earthquake duration with a very large maximum horizontal acceleration of 818 Gal. In comparison, in the case of the Tohoku Area Pacific Offshore Earthquake, the liquefactions were far from the epicenter and the maximum accelerations were not so large; however, the shaking lasted for a very long time, and it was the first case when soil with fine grains liquefied; this soil condition was considered hard to liquefy in past cases.

(4) Liquefaction did not take place in alluvial lowland formed by natural sediment

Fig. 8.3 shows a northwest to southeast geological cross-section [4] that passes through Imagawa in the City of Urayasu in Chiba Prefecture. As the figure shows, the liquefied area, with the groundwater level at 1–2 m from the ground surface, was formed with a soft landfill layer of 7–10 m with an N value of 5–10 sitting on top of the alluvial sand layer with an N value of 10–25. The figure also shows the coastline in 1945. About 74% of the City of Urayasu was landfilled in 1965 or later, and the 5–7 m

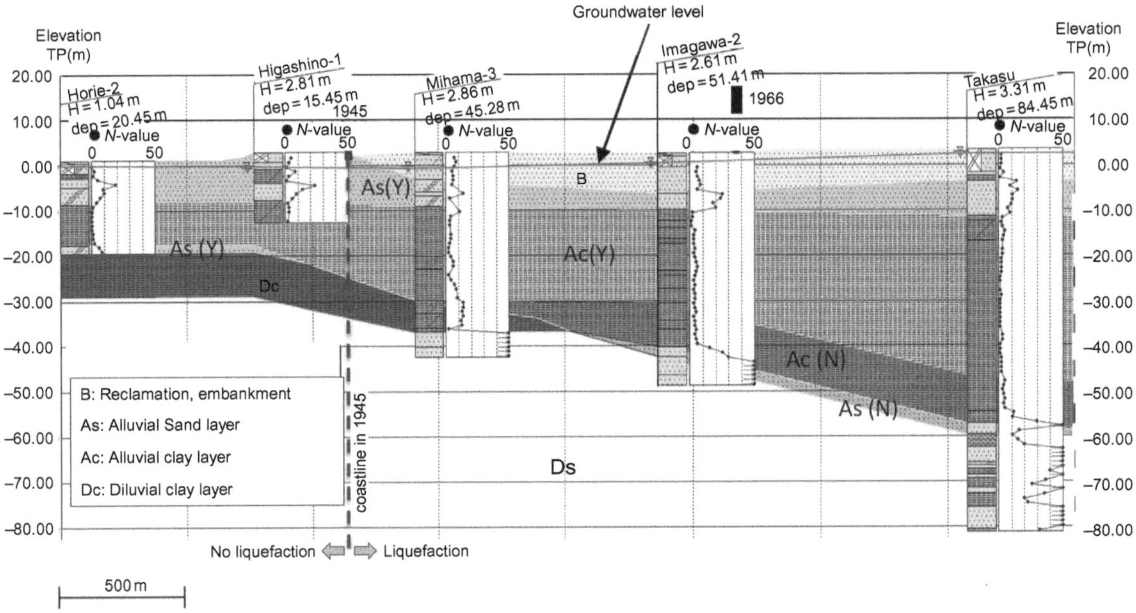

FIG. 8.3 Geological profile with liquefaction spots in Imagawa, Urayasu, Chiba (northwest to southeast cross-section). *(Based on S. Yasuda, K. Harada, Liquefaction-induced damage in the reclaimed lands along Tokyo Bay, Geotech. Eng. Mag. Jpn. Geotech. Soc. 59 (7) (2011) 38–41 (in Japanese).)*

thick part of this landfill, soaked with groundwater, liquefied. On the other hand, liquefaction did not take place in the alluvial lowland formed by natural sediment [1].

(5) Ground reliquefied in a number of places

Reliquefaction is a phenomenon in which ground, once liquefied in the past, goes through liquefaction again due to an earthquake. It was observed in a total of 91 locations around the Tokyo Bay coast, downstream banks of Tone River, Kujyukuri Flatland in Boso Peninsula and the banks of big rivers in the prefectures of Iwate, Miyagi, and Fukushima [1]. Reliquefaction occurred in limited places with specific land conditions: landfilled mud lands, reclaimed lakes and reservoirs, land where a river used to flow, lowland made of wind-carried soil, or land filled after iron sand excavation (Asahi City in Chiba Prefecture). Some land where rivers used to flow and natural riverlevees were reliquefied as well. These locations were exactly the same places liquefied in the 1987 Chiba offshore to the east, 1978 Miyagi offshore, and 2003 northern Miyagi earthquakes [6].

As we explained previously, once-liquefied ground was believed to turn solid afterwards. In the case of soil with fine grains, however, the jetted sand upon liquefaction causes stiffening deep down where the porewater dissipates, and the fine grains gather near the surface where they mix and loosely accumulate, again putting themselves in a state that is easy to liquefy.

(6) Liquefaction concentrated in artificially modified land along rivers, lakes, and reservoirs
Besides the Tokyo Bay coastline, liquefaction in the Kanto area was mostly found on the banks of the Tone and other rivers, Kasumigaura Lake, and Kitaura Lake where lands were artificially modified by reclaiming lakes or landfilling where rivers used to flow. Fig. 8.4 shows a landfill over the location where a river used to flow in Sawara in the City of Katori in Ibaraki Prefecture.

(7) River levees suffered great damage with liquefaction
Damage to riverlevee upon an earthquake includes lateral flow from the external force of the earthquake itself or ground liquefaction, and damage or destruction of levees or river structures caused by land subsidence. There are two types of riverlevee damages caused by

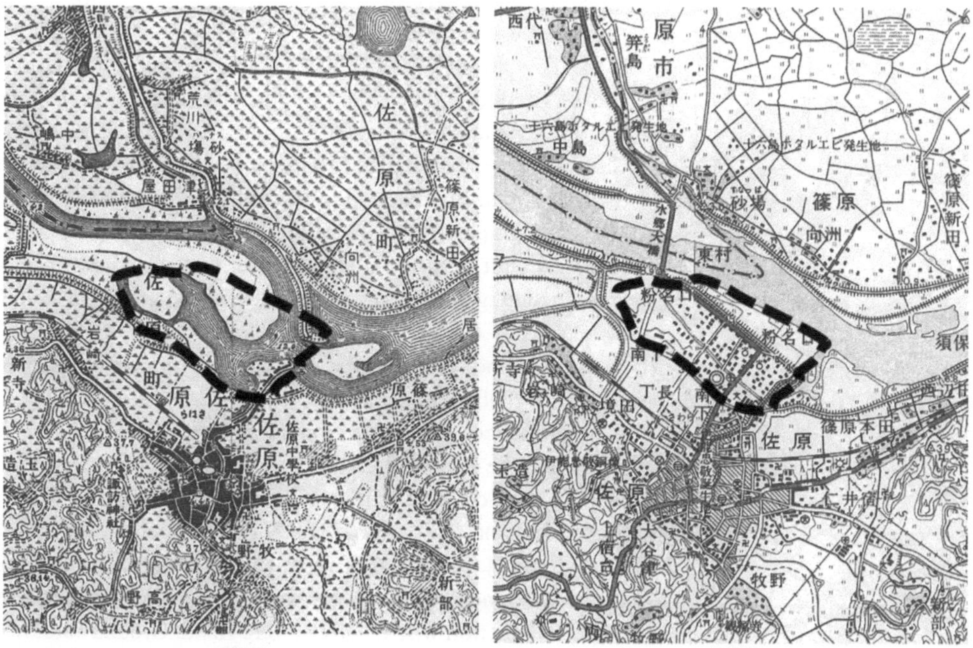

1906 1969
(from old topographical maps)

FIG. 8.4 Reclamation of an old river course in Sawara, the City of Katori in Ibaraki Prefecture.

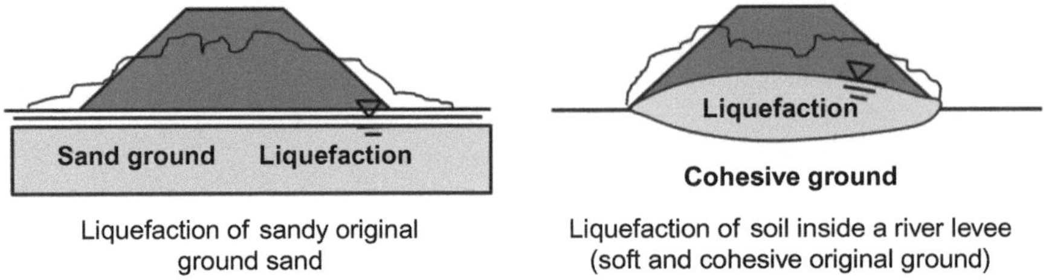

FIG. 8.5 Two different patterns of river levee liquefaction.

liquefaction (Fig. 8.5): one is large subsidence of the riverlevee by liquefaction of the sandy ground and the other is liquefaction of the levee itself (in fact, it is the sandy area surrounded by clayish base ground and groundwater inside the levee body). Up to now, countermeasures for riverlevee liquefaction have resisted the subsidence of the riverlevee. However, this time a great deal of damage was found in Tohoku on levees built on backswamp, which are cases of the latter, i.e., liquefaction of the levee structure.

(8) Liquefaction damage in Tohoku was minor except for a few places, compared to that in Kanto
In the Tohoku region, liquefaction mainly damaged riverlevees, roads, houses, lifeline, and farmland, but the level of damage to houses and the amount of jetted sand, except for a few areas like Ueda-cho in Iwaki City, were minor compared to the damage in the Kanto region. This is due to small liquefaction in landfills and reclaimed land in Tohoku. Microtopographic divisions of the liquefied areas were backswamps, lowlands built by filling the bottom of valleys, natural banks, delta areas including natural ground, hills, sand and gravel land, and loam soil developed by cut and fill.

(9) A large number of landslides and destruction of landfills were found with valley-filled areas
A large number of liquefaction sites were found inland, far from rivers and the sea, where land was developed over loam valley areas, including the cities of Ichikawa, Funabashi, Sakura, and Inzai in Chiba Prefecture, and the cities of Hokota and Tsuchiura in Ibaraki Prefecture [1]. Fushiogami in the City of Fukushima and Nankodai in Izumi-ku, Sendai City, in Miyagi Prefecture faced landslides and filling destruction where valleys were filled. Groundwater accumulates in land developed by filling valleys and local liquefaction near the bottom of the landfill (called “slip-face liquefaction”) often cause landslides.

8.5 ANTILIQUEFACTION CONSTRUCTION METHODS

A number of construction methods have been proposed to counter liquefaction. Their principles are based on the mechanism of how liquefaction occurs. Principles of countering liquefaction are divided into two groups, one to "improve the soil itself" and the other to "modify the boundary condition" [7]. Methods to "improve the soil itself" include increasing relative density by compacting the soil, changing the soil to a grain size of a hard-to-liquefy type, and solidifying the soil by injecting chemical solutions. The "modify the boundary condition" group includes lowering the degree of saturation, dispersing the excess porewater pressure, blocking the transfer of porewater pressure, and controlling the shear deformation. Table 8.1 shows the principles, methods, and constructions of these antiliquefaction methods [7].

As we pointed out earlier, the Great East Japan Earthquake caused severe liquefaction over a number of areas along the Tokyo Bay coast. Locations with sufficient antiliquefaction measures, however, did not experience

Table 8.1 Principles and Methodology for Countering Liquefaction

Principle	Methodology	Preventive Construction Method
Increase density	Compaction and/or consolidation	Sand compaction pile method Rod compaction pile method Vibro-floatation method
Improve grain size Solidification	Replacement and/or chemical treatment	Replacement method by coarse aggregate Solidifying material injection Deep mixing method
Decrease degree of saturation	Decrease groundwater level	Well point method Deep well method
Increase effective confining pressure	Compaction	Embankment
Accelerate dissipation of excess porewater pressure	Improvement in permeability	Gravel drain method Crushed stone drain method
Shutoff excess porewater pressure	Isolation	Diaphragm wall construction
Control shear deformation	Control deformation	Diaphragm wall construction, sheet piling

liquefaction even if the areas were categorized the same as the liquefied locations. Conventional antiliquefaction measures are effective to a certain extent. Liquefaction, however, took place with sandy silt and silty sand, i.e., ground that used to be understood as hard to liquefy; thus, we need to take countermeasures for these ground areas as well.

A large variety of antiliquefaction constructions have been proposed so far; however, they are not attractive economically. The cost varies with ground and construction conditions, and for ground under existing structures, the cost is JPY 50,000–100,000 for each square meter to cover approximately 100 m^2 (10 m × 10 m). Less expensive constructions (e.g., sand compaction piling method, deep mixing method, and so on) of about JPY 10,000–50,000 per square meter are available, but they need to cover a wider area (50 m × 50 m or 100 m × 100 m) in order to take effect. In any case, a person would have to spend millions of Japanese yen to cover the large cost, even with the support of a Reconstruction Grant from the central local governments. This means that countermeasures for liquefaction are still stumbling, even 5 years after the earthquake disaster. We are looking for development of inexpensive and effective (lower cost per unit area and smaller treatment area) antiliquefaction construction methods.

8.6 CONCLUSION

This subsection summarizes the lessons learned from the liquefaction damage caused by the Tohoku Area Pacific Offshore Earthquake. We believe that we should apply these as measures for preventing and reducing damage upon the highly probable direct Tokyo Metropolitan Epicentric Earthquake and the Nankai Trough Megathrust Earthquake.

(1) A trench type megathrust earthquake continues to shake the land for an extended time and is often accompanied by many aftershocks; thus, even if the acceleration amplitude is small, liquefaction can take place over a wide area. Metropolitan areas formed with alluvial lowlands and landfills in particular have to quickly take the necessary countermeasures, or otherwise enormous damage will take place.

(2) Lifelines of electricity, gas, water supply, and communication are easily damaged by liquefaction and so are small buildings like private houses. We have an urgent need to develop inexpensive and effective antiliquefaction methods, replace piping with aseismic types, and thoroughly compact landfilled soil.

(3) Large amounts of jetted sand and water discharge block roads and other transportation means, hindering emergency lifesaving and firefighting operations. Countermeasures in areas susceptible to liquefaction are important for saving roads and other transportation systems for effective evacuation, lifesaving, and support operations.

(4) A large number of energy storage facilities, like petroleum or LP gas, are found in coastal landfills that are easily affected by liquefaction. Liquefaction will cause uneven sinking of tanks and sloshing with the earthquake tremors. These phenomena can lead to major fires; therefore preventive measures are urgently required for them.

(5) In the future, when major earthquakes with aftershocks are likely, as in the case with the Tohoku Area Pacific Offshore Earthquake, or multiple earthquakes are predicted, as with the possible Nankai Trough Megathrust Earthquake, judging when to start the recovery actions is a very difficult decision.

(6) When riverlevees and river walls are severely damaged by an earthquake or ground liquefaction, they lose their original functions. Then flooding and high tides will follow, to turn the disaster into a complex one. We need to enhance the hardware so that partial loss of functions does not lead to a catastrophic disaster.

(7) Many municipalities recently provided hazard maps with liquefaction prediction to the citizens. The hazard maps, however, are not really used in preventing and reducing a disaster. We need further discussions to clarify how to use the hazard maps in the future.

REFERENCES

[1] Joint Editorial Committee for the Report on the Great East Japan Earthquake Disaster, Report on the Great East Japan Earthquake disaster, in: Fundamental Aspects 3 Geohazards, Maruzen, Tokyo, 2014. (45 pp. summary+CD-ROM) (in Japanese).

[2] Kanto Regional Development Bureau, Ministry of Land, Infrastructure, Transport and Tourism, Report on the liquefaction in the Kanto region during the 2011 off the pacific coast of Tohoku earthquake, 2011. 65 pp. http://www.ktr.mlit.go.jp/ktr_content/content/000043569.pdf (in Japanese).

[3] The Japanese Geotechnical Society (JGS), Jpn. Geotech. J. special issue on "the 2011 Off the Pacific Coast of Tohoku Earthquake" 7 (1) (2012) 1–388 (in Japanese).

[4] S. Yasuda, K. Harada, Liquefaction-induced damage in the reclaimed lands along Tokyo Bay, Geotech. Eng. Mag. Jpn. Geotech. Soc. 59 (7) (2011) 38–41 (in Japanese).

[5] Disaster Prevention Research Institute, Kyoto University, Geohazard, Section 4.3.4: Liquefaction in the Southern Hyogo prefecture earthquake in 1995, in: Practical Regional Disaster Prevention Planning, Kajima Institute Publishing, Tokyo, 1997, pp. 84–92 (in Japanese) (Chapter 4).

[6] K. Wakamatsu, Recurrent liquefaction induced by the 2011 Great East Japan Earthquake, J. Jpn. Assoc. Earthq. Eng. 12 (5) (2012) 69–88 (in Japanese).

[7] F. Oka, Liquefaction of Ground, Kinmiraisha, Nagoya, 2001. 178 pp. (in Japanese).

Chapter 9

Legal System for Supporting Disaster Victims

CHAPTER OUTLINE

9.1 CURRENT SUPPORT FOR DISASTER VICTIMS TO REBUILD THEIR LIVES

Japan has three acts to support victims in the case of natural disasters. They are the Disaster Relief Act, the Act on Support for Reconstructing Livelihoods of Disaster Victims (Act on Support), and the Act on Provision of Disaster Condolence Grant. This chapter clarifies the legal system for

The Fukushima and Tohoku Disaster. https://doi.org/10.1016/B978-0-12-814078-9.00009-1

supporting disaster victims 5 years after the Great East Japan Earthquake and also reports problems that have surfaced through these years.

9.1.1 Act on Support in Effect

The Great Hanshin Earthquake in 1995 led to the establishment of the Act on Support in 1998. The act underwent two revisions in 2004 and 2007 and is under the jurisdiction of the Cabinet Office.

The Act on Support is in effect when the number of houses totally destroyed by natural disaster exceeds the criteria for municipalities and prefectures.

The act applies to houses that were (i) totally destroyed, (ii) half-destroyed or forced to be taken down due to damage to the land, (iii) uninhabitable for extended periods due to a hazardous state, or (iv) half-destroyed and requiring major reconstruction (major half-destruction).

The Act on Support defines monetary aid as Basic Aid determined by the level of damage, and Additional Aid determined by the method of reconstruction. When added, the aids sum to a maximum of JPY 3 million (Table 9.1).

The Reconstruction Agency announced that the number of households that received Basic Aid totaled 191,607 with a sum of 153.1 billion JPY, and Additional Aid, 119,368 households with 152.6 billion JPY (as of March 31, 1015, according to the Cabinet Office) [1].

9.1.2 Independent Support by Municipal Governments

The central government established the "Reconstruction Grant Withdrawal" system with the special local grant tax and municipal governments are developing their own independent policies based on it. The independent policies of the municipal governments are different from those covered by

Table 9.1 Entitlement and Amount of Aid (in JPY Million)

Category	Reconstruction	Basic Aid	Additional Aid	Total
Total destruction	Newly built	1	2	3
Taken down	Repair	1	1	2
Uninhabitable	Rent	1	0.5	1.5
Major half-destruction	Newly built	0.5	2	2.5
	Repair	0.5	1	1.5
	Rent	0.5	0.5	1

Note: *A single member household receives 3/4 of the amount.*

policies from higher levels (central and prefectural governments), so the municipal governments can support items that are not covered by the Act on Support or the Disaster Relief Act. These policies provide aid on top of the amount covered by conventional policies (add-on), or to subjects not covered by them (from the side).

Kondo reported [2] that of the total of 21 municipalities (14 coastal municipalities in Miyagi Prefecture and 7 in Iwate), 18 (85.7%) provided direct support for building or purchasing houses (upper limit: 1–2 million JPY), 19 (90.5%) for aid in interest payments for housing purchases (upper limit: 1.46–7.08 billion JPY), 17 (81.0%) for direct aid in house repair (upper limit: 0.3–1 billion JPY), and 17 (81.0%) for land reconstruction, raising, and development (upper limit: 1–10 million JPY). The upper limits varied among municipalities; however, all of them carried out their own policies of support in one form or another.

9.2 DISASTER RELIEF ACT AND PREVENTING DISASTER-RELATED DEATHS

9.2.1 Outline of Disaster Relief Act and Special Standards

The Disaster Relief Act was established in 1947 after the 1946 Nankai Great Earthquake. The Cabinet Office has jurisdiction over it.

The act has the following provisions (Article 4, item 1):

> (1) Evacuation shelter and temporary housing, (2) Emergency supply of food and drinking water, (3) Giving or lending of clothing, bedding, and other daily necessities, (4) Medical treatment and birth assistance, (5) Rescue for victims, (6) Temporary repair of damaged houses, (7) Giving or lending of funds, facilities, and information for continuing work, (8) School supplies, (9) Burial, (10) Search for dead bodies and their disposal, (11) Removal of rubbish.

The Disaster Relief Act, thus, is designed to protect the lives of victims immediately after the disaster.

The rescue work, according to the Disaster Relief Act, is a statutory entrusted function of the Governor of each prefecture (Disaster Relief Act Article 2, Order for Enforcement of the said act Article 18), and the Mayor of each municipality carries out part of the work delegated by the Governor or supports it (Disaster Relief Act Article 13). The cost of the rescue work is shared evenly between the central and prefectural governments (Disaster Relief Act Article 18 and 21). Thus, Mayors of disaster-struck municipalities can carry out the rescue actions without having to worry about the cost.

The Governor of each prefecture sets the level, method, and extent of rescue work based on the standards set by the Prime Minister (Disaster Relief Act Article 4, Item 3, Order for Enforcement of the said act Article 3, Item 1). The General Standards of the Prime Minister are stated in "Standards of level, method, and extent of rescue according the Disaster Relief Act and actual cost reimbursement (Cabinet Office Public Notice #144, Oct. 1, 2013)" and prefectural governors follow these standards.

Although general, the standards specify the daily cost for running evacuation shelters, the extent they operate, daily meal allowance, and the cost for building temporary housing.

In cases where the General Standards set by the Prime Minister make it difficult to meet the actual needs, the prefectural Governor discusses the conditions with the Prime Minister and upon approval, modifies the level, method, and extent of rescue (Order for Enforcement of Disaster Relief Act Article 3, Item 2). These are the so-called "Special Standards."

At the time of the Great East Japan Earthquake, the Ministry of Health, Labor, and Welfare (MHLW) had jurisdiction over the Disaster Relief Act. The ministry gave a large number of notices to the municipalities affected by the disaster to promote the application of the Special Standards. They included the following:

- Evacuation shelters and food specified for 7 days shall be extended to 2 months (or longer if necessary).
- Special consideration shall be given to elderly and ill for partitions, air-conditioning, temporary laundry facilities, baths, showers, and toilets at evacuation shelters.
- Special consideration shall be given to elderly and ill for food.
- Building public welfare shelters.
- Emergency temporary housing shall be built, especially for cold regions.
- Daily reimbursement for overnight stays at private inns and hotels is 5000 JPY per person including meal allowance.
- Reimbursement for staying in private apartments and houses is about 60,000 JPY monthly for household.

These disaster reliefs reconfirmed the special ones made in the past, rather than defining the standards unapproved before. Nevertheless, the fact that so many notices were frequently sent out tells us those reliefs approved before had rarely been put into practice in the field.

Administrative officers in regions where disasters are rare are usually unaware of the details of the Disaster Relief Act. If they administer the rescue actions without the knowledge of Special Standards, they can very well take away support that the people are entitled to and cause deaths among the elderly and handicapped. During the Great East Japan Earthquake, a large number of "disaster-related deaths" were found in evacuation shelters.

9.2.2 Disaster-Related Deaths

Disaster-related deaths are those not caused directly by earthquakes or tsunami waves, but indirectly caused by exhaustion, stress, or poor living conditions while being evacuated. They were first acknowledged after the Great Hanshin-Awaji Earthquake and Hoyogo Prefecture certified 919 such deaths in 2005 [3].

In the following paragraphs, we see disaster-related deaths with the Great East Japan Earthquake.

According to the Reconstruction Agency [4], as of September 10, 2014 (within 3 years of the disaster), the number of disaster-related deaths was 3323, and during the following half-year, an additional 8 died. The age breakdown was age 7 at 20 or less, 367 at 21–65, and 2957 at 66 or over.

The disaster-related death count from the same agency [5] was 1632 as of March 31, 2012. About 90% of those were at age 66 or over and about half of them died within a month from the breakout. This report further looked into the causes of death for 1263 that were evacuated from communities with a high rate of disaster-related deaths and ordered evacuation due to the nuclear plant accident. About 30% of them died due to bodily exhaustion caused by living in the evacuation shelters.

9.2.3 For Active Application of Special Standards

Better Special Standards are needed to improve living conditions after evacuation. What would assure flexible application of the legal system? We need the administration officers in the field to better understand that these Special Standards are in place and prefectures and the Cabinet Office to explain how to set the Special Standards for the communities. Opportunities for the field officers to discuss the Special Standards with higher administrative office are also needed.

Ikuta pointed out [6] the following about the relation between General Standards and Special Standards: "The system itself is ill conditioned to have set the basic standards at low levels for administrative reasons while the reality requires discussion every time something happens" and that "at least for the level of rescue, if the standards affect the lives of the people of Japan at the time of disasters, the Cabinet Office should set a set of fine rules including exceptions."

Among all this, some municipalities declared at the ordinance level their intention to actively apply the Special Standards. The City of Minoh in Osaka Prefecture, in its Ordinance for Special Handling upon Disasters, Article 17, Item 2 states, "When the Mayor judges that rescue by the Governor of Osaka Prefecture is too late, the Mayor may initiate rescue actions. In this case, the Mayor will later request Osaka Prefecture to cover the cost needed for this effort." Administrative officers in this community are expected to take quick action [7].

Not only the administration but all people involved with disaster rescue (communities, schools, supporters, and business owners) should share knowledge about the Disaster Relief Act and we should widely spread the knowledge at opportunities such as operational drills of evacuation shelters and disaster prevention seminars (Fig. 9.1).

9.2.4 Future Measures—Hinted at by Basic Act on Disaster Control Measures

The first action to take to prevent disaster-related deaths is to improve the living conditions immediately after disasters at evacuation shelters. The Basic Act on Disaster Control Measures gives us some ideas about what to do:

The Basic Act on Disaster Control Measures was revised in 2015 and clearly states that protecting human lives and health have the highest priority

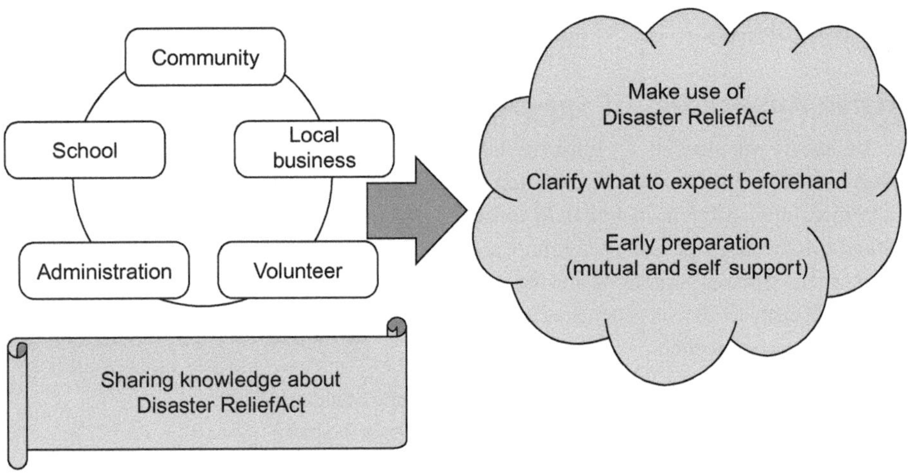

■ **FIG. 9.1** Sharing knowledge about the Disaster Relief Act.

(Article 2, Item 2, Number 4). The act calls for disaster victim support with consideration to age, gender, and handicap (Article 2, Item 2, Number 5). Elderlies, handicapped, infants, and others that need special attention are legally given the positions of "consideration required" (Article 8, Item 2, Number 15).

After Article 8, Item 2, Number 14 was set, items for organizing living conditions in evacuation shelters (Article 86, Item 6) and giving consideration to disaster victims that stay at places other than evacuation shelters (Article 86, Item 7) were added. And the Cabinet Office announced, in August of 2013, the "Guideline for targeting securement of good living conditions in evacuation shelters [8]."

The guideline discusses public nurses making rounds to evacuation shelters, arranging consultation opportunities during their visits, and sanitation and health management at evacuation shelters for the prevention of disaster-related deaths. It recommends setting up item 11, "consultation window" for consultation service and processing complaints (p. 22).

9.3 DISASTER-RELATED DEATHS AND LAWSUITS

In the case when a family member dies for reasons related to a disaster, the Act on Provision of Disaster Condolence Grant and Other Matters (Act on Provision of Disaster Condolence Grant), enacted in 1973 after the 1967 Niigata Uetsu Flooding, pays condolence grants to the family. The amount depends on the relation to the deceased, as follows:

> (Recipient)
> Spouse, child, parent, grandchild, grandparent, and sibling (living in the same household, or sharing the same account for living)
> (Amount)
> If the deceased was the primary member for supporting the household, 5 million JPY. Otherwise, 2.5 million JPY

Some lawsuits have been filed with regard to disaster-related deaths after the Great East Japan Earthquake. In some cases, the plaintiffs won the cases and received a condolence grant for disaster-related deaths. The following paragraphs explain the cases.

9.3.1 December 9, 2014, Sendai District Judgment

An 85-year-old female had her house totally destroyed. She spent 3 days in her car and returned to live in her house from the fourth day. Her condition worsened in April and she began to have a swallowing disorder. She moved into a caregiver nursing home, but developed aspiration pneumonia twice and died on August 7, 2011.

Her family filed for a condolence grant, but was turned down, saying there was no cause-and-effect relationship between the disaster and death. The family's protest was turned down, and the family filed a complaint with the court.

The judgment ruled that "the swallowing disorder was unlikely to have been caused by her dementia or aging alone. The disaster seriously damaged her lifestyle and living conditions and she suffered large burdens to her body and mind. It is reasonable to assume these burdens played large factors for her death."

9.3.2 December 17, 2014, Sendai District Judgement

A 99-year-old male, 7 days after the earthquake, developed a cerebral infarction and died due to acute respiratory failure on March 18, 2011. The man was in a nursing home, and after the earthquake, electricity, gas, and water supplies were shut down. Meals were offered only twice a day, and water intake dropped to about 40%.

His family filed for a condolence grant, but it was turned down, saying there was no cause-and-effect relation between the disaster and death. The family's protest was turned down, and the family filed a complaint with the court.

The judgment ruled that "The newly developed cerebral infarction is fairly reasonably attributed to the bodily and mental burdens caused by reduced water intake and worsened living conditions. Cause-and-effect should be acknowledged for the death and the earthquake in this case."

9.3.3 March 13, 2015, Morioka District Judgment

A 56-year-old male with chronic high blood pressure suffered worsened conditions after the earthquake. On November 22, 2011, he had an acute cardiac infarction and died of ventricular septal perforation about a month later on December 28.

His family filed twice for a condolence grant, but was turned down and the family filed a complaint to the court.

The judgment ruled that "The man worked and lived in an area largely affected by the tsunami. His house survived the disaster, but his main office building was washed away. Then after a long search for his parents-in-law, they were both found dead. While he had to make his loan payments and savings for his first daughter's education funds, there were no reconstruction plans for the main office building. He spent months without any source of income and it is clear that he had continually suffered from strong stress and anxiety caused by the disaster."

After this Morioka ruling, I joined Susumu Tsukui, Executive Director of Organization for Supporting the Building of Hanshin-Awaji Community, and Toshikazu Deguchi, Executive Director of Hyogo Research Center for Quake Restoration, and forwarded the following proposal to the Governor of Iwate Prefecture and Mayor of the City of Rikuzentakata [9].

[Proposal]

1. Based on this judgment and the framework of the judgment made on April 28, 1998 by the Osaka High Court following the Great Hanshin-Awaji Earthquake, we believe that all cases denied for disaster-related deaths by the Iwate Prefecture Committee for Provision of Disaster Condolence Grant shall be reviewed by your city.

2. If review of all cases is difficult, cases similar to this one shall not be left to the judgment by the Iwate Prefecture Committee for Provision of Disaster Condolence Grant, and your city shall judge whether the deaths were disaster related or not.

3. When you report this case to Iwate Prefecture, your city shall point out the problems in the past decisions by the Iwate Prefecture Committee for Provision of Disaster Condolence Grant and propose reviews of all cases rejected by the committee about the cause and effect between the disaster and deaths. The decisions shall be based on this judgment and the framework of the judgment made on April 28, 1998 by the Osaka High Court following the Great Hanshin-Awaji Earthquake.

I would like to point out problems with decisions on disaster-related deaths and with procedures for applying for support for them.

Yukihito Oguchi served as a committee member for providing condolence grants to the committee in the Town of Yamada in Iwate Prefecture. He said, "The decisions have to be made by committee members that experienced the disaster at where it happened [10]." The decisions were transferred to the prefectures from the municipalities and the following problems emerged [11]:

A: The local government staff lost their opportunity to understand real disaster-related death cases.

The staff would have understood what disaster related death is like by reviewing the filed applications and listening to examining experts.

B: Probability of making wrong decisions increased.

> *Only administrative workers in the disaster struck areas can know the conditions immediately after the disaster. Those in the Iwate Prefectural government, i.e., people in Morioka, will have a hard time grasping the reality and there was fear that decisions were made under wrong recognitions of facts.*

Tomomi Miyamoto who was a member of the Iwate Prefecture condolence grant approval committee agreed with Oguchi's view and wrote, "If there was a clear guideline about what investigation to make for the committee and what documents to file for the victims, a large number of decisions could probably have been made at local municipalities" [12]. She continued, "Different approval standards for the same disaster among municipalities would be unfair. A single set of standards, thus, are needed and that could probably have been set by the central government" [12].

Objections to the decisions concerning disaster-related deaths can be filed with Mayors of municipalities (request for reexamination). People can also file lawsuits to dispute decisions. The facts, however, showed that these processes were not fully functional and cases were reported in which the municipalities, responsible for teaching the victim's family about the processes of request for reexamination and taking the case to court, did not do so [13].

9.4 FUTURE OF DEEMED TEMPORARY HOUSING

Deemed temporary housing is private rental housing that a prefecture rents to provide housing for disaster victims, instead of the regular course of building temporary housing for the affected residents. It was in effect during the Great East Japan Earthquake. The Cabinet Office reported [14] the following, and we can see that this system is now well established:

> Prefectures may rent private rental housing instead of providing emergency temporary housing. In this case, to make the process of priority renting smooth, discussions and agreement with related organizations shall be made beforehand with reference to "About agreement on providing private rental housing to victims at the time of disaster (April 27, 2012)" and "About the use of private rental housing at the time of disaster (Summary of guidebook)" by MLIT and MHLW.

An investigation by the Cabinet Office reported that, as of May 2015, the number of residents in public housing was 16,163 for 6281 households,

86,720 in private rental housing for 37,206 households, and 76,347 in temporary housing for 36,390 households [15].

The advantages of deemed temporary housing are (1) disaster victims can quickly have a living space, (2) there is no initial cost compared to temporary housing and quality is above a certain level, and (3) they can meet the living needs of the tenants, e.g., commuting. On the other hand, deemed temporary housing is in general apartment houses, so it is hard to distinguish disaster victims from regular occupants. Guiding supporting organizations to meet with disaster victims in deemed temporary housing remains a problem to solve [9].

Disaster victims in deemed temporary housing are often "wide-area evacuees." The amendment of the Basic Act on Disaster Control Measures in 2012 and 2013 added items 8–13 to Article 86, about wide-area temporary housing.

Ikuta pointed out [6] that when the evacuation extends over a long period, "Maybe, a special arrangement about the extent and method of support is needed, or to discuss the establishment of a different system of disaster support when a certain time period has lapsed after the break out of the disaster." In the future, we probably need new standards for deemed temporary housing or an operation based on a system different from the Disaster Relief Act.

9.5 DISASTER VICTIMS AT HOME

The support force has a hard time reaching disaster victims at home, the same as those in deemed temporary housing. The victims are difficult to find, especially for outside supporting organizations, because they are living in their homes, as in the case with deemed temporary housing tenants. Ozeki reported that "Support and care for those disaster victims at home are the most critical for the support teams in terms of how well they can earn the cooperation from the local people in the disaster struck area."

The problem of not receiving sufficient support has existed previously, for those who did not go through the regular reconstruction process of starting from an evacuation shelter, then to temporary housing, and finally to permanent houses.

Cases have been reported [16] in which there was disparity in support between disaster victims in evacuation shelters or temporary housing and those at home, e.g., food was provided to disaster victims in shelters but not to those at home, or electrical appliances were offered to those in temporary housing but not to those at home.

The amendment of the Basic Act on Disaster Control Measures states that "efforts should be made so that at-home disaster victims also receive life support supplies, health and medical services, access to information, and other support necessary in arranging their living environment (Article 86, Item 7)." This statement means that not only supplies for living but also the living environment shall be kept at reasonable levels.

The Cabinet Office specifies in its guideline [8] the basic policy that "In managing an evacuation shelter, it is appropriate to support not only those evacuees in the shelter but also at-home victims that live in the area and the shelter shall be the hub for collecting and providing information, and supplying food, water, other supplies and services [8]." The guideline even has a section on "at-home shelter [8]."

When at-home disaster victims repair their houses, the Act on Support entitles each household to additional aid of 1 million JPY. However, the support is limited to houses that were "totally destroyed" or suffered "major half-destruction." The Disaster Relief Act offers a maximum of 520,000 JPY for emergency repair. However, the "General Standards" of this law limit those eligible to houses that are "half destructed and owned by a household without resources" or ones that had "major half-destruction" (refer to Section 9.1.2). Moreover, the number of construction workers was insufficient in the short term after the earthquake and the labor costs went up. Team-Ohkan (http://team-ohkan.net/) pointed out that the amount of 520,000 JPY paid for repair may not have been a fair amount [17]. Team-Ohkan is a corporation that has been active in providing comprehensive support, primarily for at-home disaster victims and those living in deemed temporary housing.

Some municipalities on their own offered monetary support for house repair. For example, the City of Ishinomaki in Miyagi Prefecture distributed 1 million JPY for "totally destroyed, major half-destroyed, and half-destroyed" or 1.5 million JPY for supporting interest payments.

After all, those that suffered "partial destruction" are not in the scope of governmental support. Even partial destruction costs money for house repair and purchasing appliances [16]. We may need a system for deciding the level of damage based on monetary needs for disaster victims. The conditions of some houses judged as suffering "partial destruction" were not very different from total or half destruction after ground subsidence followed by flooding [16]. We have a problem with how damage levels are determined.

9.6 DISASTER VICTIM SUPPORT, LAW INTERPRETATION, AND JUDICIAL RELIEF

9.6.1 Tsuneo Suzuki's Proposal

Discussions about the legal system for supporting disaster victims have so far developed around legal policies instead of legal interpretations. Suzuki pointed out [18] that in the field of law for disaster administration, "in Japan, not having grounds in setting laws are viewed as legal flaws and thus they are often postponed for later determination." He calls this tendency "forward interpretation" and explains it is caused by a "strong influence from ministries, agents, and proposers in interpreting disaster-related laws and the problem of 'gaps in law' being intentionally avoided." He further argues, "Having incomplete laws and pushing legal concerns to decide in the future is an act of abandoning problems in law interpretation," and, "There is much to discuss whether they could have made analogical inference or reasonable interpretation recognizing the problem as 'holes in law'."

Actually, support based on the Disaster Relief Act is emergency action and, while the government states its view [14] that "the people of Japan have no formal procedures to file complaints nor have set support processes based on such complaints," Suzuki is basing his claims on "human rights of disaster victims" to open the road for judicial relief [18]. The following sections review the actual basis of judicial relief in setting the Additional Aid in Disaster Relief Act.

9.6.2 Judicial Relief Based on Human Rights of Disaster Victims

Human rights of disaster victims, according to Suzuki, means "Legal benefit to maintain individual live, health, and quality of life under the restricted conditions during evacuation," and "legal benefit and substance under hypothetical situations that are life and health threatening caused by administration's failure to act." Suzuki further states that "These benefits are protected in Basic Act on Disaster Control Measures and Disaster Relief Act and are also legally protected by disaster related laws [18]."

Suzuki presumes "Litigation Seeking Declaratory Judgment of Public Law" when filing for judicial relief, saying "when legally protected benefit for those with legally acknowledged positions are not protected, probably leading to threats to life, health, and basic quality of life, the subjects shall seek legal affirmation of their positions before such conditions of sickness or death take place [18]."

Suzuki claims "legal benefit of life and health for the Japanese nationals and residents are benefits that are preconditions about Basic Act on Disaster Control Measures and Disaster Relief Act, thus even if people are in evacuation shelters or temporary housings, when necessary, they should request affirmation of positions presuming application of additional aid." When such requests for affirmation of positions results in "confirming illegality of not setting 'additional aid'," human rights of disaster victims will be accomplished by "local administration setting 'additional aid' by applying own discretion without relying on General Standards [18]."

Such judicial relief can apply if the evacuated life is extended over a long period.

9.6.3 Reference: Excerpts From Basic Act on Disaster Control Measures

Article 2, Item 2 (Basic Principles)

Disaster Control Measures shall take place under the following basic principles:

4. Place highest priority on lives and bodies of people by arranging workforce, supplies and other resources adequately by properly understanding the situations of the disaster, to the extent possible, even immediately after the strike when gathering of such information may be difficult.

5. While keeping attention not to interfere with the disaster victim's voluntary efforts, support the victim in timely manners given the victim's age, gender, and handicaps.

Article 8 (Consideration about disaster prevention in administration)

2. The central and local public entities shall prevent disasters and their spread by making the following efforts.

XIV. Securing health of the bodies and minds of disaster victims, their living spaces, and other matters to protect the disaster victims.

XV. Matters that are needed for disaster prevention especially for those that require special attention like elderlies, handicapped, infants, and others.

Article 86, Item 6 (Arranging living environment in evacuation shelters)

Those in charge of disaster emergency actions, upon outbreak of a disaster, shall follow the regulations and disaster prevention plans to, without delay, provide evacuation shelters, and distribute necessary safety, good living conditions, food, clothing, medicine, and other daily necessities, as well as health and medical services. They also need to make efforts to take actions to arrange the living environment of disaster victims that stay in evacuation shelters.

Article 86, Item 7 (Consideration for disaster victims staying at places other than evacuation shelters)

Those in charge of disaster emergency actions, shall distribute daily necessities, health and medical services, and information to disaster victims that cannot stay in evacuation shelters for unavoidable reasons as well. They also need to make efforts to take actions to arrange the living environment of such disaster victims.

ACKNOWLEDGMENT

The work in this chapter was supported by the Ministry of Education, Culture, Sports, Science and Technology, Project for Mitigation of Severe Disasters Caused by Urban Vulnerabilities, "(3) Study and Research about Methods for Enhancing Disaster Response upon Urban Disasters," and Japan Society for the Promotion of Science, Grants-in-Aid for Scientific Research 25300013, 25301013, 26510014, and 26560086.

REFERENCES

[1] Reconstruction Agency, Reconstruction work and related systems, June 24, 2015. http://www.reconstruction.go.jp/topics/main-cat1/sub-cat1-1/20150624_torikumi_seido.pdf (Accessed 7 September 2015).

[2] T. Kondo, The founding background and characteristics of housing rebuilding assistance subsidy programs by local governments after the Great East Japan Earthquake, Archit. Inst. Jpn. Tokyo 80 (707) (2015) 135–144.

[3] About the death toll of the Great Hanshin Earthquake, Hyogo Prefecture, Dec. 22, 2005, updated Jan. 28, 2008. https://web.pref.hyogo.lg.jp/pa20/pa20_000000016.htm (Accessed 23 September 2015).

[4] The number of disaster related deaths with the Great East Japan Earthquake (as of March 31, 2015), Reconstruction Agency, June 30, 2015. http://www.reconstruction.go.jp/topics/main-cat2/sub-cat2-6/20150630_kanrenshi.pdf (Accessed 10 September 2015).

[5] Report on disaster related deaths with the Great East Japan Earthquake, Reconstruction Agency, Aug. 21, 2012. http://www.reconstruction.go.jp/topics/20120821_shinsaikanrenshihoukoku.pdf (Accessed 7 September 2015).

[6] O. Ikuta, Laws of Disaster Prevention, Shinzansya, Tokyo, 2013.

[7] Y. Kitamura, Administration by administrative management-1, Local Gov. Study, Tokyo 91 (3) (2015) 33–51.

[8] Guideline for targeting securing good living conditions in evacuation shelters, Cabinet Office, 2013. http://www.bousai.go.jp/taisaku/hinanjo/h25/pdf/kankyoukakuho-honbun.pdf (Accessed 7 September 2015).

[9] Lawyers requested reexamination of all cases to the City of Rikuzentakata, about rejection of earthquake related deaths, Asahi Newspaper Morning edition, March 18, 2015.

[10] E. Yamasaki, Natural Disasters and Supporting the Victims, Nippon Hyoronsha, Tokyo, 2013.
[11] Y. Oguchi, Report: Actively file for decisions on "disaster related deaths." http://oguchilaw.cocolog-nifty.com/blog/2012/04/post-f1da.html (Accessed 7 September 2015).
[12] T. Iwamoto, Judgment on the disaster-related death-efforts by the Examination Board of Iwate Prefecture in the Great East Japan Earthquake, Artes Lib. 92 (2013) 67–86. http://hdl.handle.net/10140/5230.
[13] Insufficient explanation of objection procedure. 10 municipalities in the prefecture, Iwate Nippo, Feb., 22, 2012. http://www.iwate-np.co.jp/311shinsai/y2012/m07/sh1207222.html (Accessed 7 September 2015).
[14] Guideline for administrative handling of disaster support (June 2014), Cabinet Office. http://www.bousai.go.jp/taisaku/kyuujo/pdf/h26kaigi/siryo1-2.pdf (Accessed 7 September 2015).
[15] Current situation of reconstruction (June 24, 2015). http://www.reconstruction.go.jp/topics/main-cat1/sub-cat1-1/20150624_genjou.pdf (Accessed 7 September 2015).
[16] T. Takanarita, R. Fuse, At-home disaster victims, left out from reconstruction, World (No. 829), Iwanami, Tokyo, April 2012, pp. 155–164.
[17] T. Ozeki, 3.11 and supporting disaster victims, (2) Struggle with supporting reconstruction of lives – care for victims in temporary housing, at home, and at evacuation shelters, Wage Soc. Secur. Tokyo 1543 (44) (2011) 42–61.
[18] T. Suzuki, Large scale earthquake disasters and people's lives, in: T. Suzuki (Ed.), Large Scale Earthquake Disasters and Administrative Actions, Nippon Hyoronsha, Tokyo, 2015.

FURTHER READING

[1] Team-Ohkan, Report: The Great East Japan Earthquake at-home disaster victims 'Investigation of the status of house repair' – The reality after 3 years and 8 months from the Great East Japan Earthquake, Miyagi, Nov. 21, 2014.

Chapter 10

The Great East Japan Earthquake and Insurance

CHAPTER OUTLINE

10.1 NATURAL DISASTERS AND INSURANCE SYSTEM IN JAPAN

I will first summarize the insurance system covering natural disasters in Japan, but will focus only on property and casualty insurance, which is mainly to recover loss of or damage to property (including loss of profit), and cooperative insurance. First, concerning property and casualty insurance: as of September 30, 2015, there are 30 property and casualty insurance companies, including two reinsurance companies. Excluding the two reinsurance companies, the domestic direct net premium paid to the 28 insurance companies totaled 8.34 trillion JPY. The property and casualty insurance market in Japan is oligopolistic, with only three major insurance companies making up 83.8% of the total premium. In the world market, Japan ranks third after the United States and China, but the property and casualty insurance premium per capita is only 37.5% of that in the United States. The insurance density in Japan is small among developed countries.

Table 10.1 shows the coverage of the main property and casualty insurances that cover damage from natural disasters. The table shows the composition ratio of the premium for each coverage. It is an indication of the weight of

The Fukushima and Tohoku Disaster. https://doi.org/10.1016/B978-0-12-814078-9.00010-8

Table 10.1 Main Property and Casualty Insurances That Cover Damages From Natural Disasters

Insurance Type	Property Insured	Primary Contractor	Wind and Flood	Earthquake	Direct Net Premium in JFY2013 (JPY100 Million)	Premium Composition
Fire	Buildings	Ind./Corp.	O	△	11,482	13.77%
Household earthquake	Houses and personal property	Ind.	X	O	1989	2.39%
Hull	Ship hull	Corp.	O	O	847	1.02%
Cargo	Cargo during shipping on the ocean	Corp.	O	△	1326	1.59%
Transit	Cargo during domestic transit	Corp.	O	△	655	0.78%
Aviation	Airplane hull	Corp.	O	O	154	0.18%
Machinery breakdown and erection all risks	Machinery, plant under construction and after completion	Corp.	O	△	305	0.37%
Contractors' all risks	Buildings under construction and civil engineering structures under construction and after completion	Corp.	O	△	423	0.51%
Movable all risks	Movable property	Ind./Corp.	O	△	890	1.07%

O, covered by original wording; △, covered by endorsement; X, not covered.

coverage in the insurance market. Among the property and casualty insurances that cover natural disasters, fire insurance is the major one with 13.77% of the entire premium. The types of insurance that individuals can have are fire, earthquake-caused household damage, and movables all risks. Other insurance types in this table are for corporations. The household earthquake insurance for individuals is a public type supported by the government. The insurance has a policy of no loss and no profit and the private insurance companies will not gain long-term profit from selling the insurance. The insurances in the table except for the household earthquake insurance are sold by private insurance companies for the purpose of profit generation.

Next, we will discuss cooperative insurance. With cooperative insurance, the union members pay a set premium to prepare a mutual fund to pay mutual aid in the case of unexpected damage. It is a mutual-aid system to stabilize the lives of the union members and families by compensating for unexpected economic damage.

Table 10.2 Unions That Offer Fire Cooperative Insurance

Union With Cooperative Insurance	Member Organization
Japan Agricultural Cooperatives	National Mutual Insurance Federation of Agricultural Cooperatives
Japan Fisheries Cooperative	National Mutual Insurance Federation of Fisheries Cooperatives
Japanese Consumers' Cooperative Union	National Mutual Insurance Federation of Workers Cooperatives, University Cooperatives Mutual Aid Federation, Consumers Cooperatives Mutual Aid Federation, Ministry of Defense Workers Cooperatives Mutual Aid Federation
Business Cooperative Association	National Fire Insurance Federation of Cooperatives
Agricultural Insurance Scheme	NOSAI established at each prefecture and local communities

Among the types of cooperative insurance, fire cooperative insurance covers damage caused by fire, thunderstorms, explosion, and other events. Among them are some that cover earthquake, wind and water caused damage. Table 10.2 lists the main unions that offer fire cooperative insurance.

The premium paid in 2013 for cooperative insurance summed to 2.67 trillion JPY, which is greater than the total premium payment for property and casualty insurance in Table 10.1. This means that cooperative insurance plays a major role in compensating for damage to buildings and properties.

By the way, insurance companies serving individuals, such as household earthquake insurance or life insurance companies, sometimes reveal the amount of insurance paid for damage caused by the Great East Japan Earthquake as open information. Corporate earthquake insurance companies, however, do not open their files and reveal the amount of insurance paid, because such information regarding what insurance the companies prepared and the amount of insurance payments is considered confidential information related to corporate production activities.

This chapter not only discusses individual insurance, but it also analyzes the amount of corporate insurance paid and for which types of damage caused by the Great East Japan Earthquake. The payments are compared to those made at the time of the Great Hanshin Earthquake. We then evaluate the public policy of household earthquake insurance with the purposes of damage reduction and saving the disaster victims.

Methods for collecting information about insurance payments to corporations, and their disadvantages, are as follows:

(1) Questionnaires to beneficiaries of corporate insurance
   (a) What companies can answer is limited.
   (b) The research is expensive.

   (c) Companies that respond are part of an entire group and macroscopic data is unavailable.

(2) Hearings involving experts

   (a) Experts are also limited in what they can answer.
   (b) Different experts will have different opinions.
   (c) The truth cannot be found after all.

(3) Analysis reports by ranking or think-tank organizations, e.g., Standard and Poor's

   (a) Most of the analyses are based on confidential information and the credibility of such data is low.
   (b) A major cost is involved in obtaining the analysis reports.

With this evaluation, we decided to carry out our research based on widely available data. Our data source was "Insurance: Volume on statistics of property & casualty insurance." This method provides data that are more visible to readers, compared to the previously described methods (1)–(3).

10.2 INSURANCE LINKS TO FOREIGN INSURANCE COMPANIES

Insurance companies that pay for damages caused by earthquakes in Japan are those with business licenses in Japan. Thus, it may appear that insurance payment is closed within the country of Japan. However, for insurance companies to assure payment to corporations and individuals, the involvement of foreign insurance companies is needed. In general, insurance companies operate with the law of large numbers [1] in estimating insurance coverage payment. Here the law of large numbers applies when each insurance event has a probability that is independent.

The law, however, does not apply to damage caused by natural disasters such as earthquakes, typhoons, or flooding. Assume, for example, two separate buildings, one in Iwate Prefecture and the other in Fukushima Prefecture, having the same insurance company covering the buildings, each at 10 million JPY. If the insurance only covers fires, simultaneous coverage payments for both buildings are very unlikely. A fire at the building in Iwate Prefecture does not cause damage to the building in Fukushima Prefecture.

In contrast, for making a coverage payment in the case of earthquake damage, if the buildings in Iwate and Fukushima Prefectures were damaged by the same earthquake, the insurance company has to make the payment at the same time for the two buildings. That is, for an insurance that covers only damage against fire, the maximum coverage payment at a time is 10 million

JPY, whereas if it covered the risk against earthquakes, the maximum coverage payment at one time may sum up to 20 million JPY.

In this example with only two buildings, the possible maximum coverage payment was twice, but in reality, the insurance company has to estimate the maximum coverage payment for damage from an earthquake as the sum of all buildings it covers in the area that can be affected by a single earthquake. For an insurance company, the risk from natural disasters like earthquake, typhoon, or flooding can accumulate the direct claims paid; this is called accumulation risk.

Reinsurance [2] is one way of stabilizing the operations of insurance companies by moderating the effect of total coverage payment due to natural disasters. Fig. 10.1 shows the mechanism of reinsurance. Reinsurance is a mechanism for insurance companies to transfer partial or entire risks to other insurance companies.

The insurance company that covers the risk of the insured is called the direct insurance company, the contract between the insurance company and the insured is the direct insurance contract, the payment the insured makes to the insurance company is the direct premium, and the payment the direct insurance company makes to the insured is the direct claims paid. The insurance company that covers the risk of the direct insurance company, other than the direct insurance company itself, is called the reinsurance company, the contract between the original insurance company and the reinsurance company is the reinsurance contract, the payment the direct insurance company makes to the reinsurance company is the reinsurance premium, and the payment the reinsurance company makes to the original insurance company is the reinsurance claims paid [2]. Outsourcing the risk with reinsurance is called cession, and covering another insurance company's risk with reinsurance is called acceptance of reinsurance. What takes place is, by ceding the

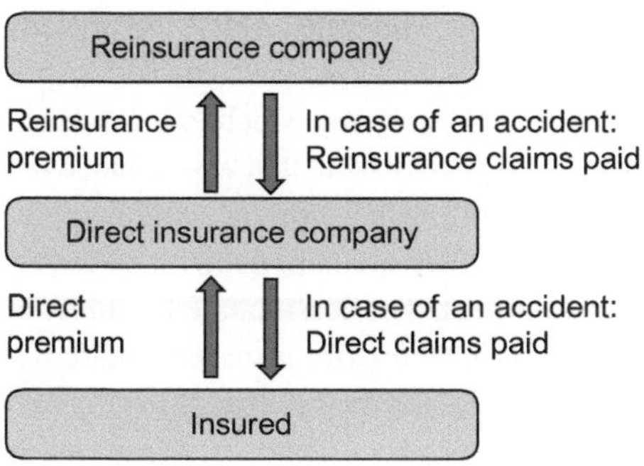

■ **FIG. 10.1** Mechanism of reinsurance.

risk of the insured from the natural disasters of earthquakes, typhoons and flooding, the direct insurance company can stabilize its operations by reducing the total net claims paid (direct claims paid less the reinsurance claims) upon damage by natural disasters.

Note here that the risk of natural disasters is an accumulation risk. Here we will discuss an automobile accident coverage for driver X. Assume a domestic insurance company A had a liability insurance agreement with the driver. Say the direct insurance coverage was 10 billion JPY and if A thought the coverage was too high, A would cede part of X's insurance risk to another domestic insurance company B. The insurance company B rarely has any other insurance contracts to cover when X has an automobile accident, and B will freely judge the portion of the coverage of 10 billion JPY it wants to accept with reinsurance.

The situation is different, however, if the direct insurance is own vehicle insurance covering property damage to X's top-of-the-line sports car. Say the own vehicle coverage was 300 million JPY, and even if A thought it was too expensive, A cannot cede the risk to B. The reason for this is that B has many other own vehicle insurance contracts and with a single hit of a typhoon, B would have to make a number of direct claims payments at the same time it covers the own vehicle coverage of X's top-of-the-line sports car. In other words, B cannot base its judgment of whether to reinsure the case on the 300 million JPY coverage.

In countering accumulation risks, entering reinsurance agreements with other insurance companies with operations in regionally nearby areas is difficult. Therefore, reinsurance companies that reinsure risks of natural disasters in Japan in practice are limited to insurance companies overseas. When we, therefore, discuss the natural disaster of the Great East Japan Earthquake and insurance, we naturally have to talk about reinsurance as well.

The total direct insurance premium, excluding the household earthquake insurance and compulsory automobile liability insurance, of direct insurance companies in Japan Fiscal Year (JFY) of 2013 (April 2013 to March 2014) was 7189.3 million JPY, among which 952.7 million JPY were ceded with a cession rate of 13.25% [3]. Note that we excluded the direct premium payment for household earthquake insurance and compulsory automobile liability insurance because they are both public insurances with a high rate of cession by the government. The cession rate for direct insurance companies in JFY 2013 was 30.52% for fire insurance that covers accumulation risks by natural disasters. Fire insurance, in addition to building fires, covers wind and water damage from, e.g., typhoons. Corporate earthquake insurance, furthermore, has special agreements to cover earthquake damage.

In contrast, the cession rate for liability insurances that do not cover accumulation risks by natural disasters was 11.71% (including reinsurance to overseas and domestic insurance companies because separately counting the two is difficult). That is, the cession rate is largely affected by whether the insurance covers damage from natural disasters or not.

10.3 **INSURANCE COVERAGE PAID**

In estimating the amount of paid coverage, we looked into "Statistics of non-life Insurance [3]." Our estimate started by calculating the loss ratio of each insurance type of fire, hull, cargo, transit, personal accident, aviation, machinery breakdown, erection all risks, contractors' all risks and movable all risks, to see they were notably large in JFY 2011. For insurance with large differences, we assumed the difference to be direct claims paid toward damage caused by the Great East Japan Earthquake (earthquake claims).

Fig. 10.2 shows the earthquake claims for property and casualty insurance and life insurance. The figures do not include cooperative insurance. The earthquake claims of life insurance in JFY 2011 were only 1.02% of the total direct claims paid during the fiscal year [4]. The earthquake claims of property and casualty insurance, on the other hand, amounted to 30.72% of the total. The figure shows there was a large influence on property and casualty insurance covering the Great East Japan Earthquake damage.

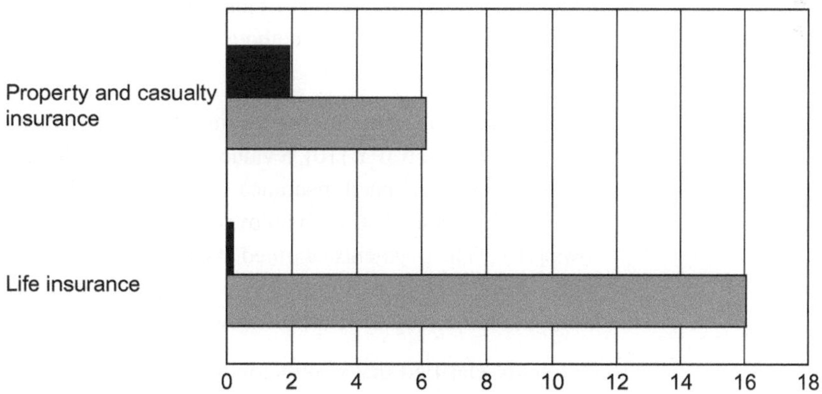

	Life insurance	Property and casualty insurance
■ Earthquake claims	0.16	1.9
□ 2011 Direct claims paid	16.08	6.17

■ **FIG. 10.2** Claims paid for the Great East Japan Earthquake (unit: trillion JPY).

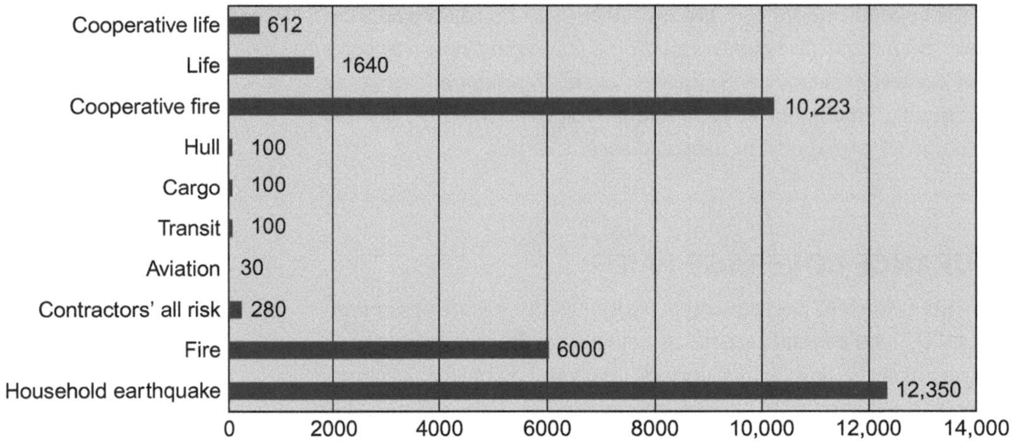

■ **FIG. 10.3** Claims paid for the Great East Japan Earthquake (unit: 100 million JPY).

Fig. 10.3 shows the breakdown of claims paid by cooperative insurance [5–8] and of property and casualty insurance [9]. Property and casualty insurance, life insurance, and cooperative insurance sum to about 3143.5 billion JPY. Of this sum, property and casualty insurance and fire cooperative insurance are paid for recovery of property damage and their sum was 2918.3 billion JPY.

Ishii [10] also calculated insurance claims paid for damage from the Great East Japan Earthquake at 3.1 trillion JYP. The small difference is due to differences of classes in each type of insurance, when the claims were paid, and differences in calculation methods. The results, however, were about the same.

Ishii's calculation of the sum of insurance claims paid for recovery resulted in about 2.9 trillion JPY [10], a value that also matches our calculation. He, thus, correctly evaluated insurance claims paid (2.9 trillion JPY above) in terms of their contribution to recovery from the earthquake disaster as follows [10]: "The payments summed to only 15% of a conservative total economic damage of 16.9 trillion JPY, and each household and company had to cover the remaining 85% on their own."

Our study clarified that contractors' all risks, aviation, transit, cargo, and hull for corporations are likely to have paid to cover earthquake damage. Application of our analysis method to the Great Hanshin Earthquake resulted in a small possibility of such insurance payment. Not just these types of insurance but also earthquake insurance for corporations have small possibilities of having made large payments.

The wide area of damage from the Great East Japan Earthquake compared to those from the Great Hanshin Earthquake may have caused these results. Furthermore, the trend of more corporations entering insurance contracts to cover earthquake damage after the Great Hanshin Earthquake also contributed to these results.

10.4 RELATION BETWEEN THE NUMBER OF VICTIMS AND INSURANCE PAYMENT

Some insurance companies have premium payments that are adequate for the risk covered by the insurance, but some premiums are out of balance. Private insurance offered by private insurance companies are the former, with customers purchasing them to cover themselves. In these cases, the premium is strongly linked to the risk and the variance in the premium encourages the insured to take disaster prevention and mitigation activities. This efficient system enhances the economic efficiency of society, with investment in disaster prevention and mitigation, and the overall benefit to society is optimized. This type of insurance is "market driven."

Those that are facing high risk can purchase market-driven insurance at the cost of a high premium. If we, however, think about regional differences among places of birth, people born in regions with frequent earthquakes or tropical depressions of typhoons and cyclones are facing much higher risk. Some people are born with genetic disturbances accompanied by higher risk. Some are born sensitive to the environment, while others are more resistant. The risk of illness greatly depends on the individual.

The place of birth and genetic differences are out of people's control. "Mutual aid insurance" is based on the concept of avoiding high premiums for such people with high risk and sharing the risk among the insured. These types are social insurance with mandatory participation, e.g., national health insurance, long-term nursing insurance, social security, or unemployment. The premium does not depend on the risk covered by the insurance. Social insurance is a type of public insurance. Some privately operated insurance companies, e.g., compulsory automobile liability insurance or nuclear energy insurance, are also categorized as public.

This chapter has so far reviewed insurance from the viewpoint of market-driven insurance. We will change our viewpoint to mutual aid, i.e., insurance as a tool for public policies, or the viewpoint of helping disaster victims. Our further analysis calls for a new variable, the number of victims. Tables 10.3 and 10.4 show worst rankings in the journal *sigma* published annually by Swiss Re [11,12], the second largest reinsurance company.

Table 10.3 The 40 Worst Catastrophes in Terms of Victims (1970–2014)

Rank	Dead and Missing	Insured Loss[a] (US$M)	Date (Start)	Event	Country
1	300,000	–	11/14/1970	Storm and flood catastrophe	Bangladesh
2	255,000	–	7/28/1976	Earthquake (M7.5)	China
3	222,570	109	1/12/2010	Earthquake (Mw7.0)	Haiti
4	220,000	2506	12/26/2004	Earthquake (Mw9), tsunami in Indian Ocean	Indonesia, Thailand et al.
5	138,300	–	5/2/2008	Tropical cyclone Nargis	Myanmar (Burma), Bay of Bengal
6	138,000	4	4/29/1991	Tropical cyclone Gorky	Bangladesh
7	87,449	403	5/12/2008	Earthquake (Mw7.9) in Sichuan, aftershocks	China
8	74,310	–	10/8/2005	Earthquake (Mw7.6); aftershocks, landslides	Pakistan, India, Afghanistan
9	66,000	–	5/31/1970	Earthquake (M7.7) triggers massive avalanche and floods	Peru
10	55,630	–	6/15/2010	Heat waves with temperatures up to 40 degrees Celsius	Russia, Czech Republic
11	40,000	208	6/21/1990	Earthquake (M7.7); landslides	Iran
12	35,000	1622	6/1/2003	Heat wave and drought in Europe	Spain, Italy, Germany et al.
13	26,271	–	12/26/2003	Earthquake (M6.5), destroys 85% of Bam	Iran
14	25,000	–	12/7/1988	Earthquake (M7.7) in Tabas	Armenia, ex USSR
15	25,000	–	9/16/1978	Earthquake (M6.9)	Iran
16	23,000	–	11/13/1985	Volcanic eruption on Nevado del Ruiz and avalanches	Colombia
17	22,300	312	2/4/1976	Earthquake (M7.5)	Guatemala
18	19,737	134	1/26/2001	Earthquake (Mw7.6) in Gujarat	India, Pakistan
19	19,118	1421	8/17/1999	Earthquake (M7.0) in Izmit	Turkey
20	18,520	36,828	3/11/2011	Earthquake (Mw9.0) triggers tsunami (The Great East Japan Earthquake)	Japan
21	15,000	142	10/29/1999	Tropical cyclone 05B	India, Bangladesh
22	14,204	–	11/20/1977	Tropical cyclone in Andhra Pradesh	India
23	11,069	–	5/25/1985	Tropical cyclone in Bay of Bengal	Bangladesh
24	10,800	–	10/26/1971	Odisha cyclone, flooding in Bay of Bengal and Orissa state	India
25	10,000	313	12/12/1999	Floods, mudflows and landslides	Venezuela
26	9500	1041	9/19/1985	Earthquake (M8.1)	Mexico
27	9475	–	9/30/1993	Earthquake (M6.4)	India
28	9000	726	10/22/1998	Hurricane Mitch in Central America	Honduras, Nicaragua et al.

Table 10.3 The 40 Worst Catastrophes in Terms of Victims (1970–2014) *Continued*

Rank	Dead and Missing	Insured Loss (US$M)	Date (Start)	Event	Country
29	8135	518	11/8/2013	Typhoon Haisan, storm surge	Philippines, Vietnam, China, Palau
30	7079	–	8/17/1976	Earthquake (Mw7.9), tsunami in Moro Gulf	Philippines
31	6425	3839	1/17/1995	Great Hanshin earthquake (M7.2) in Kobe	Japan
32	6304	–	11/5/1991	Typhoon Thelma (Uring)	Philippines
33	6000	–	12/2/1984	Accident in chemical plant; methyl isocyanates released	India
34	6000	–	6/1/1976	Heat wave, drought	France
35	5749	47	5/27/2006	Earthquake (M6.3); Bantul almost destroyed	Indonesia
36	5748	508	6/14/2013	Floods caused by heavy monsoon rains	India
37	5422	–	6/25/1976	Earthquake (M7.1)	Indonesia
38	5374	–	4/10/1972	Earthquake (M6.9) in Fars	Iran
39	5300	–	12/28/1974	Earthquake (M6.3)	Pakistan
40	5000	–	12/23/1972	Earthquake (M6.2)	Iran

[a]Property and business interruption, excluding liability and life insurance.
From Swiss Re, sigma No 2/2015.

The insurance damage in these tables is the sum of insurance payments for physical damage covered by property and casualty insurance, such as fire and earthquake insurances. The amounts were adjusted with the level of prices. It may be questionable whether insurance payments to cover physical damage can directly save disaster victims, but it is an index to judge how well the mechanism of insurance covers damage from a disaster.

Table 10.3 shows the worst ranking of the number of victims from 1970 to 2014. Disasters with large numbers of victims were natural disasters in developing countries. The only man-caused social disaster in the ranking is the chemical factory disaster at Bhopal, India at 33. Disasters in Japan that are listed are the Great East Japan Earthquake at 20 and the Hanshin Great Earthquake at 31.

Table 10.4 shows the worst insurance payment ranking from 1970 to 2014. Disasters that resulted in large insurance payments were natural disasters in advanced countries in Europe and America. The only human-caused social disaster was the 9/11 terrorist attack in the United States alone. Disasters in

Table 10.4 The 40 Most Costly Insurance Losses (1970–2014)

Rank	Insured Loss[a] (UD$M)	Dead and Missing	Date (Start)	Event	Country
1	78,638	1836	8/25/2005	Hurricane Katrina; storm surge, damage to oil rigs	United States, Gulf of Mexico, Bahamas
2	36,828	18,520	3/11/2011	Earthquake (Mw 9.0) triggers tsunami (The Great East Japan Earthquake)	Japan
3	36,079	237	10/24/2012	Hurricane Sandy, massive storm surge	United States
4	26,990	43	8/23/1992	Hurricane Andrew; floods	United States, Bahamas
5	25,104	2982	9/11/2001	Terror attack on WTC, Pentagon, other buildings	United States
6	22,355	61	1/17/1994	Northridge earthquake (Mm 6.6)	United States
7	22,258	136	9/6/2008	Hurricane Ike	United States, Caribbean, Gulf of Mexico et al.
8	16,836	181	2/22/2011	Earthquake (Mw 6.3), aftershocks	New Zealand
9	16,157	119	9/2/2004	Hurricane Ivan; damage to oil rigs	United States, Caribbean, Barbados et al.
10	15,783	815	7/27/2011	Floods caused by heavy monsoon rains	Thailand
11	15,234	35	10/19/2005	Hurricane Wilma; torrential rain, floods	United States, Mexico, Jamaica, Haiti et al.
12	12,240	34	9/20/2005	Hurricane Rita: floods, damage to oil rigs	United States, Gulf of Mexico, Cuba
13	11,339	123	7/15/2012	Drought in the Corn Belt	United States
14	10,087	24	8/11/2004	Hurricane Charley	United States, Cuba, Jamaica et al.
15	9813	51	9/27/1991	Typhoon Mireille	Japan
16	8730	71	9/15/1989	Hurricane Hugo	United States, Puerto Rico et al.
17	8682	562	2/27/2010	Earthquake (Mw 8.8) triggers tsunami	Chile
18	8458	95	1/25/1990	Winter storm Daria	France, United Kingdom et al.
19	8241	110	12/25/1999	Winter storm Lothar	Switzerland, United Kingdom, France et al.
20	7681	321	4/22/2011	Major tornado outbreak; 343 tornadoes, hail	United States
21	7418	177	5/20/2011	Major tornado out break (180 tornadoes)	United States
22	6959	54	1/18/2007	Winter storm Kyrill, floods	Germany, United Kingdom et al.

Table 10.4 The 40 Most Costly Insurance Losses (1970–2014) *Continued*

Rank	Insured Loss (UD$M)	Dead and Missing	Date (Start)	Event	Country
23	6456	22	10/15/1987	Storm and floods in Europe	France, United Kingdom et al.
24	6449	38	8/26/2004	Hurricane Frances	United States, Bahamas
25	6134	50	8/22/2011	Hurricane Irene, torrential rainfall, flooding	United States, Canada, Bahamas et al.
26	5780	64	2/25/1990	Winter storm Vivian	Switzerland, Germany
27	5740	26	9/22/1999	Typhoon Bart	Japan
28	5426	–	9/4/2010	Earthquake (Mw 7.0), over 300 aftershocks	New Zealand
29	5125	600	9/20/1998	Hurricane Georges; floods	United States, Caribbean
30	4818	41	6/5/2001	Tropical strom Allison; heavy rain, floods	United States
31	4765	3034	9/13/2004	Hurricane Jeanne; floods, landslides	United States, Caribbean, Haiti et al.
32	4492	45	9/6/2004	Typhoon Songda	Japan, South Korea
33	4200	25	5/27/2013	Floods	Germany, Czech Republic et al.
34	4123	51	5/2/2003	Thunderstorms, tornadoes, hail, flash floods	United States
35	4010	70	9/10/1999	Hurricane Floyd; heavy rains, floods	United States, Bahamas
36	3899	–	7/27/2013	Hailstorms	Germany, France
37	3882	59	10/1/1995	Hurricane Opal; floods	United States, Mexico, Gulf of Mexico et al.
38	3839	6425	1/17/1995	Great Hanshin earthquake (M7.2) in Kobe	Japan
39	3501	25	1/24/2009	Winter storm Klaus	France, Spain
40	3410	45	12/27/1999	Winter storm Martin	Spain, France, Switzerland, Italy

[a]Property and business interruption, excluding liability and life insurance losses; United States natural catastrophe figures based on Property Claim Services, including National Flood Insurance Program losses.
From Swiss Re, sigma No 2/2015.

Japan that are ranked were the Great East Japan Earthquake at number 2, the 1991 typhoon number 19 ranked 15th, the 1999 typhoon 18 at 27th, the 2004 typhoon 18 at 32nd, and the Great Hanshin Earthquake at number 38. From these facts, we can conclude the following:

(1) The number of victims is small with disasters with large insurance payments.

(2) Insurance payment is small for disasters with a large number of victims.
(3) Natural disasters cause either a large number of victims or large insurance payments, but hardly any marked large numbers in both aspects.
(4) Japan is unique with some natural disasters with both large insurance payment and a large number of victims.

From the first point (1), we can see that advanced countries have many insured (high insurance density) with high income who can afford the high premium for market-driven insurance and when struck by a natural disaster, large insurance payments are made.

From point (2), regions with frequent natural disasters (in most cases, developing countries) have market-driven insurance with high premiums that are rarely purchased; thus, when a disaster with a large number of victims strikes, the insurance payments are low, causing a delay in the recovery and the economy to stagger. As a result, a negative spiral further lowers people's income levels and no one can afford the premium payment for insurance.

The reason behind point (4) is the household earthquake insurances in Japan. Natural disasters in Japan that result in large numbers of victims are earthquakes. When one happens, residential earthquake insurance makes large insurance payments. The next section will discuss household earthquake insurances from the viewpoint of their links to overseas insurances, i.e., reinsurance.

10.5 OVERSEAS LINKS OF RESIDENTIAL EARTHQUAKE INSURANCES

Private insurance companies have household earthquake insurance policies, so they are market driven. The government, however, assumes the role of reinsurance companies so more people will purchase earthquake insurance. The target is to stabilize the lives of earthquake victims. Earthquake insurance, therefore, serves also as public insurance as a tool of public policies.

The amounts of individual premium payments for public insurance are not adequately calculated, and the policies are more like mutual aid insurance. Among all social insurance, only worker's compensation is designed to reflect an appropriate amount of payment for their premiums and this is insurance for which corporations make premium payments.

Other social insurance is for individuals and the premium payments do not reflect the risk covered by their policies. Compulsory automobile liability insurance is public insurance for individuals, but their premium payments are more like mutual aid insurance, without reflecting adequate amounts compared to voluntary automobile liability policies.

Household earthquake insurance is public insurance designed for individuals and is a type of mutual aid insurance. The Japanese government functions as the reinsurance company, so these household earthquake insurance policies are not linked to overseas reinsurers discussed in Section 10.2. In other words, they are internationally isolated. The long-term balance that the government targets makes such arrangements possible, compared to the short-term balance that private reinsurance companies look for. Fifty years have passed since household earthquake insurance came on the market and 1994 and 2011, i.e., the years of the Great Hanshin Earthquake and the Great East Japan Earthquake, have been the only years when the annual loss ratio exceeded 25%. Private reinsurance companies cannot operate to target a balance with major earthquakes, with a frequency of only twice in 50 years.

If private reinsurance companies had handled policies for household earthquake insurance, they would have raised the reinsurance premium to quickly recover their loss after major earthquakes. In fact, corporate earthquake insurance companies raised their premium payment in 2012 following the Great East Japan Earthquake.

Fig. 10.4 shows the time history of cession rate of both premium and claims paid for fire insurance including earthquake coverage for corporations. The figure shows a large cession rate of claims paid for the years 2011 and 2012. This indicates large reinsurance payments were made by reinsurance companies to the insurance companies to cover their large direct claims paid. The cession rate of premiums for 2012 and 2013 were also high, showing the increase in reinsurance premiums. The rise in earthquake insurance premiums for corporations in 2012 was partly due to the increase in reinsurance premiums.

Household earthquake insurance has its doors closed to overseas reinsurers to avoid this type of sudden change in the insurance premium. Household earthquake insurance reviewed their earthquake source model and raised their premium payments in July of 2014. This increase was much later than other types of insurance with links to overseas reinsurance companies. The Great East Japan Earthquake confirmed the advantage of international isolation of household earthquake insurance.

It is said that calculation of the premiums for insurance that covers earthquake risks is different with overseas private reinsurance companies and

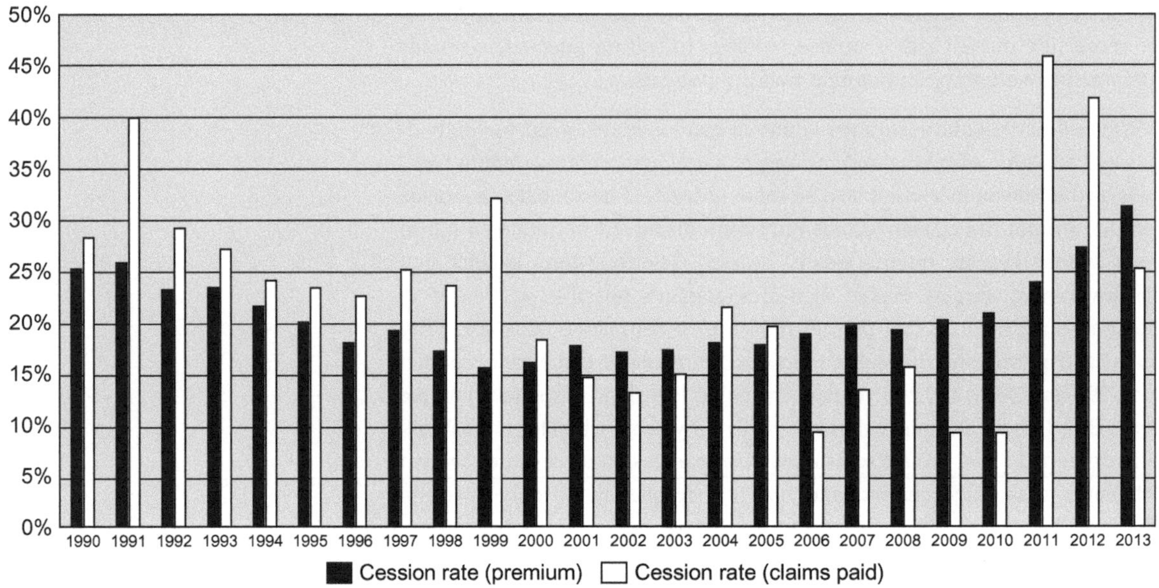

■ **FIG. 10.4** Cession rate of fire insurance. *(Based on Statistics of Non-life Insurance, Insurance Research Center, Tokyo, 1966–2014.)*

Table 10.5 Business of Residential Earthquake Insurance (Unit: 100 million JPY)

	Insurance Premium	Insurance Payment	Loss Ratio	Balance
Private insurance company	19,152	9231	48.2%	9921
Government	10,598	5360	50.6%	5238
Total	29,750	14,591	49.0%	15,159

Based on Statistics of Non-life Insurance, Insurance Research Center, Tokyo, 1966–2014.

the Japanese government, so premiums in Japan for household earthquake insurance are low [10]. In other words, isolating household earthquake insurance from foreign reinsurance companies allows us to buy earthquake coverage at low premiums. It is an advantage of international isolation of household earthquake insurance as a form of mutual aid insurance.

Table 10.5 shows how the household earthquake insurance performed during the 48 years of 1966–2013. The insurance premium may be lower compared to the overseas reinsurance market, but both private insurance companies and the government have not posted a loss in their operations.

10.6 IMPORTANCE OF INSURANCE UPON NATURAL DISASTERS

As we have seen, large insurance payments were made against damage from the Great East Japan Earthquake. Payments made to corporations were, in fact, much less than their business operations. For individuals, however, the insurance payments were important in their recovery. Mori, for example, published comments of the disaster victims, such as, "Earthquake insurance was a shed of sunlight." Insurance payments were important, not only in reconstructing individual buildings and recovering household articles, but also in maintaining the victims' lives immediately after the disaster.

We will now discuss important factors of insurance especially for individual victims. The loss ratio of contractors' all risks insurance to cover damage from the Great East Japan Earthquake show that insurance payments were made from 2011 to 2013. This insurance paid for physical damage to structures like buildings under construction to cover shortages of recovery cost. Oftentimes the amount of insurance payment cannot be determined until the recovery plans are settled and the cost identified. Thus, insurance payments were delayed. The insured were corporations and such delays were likely acceptable.

This type of insurance contract for individuals, however, would significantly damage the actual value of insurance, since individuals do not typically have such large buffers of funds. Household earthquake insurance has simplified the claim assessment so insurance payments are quickly made to the victims. After the Great East Japan Earthquake, insurance payments in excess of 1 trillion JPY were made within 3 months. Claim assessments have been made simpler through the long history of household earthquake insurance, and the advantages were proven with the Great East Japan Earthquake.

Household earthquake insurance is voluntary. For a voluntary insurance to play an important role at the time of natural disasters, its premium payment must be affordable. Some argue that, after the Great East Japan Earthquake [13], premium payments for household earthquake insurance should properly reflect the risk to raise the awareness of the insured. This claim has a point in continuing the system of household earthquake insurance, but we should not close our eyes to the fact that such a move would weaken the system for victim relief. Its function as mutual aid insurance would be lost.

Victims of automobile accidents in Japan are covered by two layers of insurance: compulsory and voluntary automobile liability insurance. As we stated earlier, the premium for compulsory insurance reflecting the driver's risk is much lower than the voluntary automobile liability insurance and it is more of a mutual aid insurance.

We might be able to apply this insurance system for traffic accident victims to earthquakes as well. We could continue the conventional household earthquake insurance and add a new earthquake insurance system on top of it. The added earthquake insurance could open its doors to foreign reinsurers to better reflect the risk and the variation in its premium would encourage the insured to promote disaster prevention and mitigation. In any case, it is desirable to clearly separate mutual aid insurance from market-driven types and the design of our insurance systems.

REFERENCES

[1] K. Hiroumi, Y. Hanawa, Dictionary of Insurance Words, Nikkei, Tokyo, 1997. p. 224.
[2] M. Otani (Ed.), Toa Reinsurance, Its Theory and Application Rev, Nikkei BP consulting, Tokyo, 2011, p. 601.
[3] Insurance Research Center (Hoken Kenkyujo), Statistics of Non-Life Insurance, Insurance Research Center (Hoken Kenkyujo), Tokyo, 1966–2014.
[4] Insurance Research Center (Hoken Kenkyujo), Statistics of Life Insurance, Insurance Research Center (Hoken Kenkyujo), Tokyo, 2012.
[5] H. Wakamatsu, Zenkyoren's response to the Great East Japan Earthquake, J. Insur. Sci. 619 (2012) 83–98.
[6] J.F. Kyosairen, 2013 Disclosure Data, JF Kyosairen, Tokyo, 2014. p. 74.
[7] Japanese Consumers' Co-operative Union. http://www.kyosai-cc.or.jp/info/ (Accessed 31 August 2015).
[8] Zenrosai, Fact Book 2013, Zenrosai, Tokyo, 2014. p. 127.
[9] Y. Kuriyama, General insurers' response to the Great East Japan Earthquake and the issues of the earthquake insurance, J. Insur. Sci. 619 (2012) 63–82.
[10] T. Ishii, The last to take the risk 2, the last resort for Japan's safety assurance, in: Huge Natural Disaster and Reinsurance, Hokenmainichi, Tokyo, 2013, p. 215.
[11] Swiss Re, *Sigma* World Insurance in 2013, Swiss Re Ltd., Zurich, 2014. p. 44.
[12] Swiss Re, *Sigma* Natural Catastrophes and Man-Made Disasters in 2014, Swiss Re Ltd., Zurich, 2015. p. 47.
[13] M. Onzo, Risk control and earthquake insurance in Japan, J. Insur. Sci. 619 (2012) 147–162.

FURTHER READING

[1] T. Mori, Testimony, The Great East Japan Earthquake, 1.2 Trillion Yen of Earthquake Insurance Money, Hoken Mainichi Newspaper, Tokyo, 2014.

Chapter 11

Problems With "Disaster Relief Volunteers"

CHAPTER OUTLINE

11.1 THE GREAT EAST JAPAN EARTHQUAKE AND DISASTER RELIEF VOLUNTEERS

The Great East Japan Earthquake was the most difficult disaster ever for disaster relief volunteers in carrying out their volunteer work. First of all, immediately after its outbreak, volunteers could not reach the disaster-hit areas. The enormous earthquake and tsunami devastated areas from Tohoku to Northern Kanto and the Tokyo Metropolitan area along the Pacific coast, destroying the traffic network, and on top of this, gasoline shortages impeded people from reaching the area. Even if the volunteers were able

The Fukushima and Tohoku Disaster. https://doi.org/10.1016/B978-0-12-814078-9.00011-X

to reach the affected regions, the frequent earthquake aftershocks and the nuclear power plant accident made securing a safe environment for activities difficult. The earthquake and tsunami affected areas, however, had shortages of supplies and the demand for manpower and emergency relief goods was serious. The disaster relief volunteers were needed, but they could not reach the disaster-affected regions for a while.

In addition to the physical obstacles, a number of media outlets sent out the message that disaster relief volunteers should not go to the disaster-affected areas. The situation was, of course, difficult, as explained, but moreover the serious shortages of supplies hampered preparations in the affected regions for readying themselves to accept volunteers [1], and there were strong opinions that volunteers would "impose burden on the affected regions with shortage of supplies" or "volunteers being there would just add to the confusion," so "they should not go."

The Japan National Council of Social Welfare (JCSW) published a report on the Web [1] that stated: "As of March 18, today, the Disaster Relief Volunteer Centers (DRVC) are preparing for their operations, however, emergency support, collecting information about the affected areas, roadblocks, and shortage of fuel and supplies are causing delays and most of them are not ready to accept volunteers. Currently the shortage of gasoline is serious and public transportation are still in the process of recovering. Volunteers will need to find their own ways of transportation, prepare their food supply, and secure places for overnight stay for themselves."

These opinions were sent out not only from such organizations but also on SNS as individual voices, and as they continued, they created the feeling that, for whatever reason, going to the disaster-struck regions was wrong. We presume that there were some that gave up the idea after hearing or reading "disaster relief volunteers should not go." Then other voices were frequently heard, saying things like, "The activity level at the Great East Japan Earthquake is low" or "The efforts are smaller compared to the time of Great Hanshin-Awaji Earthquake." The actual regions were in great need of support and some even said that they "should go even if they may appear they are barging in" [2].

Many discussions on activities of "disaster relief volunteers" took place, but the number of people who actually went to offer their aid is not clear.

On the other hand, it is said that the difficulty in reaching the area led to the largest amount of "earthquake donations" ever [3]. Just for this earthquake disaster, separately from the "public donations" that were directly distributed to those affected, an "activity support fund" was widely announced to support activities by the disaster relief volunteers. Although there were some negative opinions, we can say at least that social expectations of the volunteer work were high.

Now, did the "disaster relief volunteers" meet these expectations? After the Great Hanshin-Awaji Earthquake, the Social Welfare Council (SWC) worked on organizing DRVCs to accept individual volunteers to provide help in disaster-struck areas. The Great East Japan Earthquake, however, required long travel distances and times to reach the disaster locations as well as huge amounts of supplies, which were out of the hands of DRVCs, prepared only for 1-day-trip volunteers. New frameworks in the civilian sector were needed to support disaster-struck areas and victims. In fact, this disaster initiated a number of projects in the civilian sector. Note that by "civilian sector" we mean the third sector of volunteer groups, NPOs, coops and the like. They are the third sector after the first, the government, and the second, private corporations. Yamaoka says that, "As long as the group is called a sector, it is not just a collection of individual groups but they have a position in the society and each member has to acknowledge that the group is part of that sector [4]."

In the rest of this chapter, we will discuss what was expected, what was done, and what was not done in the social area of "disaster relief volunteering" during and after the Great East Japan Earthquake.

First, we will refer to some studies on the reality of disaster relief volunteers and confirm the gaps between expectation and reality.

Then, under disaster donations, we will review the "activity support fund," which was an indicator of people's expectations of the "civilian sector," and through some research, we will see how the funds flew into the civilian sectors and the mechanism for applying the funds to support activities in the field. We will then discuss a trial in responding to a wide-area disaster that was hard to handle through the conventional response system and introduce a new system of disaster response for cases of megasize disasters in the future. Expectations and concerns with this system are evaluated. Our study reveals what studies we need to pursue and the problems in real application. We would like to put this forward as an interim discussion to reevaluate the conventional concepts of "disaster relief volunteers."

11.2 HOW WE CONCEIVE "DISASTER RELIEF VOLUNTEERS" AND THE REALITY

11.2.1 The Number of Volunteers Remains Unclear

The number of volunteers during the 1 month period following the Great Hanshin-Awaji Earthquake was 20,000 a day, and during the 2-month total was 1 million man-days [5]. These numbers were based on interviews with administration workers that patrolled the evacuation shelters and telephone interviews with municipalities and volunteer organizations; as such, they are likely to have multiple counts and dropped counts. The number announced as "the number of disaster relief volunteers" is the number of people registered at DRVCs opened by the SWC. The number of volunteers during the Great East Japan Earthquake was also counted by SWC. JCSW reported that the number was about 926,200 during 1 year. (This is the number of people who volunteered and worked in the three severely disaster-struck prefectures since the outbreak until the end of February 2012. The numbers added were rounded to the nearest 100.) [6]. When the media said "the number of volunteers was less than the time of the Great Hanshin-Awaji Earthquake," they were probably basing their words on these numbers. This JCSW report also stated "there were probably a number of volunteers that worked in NPOs and so without going through DRVCs, thus, numbers in this report are not the total count of volunteers during the Great East Japan Earthquake." In fact, Japan NPO Research Association reported [7] that most people replied that they went to volunteer through windows at their companies (29.5%). The City of Ishinomaki, which accepted individual volunteers through DRVC and group volunteers through a different organization, reported 121,969 volunteers through DRVC while group volunteers reached 158,298 [8], also hinting that a larger number of group volunteers did not go through DRVC.

Okamoto [9] tried, by referencing a number of studies, to count the number of volunteers for the Great East Japan Earthquake and he concluded that "the count possibly was 2 to 6 times" the number accumulated by DRVC of SWC, and that "the SWC's way of accepting volunteers only through DRVC" was "not adequate for this case of a mega disaster." Activities in Ishinomaki City started mostly with removal of mud and debris for half a year [8]. This work required intense labor and groups with heavy machinery handled it in many cases. Another reality was that it was hard to find places to stay overnight in the field, and rather than accepting individual volunteers in the field, like DRVC, and then looking for places for them to stay, it was more effective to secure transportation and locations to stay off site and dispatch the volunteers in the morning.

As Okamoto pointed out, the problem was the tacit condition that only SWC's DRVCs were to handle coordination of the needs and supply of volunteers. The delays in opening DRVCs and restricting the acceptance of volunteers should probably be blamed on SWC.

Next we will see how this system of DRVC-managed "disaster relief volunteer" work developed in Japanese society.

11.2.2 Formation of "Disaster Relief Volunteers"

The year the Great Hanshin-Awaji Earthquake broke out, 1995, was called "the first year of volunteers." Voluntary activities had already started in the 1960s as social activities by civilians. In the 1980s, social issues that were difficult for the government or corporations to handle (e.g., need for nursing care with the aging of the population, or the concerns with foreign labor that came with globalization) turned to the "volunteers" in hopes of finding solutions. In the 1990s the social systems started to take shape so the voluntary organizations could take responsibilities in their efforts in solving social issues (e.g., by acquiring legal status or receiving tax benefits). In fact, it was this group of people who worked on establishing the social base who developed the system of coordinating individual volunteers in disaster-struck regions, to support the activities in "the first year of volunteers" [10,11].

The reason why the year after the Great Hanshin-Awaji Earthquake was called "the first year of volunteers" was probably because the attitude of the people changed at that point in their acceptance of "volunteers." Before then, the word "volunteer" had a negative connotation of being a servant or of self-sacrifice. However, in disaster-struck areas, people naturally started to help one another. Also, another factor that contributed to raising the social recognition of volunteers was that the actual outcome of the activities (e.g., cleaning up the debris) were visible to the people. Many people who experienced volunteering found new "networks" that were different from communities or workplace relations and such new relations contributed to forming the system of new support.

The disaster relief activities after the earthquake, thus, greatly stimulated two regions of society and urged them to make changes [12]. One was "civilian activities" by volunteers and NPOs. Three years after the Great Hanshin-Awaji Earthquake, the Act on Promotion of Specified Nonprofit Activities was enacted to allow legal status and intermediaries were built to support NPO activities and operations to promote the forming of the civilian sector. The volunteer activities after the earthquake must have pushed this move.

The other was social recognition of disaster prevention. The Great Hanshin-Awaji Earthquake forced Japan to seriously review its system of disaster management and prevention, and to reform its fundamental concept of "disaster prevention." The principle of action of volunteers who can flexibly handle the individual problems in front of them was evaluated highly, compared to administrative actions, which could not make a move until it grasped the overall picture, under the principle of fairness. This difference triggered

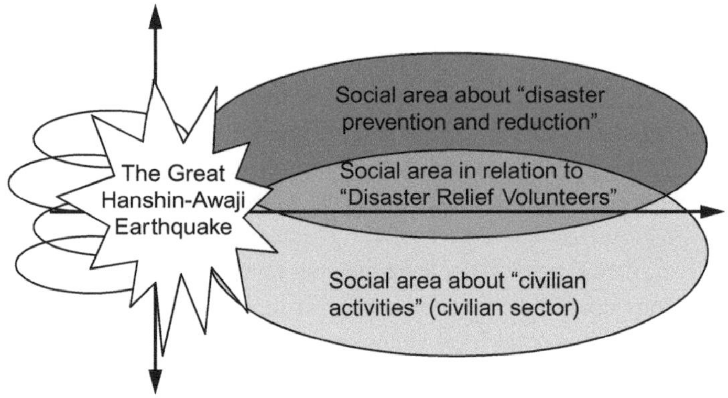

■ **FIG. 11.1** Formation of the social sector "Disaster Relief Volunteers."

volunteers to take firm positions in the social area of "disaster prevention." First, after 1995, regional disaster prevention plans that were undergoing revision at the time clarified departments that accept disaster relief volunteers, and at the same time, a system was formed to register "specialist volunteers" with special skills at the time of disasters. Later, the system of accepting disaster relief volunteers took the form of the SWC extending its regular DRVC functions to open DRVCs. SWCs also strengthened their bases of operation both in the budget (access to reserve fund of chest society) and resource (e.g., systematic dispatch of SWC workers, or training of managers). Then more municipalities entered into agreements with SWCs and placed their DRVCs in the municipality plans [13].

This is how the activities by disaster relief volunteers after the Great Hanshin-Awaji Earthquake gained positions for volunteers in the social areas of "disaster prevention" and "civilian activities" as they were reformed. We can say they created a social position of "disaster relief volunteer" at the crossing of the two social areas (Fig. 11.1).

11.3 EXPECTATIONS TOWARDS THE CIVILIAN SECTOR AND THE REALITY

11.3.1 Flow and Distribution of the Activity Support Fund

The Japan Fundraising Association started its studies on donations in 2010 based on available statistics reports and its own research, and the results have been published in a "Donation Whitepaper." Its 2012 issue posted a special article on "earthquake related donations" and summarized donations made in relation to the earthquake for 1 year after its breakout [3].

We also referenced the fact book [7], a publication by Japan NPO Research Association that consolidated its 3½ years of studies on NPO activities after the earthquake. "Earthquake donations," however, were concentrated in the short period immediately after the earthquake and our primary source of information on donation was the previously mentioned whitepaper. Both the whitepaper and fact book categorized "earthquake donations" in the form of monetary donations of (1) Charity Fund, (2) Activity Support Fund, and (3) One-Time Donation and Educational Fund, and nonmonetary donations of (4) Supplies, and (5) Volunteering. The two references, in addition to their own work, summarized past research and published data.

The "Donation Whitepaper [3]" revealed that as many as 85 million people made donations, either monetary or supplies, for the victims of the Great East Japan Earthquake, and the total amount was about 600 billion JPY (Fig. 11.2).

We then turned our attention to the "Activity Support Fund" which is a sign of people's expectations towards disaster relief volunteers. The delay in

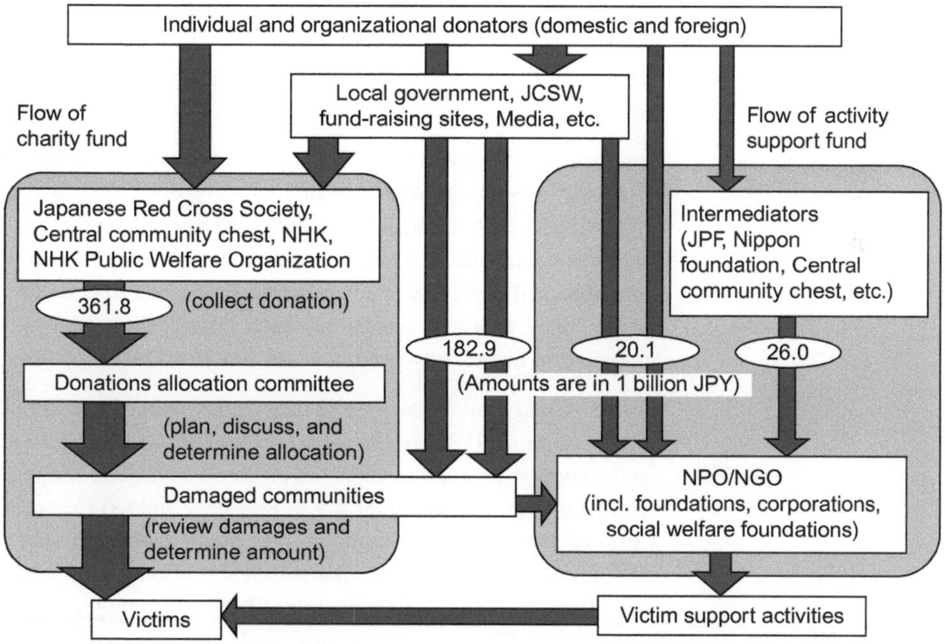

■ **FIG. 11.2** Flow of charity fund and activity support fund. *(Based on Japan Fundraising Association, Giving Japan 2012, Tokyo, Keidanren Publication, Tokyo, 2012; The Great East Japan Earthquake Civilian Support Fact Book, Japan NPO Research Association, Tokyo, 2015. http://www.janpora.org/shinsaitokubetsuproject/seika/seika_fact_150725.pdf (Accessed 20 September 2015); N. Hayase, Summarizing problems with charity funds and activity support fund, J. Volunt. Res. 2 (2013), Osaka Voluntary Action Center (OVAC).)*

distributing charity funds moved the attention of the victims to activity support funds that quickly reached the disaster-struck areas.

Some people made monetary donations, however, without marking the distinction of charity fund or activity support fund, and there were also some receiving organizations that did not clarify the difference. As the funds were passed on to multiple intermediators, some changed their categories and there was an issue in figuring out the accurate amount [14], but some efforts were made to estimate the amounts.

Provisional calculations in the whitepaper reported the total amount of activity support fund collected by intermediators was about 26 billion JPY and about 20 billion by NPO/NGO. The Japan Fundraising Association states that the two amounts cannot be simply added to compute the total activity support fund, because of possible double counting [3]. Although a digit less than the total charity fund of 362 billion JPY, there had been no past disaster that received such a large amount of monetary donations to support activities by "disaster relief volunteers."

One of the recent disasters that collected an activity support fund was the 2004 rainstorm and flooding disaster in Niigata Prefecture. Niigata NPO Association collected a total of 956 million JPY. At this time, some people raised concerns about donations that do not go directly to the victims, but the significance of this type of donation needs more recognition [15].

Now we will review how the civilian sector prepared a system to use to distribute these large amounts of monetary donations to the supporting groups. With this earthquake, a number of intermediate organizations and supporting foundations served as "fund intermediators" to receive monetary donations and distribute them as supporting funds to activity groups. The amount they received was large. Each intermediator showed its support guideline to the activity groups when they distributed the funds [15].

For the Great East Japan Earthquake, funds were concentrated into three intermediators; (1) Japan Platform (JPF) with The Great East Japan Earthquake Supporting Fund of about 6.8 billion JPY, (2) The Nippon Foundation with ROAD Project of about 4.9 billion JPY, and (3) Central Community Chest of Japan with a Volunteer NPO Support Fund of about 3.6 billion JPY. The three intermediators had different support guidelines in their offering conditions of funds. For example, JPF provided major support in units of 100 million JPY to support large-scale activities of transporting large amounts of supply. In contrast, the Central Community Chest of Japan supported smaller-scale activities and distinguished between short-term activities of less than a month and mid- to long-term activities of more than a month and set the upper limits for them to 500,000 and 3 million JPY, respectively [16]. The Nippon Foundation, in addition to regular support, arranged their own programs such as Footbath Volunteers or Student Volunteers, organized the volunteers and dispatched them to the disaster-struck areas. The activity support funds collected in this manner were provided to the disaster areas and victims in these forms through supporting groups, along the supporting guidelines the fund intermediators determined.

11.3.2 Participation of New Players—Private Companies and International NGOs

As we saw above, supporting the disaster areas of the Great East Japan Earthquake required huge amounts of supplies spread to wide areas, a form of support different from the DRVC with only plans of individual service for disaster victims. Within the civilian sector, social support activities by corporations and international NGOs provided major support. Separately from DRVCs, they gathered funds through their own routes, collected supplies, and systematically distributed them to the disaster areas, and at the same time, made contributions to introduce large activity support funds to the civilian sector; they played major roles in both collecting and distributing supplies.

11.3.2.1 Social Contribution Activities by Private Corporations [17,18]

The Japan Business Federation (Keidanren) made a call, in 1990, to form the 1% Club with members who agreed to make a societal contribution in the amount of 1% of after-tax income in the case of individuals and 1% of income before taxes in the case of corporations. With its joint operation with NPOs and NGOs, the club has made efforts to gather voluntary business people and corporate donations. It has records of disaster support activities after The Great Hanshin-Awaji Earthquake (1995) and Niigata Chuetsu Offshore Earthquake (2004). In 2005, with the Central Community Chest of Japan and JCSW, it established the "Disaster Relief Volunteer Activity Support Project Council (Support P)," so companies can write off their donations of supplies and cost in coordinator dispatching. The club has been active in promoting disaster support by private companies.

Keidanren worked with the 1% Club and started "The Great East Japan Earthquake Response Headquarters" immediately after the disaster hit. Through its homepage and public relations, it broadcast information about support for affected regions and victims. At the same time, it joined forces to start supplying aid supplies with JPF it had been supporting, and NPOs and NGOs it had built good relations with, through joint work in Support P. It also joined the Japan Civil Network for Disaster Relief in the East Japan (JCN) from its very start. We will discuss JCN activities later.

The DRVC activity style of coordinating individual needs with matching volunteers worked effectively in the case of earthquakes with diverse needs. However, this time, when the activity environment was destroyed and huge manpower was needed, the coordination among the needs and volunteers faced difficulties.

Among the societal contribution efforts by companies, the business person volunteer program by joint efforts of 1% Club and Support P was noteworthy. The areas attacked by tsunami had a need for major manpower to remove the mud and debris. However, it was hard for the volunteers to reach the areas. The program by the 1% Club and Support P first secured transportation and lodging for the volunteers to systematically dispatch people into the affected areas.

The NGO Peace Boat was early in starting this method of securing a local center and a means of transportation. The "Tono Magokoro (Magokoro means sincere heart) Network," which functioned as the hub for sending relief supplies to the disaster-struck communities along the coast, also organized volunteers to send them in buses that took 30 minutes to an hour to reach coastal disaster areas.

The 1% Club and Support P program sent a total of 2101 man-days of corporate employees over 20 times during the period from April to August. They carried mud out of houses and street gutters, removed debris, cleaned photographs, and delivered relief supplies to evacuation shelters. It was the time of year difficult for students to participate in the work and dispatching groups of business persons was highly appreciated in the disaster areas.

The 1% Club documented corporate support activities during the half year from the earthquake [17]. The document subtitled "Mutual Aid Efforts by the Business World" shows they carried out a variety of support activities. The document reports 259 companies encouraged their employees to participate in the support activities for the victims and disaster areas. Corporate employees that participated in the program and replied to Keidanren's survey totaled about 180,000 man-days in the disaster areas.

In addition, 170 companies started their own support activity programs and some report that their support activities triggered policies of volunteer vacation and expense support [18].

We should further note that the corporate volunteers gave away their intellectual properties of specialized knowledge, technology, and know-how of operations and management. They also took advantage of their affiliated company and customer networks for the support activities [19]. We would like to also note that cooperation between the business sector and the civilian sector has started new activities.

11.3.2.2 International Cooperation NGO

International Cooperation NGOs are established for the purpose of support activities in foreign countries. Some have articles of incorporation that do not specify domestic activities, and many did not have experience in supporting disasters in their own countries. However, the magnitude of this disaster triggered many organizations to start support activities. An intermediate support organization, Japan NGO Center for International Cooperation (JANIC), reported that research involving its 157 organizational members showed that 59 (37%) of them took parts in support activities and 18 of them started within 3 days from the earthquake [20].

In addition to their swift action, their resource power captured attention. Results of JANIC's 3-month surveys of activities showed that most efforts spent during the first three months were on "relief supply distribution" followed by "meal preparation and distribution" and "medical support and caring for the minds." Disaster areas that were devastated over a wide area faced prolonged periods of shortages of everything, and these results reveal that supplies of goods and meals were needed for extended times.

In the disaster areas where the living environment was devastated, the activities for distributing supplies and services were "greatly helped by the daily experience of development programs and the staff of international cooperation NGOs in developing countries" [21].

Yamaguchi reported [21] that the background of international cooperation NGOs in their ability to quickly send resources in large quantities were; (1) many organizations had experience in emergency support in foreign countries, (2) the staff was skilled in project management, and (3) they had initial funds to get started. Especially concerning (3), the aforementioned JPF, established in 2000 as a funding mechanism to aid humanistic support activities for overseas disasters and war zones, from its experience contributed to the initial moves by NGOs that joined the movement immediately after the earthquake. As we saw, large funding by JPF allowed NGO activities with major resources, and JANIC, in addition, started "Compiled Donation for Relief Support Activities of The Great East Japan Earthquake" and intermediated funds to NGOs in support activities.

11.3.3 Creating a New System of "Connection"

11.3.3.1 Japan Civil Network for Disaster Relief in the East Japan

With this earthquake, from immediately after its breakout, efforts were made to establish a system of coordination and cooperation with the entire civilian sector in the scope. This was the Japan Civil Network for Disaster Relief in the East Japan (JCN). People that worked in establishing this network were those that had been working as "disaster relief volunteers" at the frontline in the field while struggling to form a foundation for their activities. From the beginning, they recognized the limits of the response organization centered around DRVCs and immediately after the strike of the disaster, they posted the slogan "Never let any area or anyone be left behind" and targeted building a fully civilian system that spread over boundaries of specialties and regions over the entire nation [12,22,23].

With 855 organizations that participated at its peak in April 2013, this network (1) represented the civilian sector in coordinating roles with other sectors (by organizing regular meetings with the governments), and (2) aimed at building coordination and relations among multiple volunteers and among volunteers and victims by supplying timely and visible information on the statuses of disaster areas, victims, and supporting groups through coordinated examinations with other organizations (Keidanren, universities and research organizations, SWC, and so on).

Table 11.1 shows the primary activities of JCN based on its website information. When it comes to support that requires nationwide efforts (e.g., supplying information on support that requires long-distance transportation, such as arranging volunteer buses or recreational projects) it has to revert to support in the form of collecting and sending out information. However, it is also actively involved with support efforts creating opportunities for "people to meet people" by dispatching office staff directly into the field.

11.3.3.2 Long-Distance Evacuee Support Business by JCN

Among the activities of JCN, local meetings for support organizations with activities in the three disaster-struck prefectures and long-distance evacuee support meetings for support organizations to promote acceptance of long-distance evacuees in prefectures over the nation were especially noteworthy. When it organized these meetings, JCN office staff visited the area beforehand to check on the activities of the support organizations in the area, communicated with the organizations that were playing central

Table 11.1 Activities by Japan Civil NW for Disaster Relief in the East Japan (JCN)

Year/Month	(Overall)	(Information)	(Field Support)	Activity type			Number of Groups
				LDE: LDE	NW: Network	Pref: Prefecture	
				SI: SI	Mtg: Mtg		
				(Long-Distance Evacuation Support)			
2011							
March	Emergency gathering (14th) → Set mission statement, started (16th) → inauguration Mtg (30th)						193
April		Started official website and mailing list					443
	Started regular Mtgs with ministries and agencies						
		Announced support status map (preliminary), and volunteer bus time table					
		Issued "the Great East Japan Earthquake Disaster Relief Volunteer Activity Guideline"					
May–June			First local meetings in Miyagi, Fukushima, and Iwate Prefs				577
July		Issued "Tips on safety and sanitation for disaster relief volunteer activities"					
August	Stopped regular meetings with ministries and agencies (had 6 total)						
September	First JCN general Mtg (Tokyo)						657
October		Announced support status of LDEs					
November–December			Second local Mtgs in Miyagi, Fukushima, and Iwate Prefs				709
2012							
January		Opened "Good reconstruction support! Cases"					
February–March			Third local Mtgs in Miyagi, Fukushima, and Iwate Prefs				753
	Second JCN general Mtg (Tokyo)						
				First LDEs SI Mtg (Tokyo)			
April				Research on field activities for building NWs of LDE supporting groups			
				Second LDEs SI Mtg (Tokyo)			
		Opened recreational program information					
	JCN annual conference (Tokyo)						
June				LDEs SI Mtg in Tokai			795
July	Hosted a side event in "World Ministerial Conference on Disaster Reduction in Tohoku"						
July–August			Fourth local Mtgs in Miyagi, and Iwate Prefs				
				LDEs SI Mtgs in Kinki (Osaka) and Yamagata Pref			
September		Closed recreational program information					824
October				LDEs SI Mtg in Shikoku (Matsuyama)			

Continued

Table 11.1 Activities by Japan Civil NW for Disaster Relief in the East Japan (JCN) *Continued*

Year/Month	(Overall)	(Information)	(Field Support)	Activity type			Number of Groups
				LDE: LDE	NW: Network	Pref: Prefecture	
				SI: SI	Mtg: Mtg		
				(Long-Distance Evacuation Support)			
November–December			Local Mtgs fifth in Miyagi, and Iwate, fourth in Fukushima Prefs				826
			Fourth local Mtg in Fukushima Pref				
				LDEs SI Mtg in Chugoku (Hiroshima)			
2013							
January		Opened information on national NWing Mtg					835
February			First youth conference in Iwate Pref				
				LDEs SI Mtg in Kyushu (Fukuoka)			
March			Local Mtgs sixth in Miyagi, and Iwate, fifth in Fukushima Prefs				854
				LDEs SI Mtgs in Tokyo and Hokusinetsu (Niigata)			
April	(recorded maximum number of registered groups)						855
May			Started case reviews (Iwate, Miyagi, and Fukushima Prefs)				
	First NW promotion conference (Tokyo), JCN annual conference (Tokyo)						
June			Sixth local Mtg in Fukushima Pref				803
July–August			Seventh local Mtgs in Miyagi, and Iwate Prefs				
				2nd LDEs SI Mtgs in Tokyo and Chugoku (Hiroshima)			
				1st LDEs SI Mtg for all nation (Tokyo)			
	Second NW promotion conference (Tokyo)						
				Published "Report on research in the field of SWC LDE support"			
September			Seventh local Mtg in Fukushima Pref				797
	First regular discussions with related ministries and agents (Tokyo)						
October			Eighth local Mtg in Iwate Pref				
November				2nd LDEs SI Mtg in Shikoku (Kochi)			
				3rd LDEs SI Mtg in Tokyo (joint Mtg)			
November–December			Eighth local Mtgs in Miyagi, and Fukushima Prefs				800
				2nd LDEs SI Mtg in Yamagata			

2014					
January			Ninth local Mtg in Iwate Pref		
	Second regular discussions with related ministries and agents (Tokyo)				
February			Ninth local Mtg in Miyagi Pref		
March				2nd LDEs SI Mtg for all nation (Tokyo)	796
	Third JCN general Mtg (Tokyo)				
April				2nd LDEs SI Mtg in Tokai (Nagoya)	
	Third regular discussions with related ministries and agents				
June				LDEs SI Mtgs in Tokyo and Kyushu (Fukuoka)	782
	JCN annual conference (Tokyo)				
July	Fourth regular discussions with related ministries and agents				
August			Tenth local Mtg in Iwate Pref		
September		Stopped announcement of volunteer bus time table			
September–October				LDEs SI Mtgs in Kinki (Osaka) and Hokkaido	
October	Fifth regular discussions with related ministries and agents				
October–November			Local Mtgs tenth in Miyagi, and ninth in Fukushima Prefs		
2015					
February–March			Local Mtgs tenth in Fukushima, and eleventh in Miyagi and Iwate Prefs		636
				3rd LDEs SI Mtg for all nation (Tokyo)	

Based on information from JCN website http://www.jpn-civil.net/2014/about_us/history/ [23a].

roles in the areas, and provided opportunities for those involved with the support activities to meet. The preparations and administration by JCN kept the meetings highly effective. It also continued support from the side so the connections among organizations continued even after the meetings.

For these long-distance evacuee support meetings, JCN cooperated with the support organizations to respond to each evacuee's individual concerns that were sometimes difficult to uncover, brought to light the difficult evacuee realities, and developed movements to appeal to the society at large for needs. The realities that evacuees were coping with, especially, were not well known even to the government, and JCN worked hard with practitioners (lawyers) that supported research organizations and evacuees in participating in the editing and publication of the "Nuclear Plant Evacuation Whitepaper" [24]. The concerns with the nuclear plant evacuees were something that Japanese society had never experienced in the past and efforts in solving these concerns were made by trial and error. JCN, in addition to its connection building through the nationwide Long-distance Evacuee Support Meetings, had contracted work from Fukushima Prefecture to support building people's networks in their new locations with local connection coordinators [24].

11.3.4 Expectations and Concerns Identified Through New Systems

11.3.4.1 Significance and Problems With Building Support Systems With the Entire Civilian Sector in the Scope

Sharing the grand purpose of "supporting the disaster victims" is easy, but the variety of civilian organizations with different activity guidelines and senses of values can quickly bring about disagreements. The JCN efforts are greatly appreciated in creating loose connections among the civilian organizations and managing the groups without much conflict, sharing information, coordinating the flow of activity resources, negotiating with other sectors, and bringing in benefits to the entire civilian sector. The outreach from each prefecture contributed to offering opportunities for face-to-face connections among the support personnel, disaster-struck areas, and the victims. The efforts were outstanding, especially as a new support arrangement tried for the first time.

On the other hand, the network organization supported by people's voluntary motivation had its limits. In the early stages, a "team system" in which management was left up to anyone who wanted to take on the role, encountered difficulties in having those managers take responsibility for completing the work to the end, when there were plans for the administration office to take control. It was also difficult to form agreement throughout the entire network.

11.3.4.2 Expectations and Concerns as Responsible Party for Emergency Response

Disaster response is a logistics process to revive a destroyed society by repairing its functions [25]. First, a public organization "substitutes" for the destroyed social system and takes actions to "recover" it. Expectations of the disaster relief volunteers from DRVC were: to offer additional manpower to public services at the time of the disaster (e.g., support the limited number of clerks to visit the evacuation shelters); to provide functional service (e.g., prepare side dishes for evacuees to prevent poor nutritional balance from eating just the plain rice balls distributed in the evacuation shelters); or to help victims trying to rebuild their own lives by removing obstacles in their private home (e.g., help clean up their houses). The Great East Japan Earthquake, however, greatly damaged the public organizations and at times work usually performed by the public workforce had to be made by civilian support personnel. The major devastation over a wide area called for huge amounts of aid supplies and food. International cooperation NGOs and private companies showed their power in leading the efforts and these were highly appreciated; however, some concerns were pointed out.

The aforementioned JANIC report [20] emphasized in its recommendation section the relationship between administrative offices and the support organizations. Many administration offices saw disaster relief volunteers as amateur individuals and did not view the international cooperation NGOs as partners to collaborate with in their efforts, although they had experience and resources for humanistic support in foreign countries. For emergency aid overseas, once the United Nations (UN) takes the central role in orchestrating support, NGOs, the UN, and the local administration sit at the same table to discuss how to split the administration and support activities. This time in the field, however, there was no coordination allowing NGOs to take a share in administrative support. The report pointed out that the NGOs had the capability to substitute for and support administrative services, but were not given the opportunities to put it to work.

The organization with high expectations of the civilian sector during this earthquake disaster was, in fact, the government. The next section outlines the coordination between civilian support by the government and the volunteers, introduces the new system born from those efforts, and explains the expectations and concerns about the system.

11.4 BEYOND DISASTER VOLUNTEERING—FROM SEPARATE ACTIVITIES TO COORDINATED SUPPORT

11.4.1 Government Expectations and Concerns About the Civilian Sector

The government of Japan supported the civilian sector from immediately after the breakout of the disaster and sought coordination throughout. The government had not made such movements during past disasters. Table 11.2 shows the sequence of its support.

First, on the day after the earthquake, the Cabinet Secretariat opened a contact point for the civilian sector and arranged periodic liaison meetings with JCN and implemented some of the requests from the civilian sector, e.g., opening highways in the disaster area for free, for easier transportation.

Table 11.2 Government Actions for Supporting Civilian Activities

Date	Actions
3/12/2011	Cabinet Secretariat: Set "The Cabinet Secretariat Office for Volunteer Cooperation in response to the Earthquake." Hired staff from the civilian section. Appointed House of Representatives Member Kiyomi Tsujimoto as the prime ministerial aide in charge of earthquake relief volunteer activities
3/15/2011	Ministry of Finance: Adopted tax deduction for contributions to promote giving of charity funds and support funds. Appointed Central Community Chest of Japan's "Disaster Relief Volunteer & NPO Support Fund" as specified contribution. Individuals could write off their contributions as deductions and corporations report the full amount as expense
3/1/2011	Government of Miyagi Prefecture: Had a four-party meeting of Central and Miyagi Governments, Self-Defense Force, and NGO (JCN Liaison) to set roles of each party (e.g., NGO boards Self Defense Force vehicles to distribute food, and the administration purchases food materials)
4/1/2011	Ministry of Finance: Applied rules of specified contributions to fundraisings by approved NPOs
4/1/2011	JCN started (3/30): JCN served as administration office to arrange regular meetings with civilian volunteer groups and ministries and agencies to exchange information and opinions. A total of six meetings were held
5/1/2011	Ministry of Finance: Applied rules of specified contribution to fundraisings by Public Interest Incorporated Associations and Incorporated Foundations
9/1/2011	Cabinet Secretariat: Functions of "The Cabinet Secretariat Office for Volunteer Cooperation in response to the Earthquake" transferred to "Great East Japan Earthquake Restoration Support Headquarters, Disaster Relief Volunteer Section"
	Reconstruction Agency: With the establishment of the Reconstruction Agency, the interface with volunteers and NPOs was transferred to "Reconstruction Agency, Public Interest Civilian Liaison Section"
4/1/2012	Reconstruction Agency: Started activities with publishing "Financial government support for future NPO activities," and "Roadmap for various organizations for reconstruction support"

Based on Cabinet Office website, Reconstruction Agency website (http://www.reconstruction.go.jp/topics/main-cat4/sub-cat4-2/) and Japan Fundraising Association, Giving Japan 2012, Tokyo, Keidanren Publication, Tokyo, 2012.

The Ministry of Finance also specified tax-exempt charities to promote disaster donations. Specifying tax-exempt charities is said to have extremely strong promotive effects and the amount of tax exemption totaled 126 billion JPY. The Japan Fundraising Association [3] estimated the effect based on results of the 2011 tax returns. It reported that of the 1,360,000 citizens that filed charity exemptions, 819,000 filed for the earthquake-related charity, and adding the roughly 250,000 that filed the special exemption introduced with the new charity tax system, a total of about 5 billion JPY was filed as tax exemptions. The report concluded that, compared to the previous 2010, "Both the number of people that filed charity exemption and the amount doubled. Donations in response to The Great East Japan Earthquake was a big push."

The government also prepared multiple grants to support civilian support activities. For example, "Building New Public Areas (Cabinet Secretariat)" or "Support for Building Connections (Ministry of Health, Labor, and Welfare)." Many of these grants had conditions to work with local governments, and some criticized that the grants could hinder independence and motivation of the civilian organizations that could "turn into subcontractors," or that the limited 1-year budget would cause instability in the civilian organization finances [26,27]. For earlier disasters, however, there had never been such a wide selection of grants and this fact shows the government's high expectations towards the civilian sector.

11.4.2 For the Establishment of a New System—Possibilities and Problems

The amendment of the Basic Act on Disaster Control Measures in June 2013 added the following sentence: "(Coordination among the central and local governments and volunteers) Article 5, 3 The central and local governments shall take into account the importance of the role volunteers play in disaster prevention activities at the time of disasters, and respect their autonomy and endeavor to collaborate with them."

The Cabinet Office had organized a "Disaster Prevention Volunteer Activity Discussion Group" and had worked for 6 years building relations with civilian groups. However, a coordination interface to make use of the experience, accumulated knowledge and know-how in postdisaster activity coordination was not set up with this earthquake, and instead the Cabinet Secretariat set up a Volunteer Coordination Group. The government thus ended up opening two windows for coordination with civilian groups and we list this as a problem.

On the other hand, the relation of mutual trust built through the Discussion Group functioned well in coordination between the Miyagi Prefecture Local Disaster Response Headquarters and NGOs. A new system of coordination is now being built out of these experiences and lessons.

In 2013, the Cabinet Office built the "Opinion Exchange about Wide-Area Coordination of Volunteer Activities upon Large-Size Disasters" to exchange opinions among people with experience and knowledge about disaster relief volunteers such as international cooperation NGOs, disaster relief NPOs, JCSW, and corporate employees in charge of social contributions that had experience in support activities during the Great East Japan Earthquake [28].

At the same time, some civilian organizations with experience in operating JCN and support activities during the Great East Japan Earthquake acknowledged the need for building response systems in bracing against future disasters. They are working on enhancing their own activity basis (funding and effective information broadcasting) and building a full-time institution to promote coordination with the governments [29].

In the past, people assumed DRVC would control "disaster relief volunteers" with a single local resource interface with staff with little experience in disaster response. This concept is still around on the administration side. This time, however, NGOs with special skills in disaster fields and ample resources, or disciplined organizations like private companies, made outstanding contributions. Public organizations now need to enter matching relationships with these organizations and, for that purpose, the government needs to renew its framework of disaster response. On the other hand, the civilian sector side also needs to prepare its funding capacity and resources for this mutual relationship.

The most important assignment, over enhancing their activity bases, for the citizen groups in the civilian sector is to encourage participation of a variety of people and build systems to take advantage of the variety. This move will strengthen the citizen society to accept a variety of citizens at the time of the disaster and enhance the adjustment capability in demanding supplies [9,30]. Thus, the requirements of citizen groups in the civilian sector and their networks are to open their doors wide so people can better participate in activities to respond to social problems, and to function as a means of solving such problems.

ACKNOWLEDGMENT

The author would like to acknowledge Mr. Okasaka and Mr. Tsuga at JCN office for providing information for this work.

REFERENCES

[1] Tohoku Area Pacific Offshore Earthquake, Report 5, Japan National Council of Social Welfare, Tokyo, March 18, 2011.

[2] M. Murai, Mindset of Disaster Relief Volunteers, Softbank Shinsho, Tokyo, 2011.

[3] Japan Fundraising Association, Giving Japan 2012, Keidanren Publication, Tokyo, 2012.

[4] Y. Yamaoka, Trend of civilian sector, in: Let the Circle of Volunteering Expand, Tutui Shobo, Liaison Committee, Tokyo, 2014. Volunteer white paper.

[5] Journal of Recovery from the Great Hanshin-Awaji Earthquake, vol. 1, Association of the 21st Century Hyogo Creation, Hyogo, 1997.

[6] Report on the Great East Japan Earthquake Disaster Relief Volunteer Centers, National Social Welfare Council, Tokyo, 2012. http://www.shakyo.or.jp/research/2011_pdf/11volunteer.pdf (Accessed 30 September 2015).

[7] The Great East Japan Earthquake Civilian Support Fact Book, Japan NPO Research Association, Tokyo, 2015. http://www.janpora.org/shinsaitokubetsuproject/seika/seika_fact_150725.pdf (Accessed 30 September 2015).

[8] 2 years from the Great East Japan Earthquake, Ishinomaki Disaster Recovery Support Chamber Activity Report, Ishinomaki Future Support Association, Miyagi, 2013. http://ishinomaki-support.com/category/ishinomaki_cat/idrac_report/ (Accessed 9 January 2017).

[9] M. Okamoto, From the question that 'How many people volunteered for the Great East Japan Earthquake?', J. Volunt. Res 2 (2013) 3–14 Osaka Voluntary Action Center (OVAC).

[10] Earthquake relief volunteers, Citizen's meeting to support disaster victims of the Great Hanshin-Awaji Earthquake, Osaka, 1996.

[11] N. Hayase, The Theory of Volunteers, Questioning How to Build Communities (Annual Report of Study of Municipalities 9), Good Book Distribution Society, Tokyo, 1996.

[12] M. Suga, Disaster Relief Volunteers, Sociology Before 3.11, Seikatsushoin, Tokyo, 2014 (Chapter 3).

[13] M. Suga, Development of disaster relief activities, Introduction to the Theory of Disaster Volunteering, Kobundo, Tokyo, 2008 (Chapter 4).

[14] T. Nakajima, Disaster support giving to March 11 Japan Earthquake Relief: with a focus on support funds on recovery, OSIPP discussion paper, Osaka, DP-2014-J-007, 2014. http://www.osipp.osaka-u.ac.jp/archives/DP/2014/DP2014J007.pdf (Accessed 11 September 2015).

[15] N. Hayase, Summarizing problems with charity funds and activity support fund, J. Volunt. Res. 2 (2013) 106–113 Osaka Voluntary Action Center (OVAC).

[16] Central Community Chest of Japan, Interim report: Red Feather Disaster Relief Volunteers and NPO activities support fund, Tokyo, 2015. http://www.akaihane.or.jp/er/pdf/20150115_borasapo_cyukan.pdf (Accessed 20 September 2015).

[17] Japan Business Federation Societal Service Promotion Committee 1% Club, Report on support activities for business world victims and regions struck by The Great East Japan Earthquake: mutual aid efforts by the business world. 3 Tokyo, 2012. https://www.keidanren.or.jp/japanese/policy/2012/011.pdf (Accessed 26 September 2015).

[18] H. Saitou, Support Efforts Involving Corporations and a Variety of Individuals and Organizations, Tutui Shobo, Tokyo, 2014. Volunteer whitepaper.

[19] Nippon Foundation, Knot Generated 25 Stories, Kirakusya, Tokyo, 2012.
[20] Japan NGO Center for International Cooperation (JANIC), The Great East Japan Earthquake and International Cooperation NGO, Tokyo, 2012. http://www.janic.org/MT/img/shinsai_ngo_report.pdf (Accessed 20 September 2015).
[21] M. Yamaguchi, Roles performed by NGOs and future concerns, in: Let the Circle of Volunteering Expand, Tutui Shobo, Liaison Committee, Tokyo, 2014. Volunteer white paper.
[22] M. Suga, Problems related to disaster relief volunteers, in: Review of the Great East Japan Earthquake, Minerva Shobo, Kyoto, 2012 (Chapter 11).
[23] M. Suga, Expectations and concerns in civilian activities for disaster victim support, in: Social Safety for Disaster Prevention and Mitigation, Minerva Shobo, Kyoto, 2014.
[23a] Japan Civil Network (JCN) website. http://www.jpn-civil.net/2014/about_us/history/ (Accessed 11 September 2015).
[24] Kwansei Gakuin University Institute for the Research of Disaster Area Reconstruction, The Japan Civil Network for Disaster Relief in the East Japan, Save Fukushima Children Lawyers' Network (SAFLAN), Nuclear Plant Evacuation Whitepaper, Jinbun Shoin, Kyoto, 2015.
[25] S. Tanaka, Perspective of Sociology of Disasters, Introduction to Sociology of Disasters, Kobundo, Tokyo, 2007.
[26] Y. Matsuda, T. Tsuga, Analysis of Government Subsidies for Organizations Supporting Nation-wide Evacuees from the Accident of Fukushima Nuclear Plant, Studies in Disaster Recovery and Revitalization, Kwansei Gakuin University, Institute for the Research of Disaster Area Reconstruction, Hyogo, 2014.
[27] T. Sugano, Analysis of the revenue structure of support groups for victims of the Great East Japan Earthquake, J. Soc. Saf. Sci. 24 (2014) 263–271.
[28] Cabinet Office, Opinion exchange and recommendations about wide-area coordination of volunteer activities upon large-size disasters, 2015. http://www.bousai-vol.go.jp/product/proposal.pdf (Accessed 11 September 2015).
[29] T. Myojo, Concerns in disaster response in Japan and coordination function in need, in: The 3rd World Conference on Disaster Risk Reduction Public Forum, Miyagi, 2015.
[30] Y. Tanaka, Theory of Civilian Society Policy, Akashi Shoten, Tokyo, 2011.

FURTHER READING

[1] Japan Platform, Japan platform report on The Great East Japan Earthquake Victim Support (March 2011 to March 2012), Tokyo, 2012. http://tohoku.japanplatform.org/lib/data/2012tohoku_rpt.pdf (Accessed 20 September 2015).
[2] T. Tsuga, Efforts in Solving Concerns of Long-distance Evacuees, Tutui Shobo, Tokyo, 2014. Volunteer whitepaper.

Chapter 12

How the Great East Japan Earthquake Affected Corporate Risk Management

CHAPTER OUTLINE

12.1 THE GREAT EAST JAPAN EARTHQUAKE AND CORPORATE OPERATION

The Great East Japan Earthquake on March 11, 2011 forced Japanese corporations into a number of "unexpected" situations. This complex disaster was something Japanese corporations had never experienced before, but they not only recovered and reconstructed themselves but also modified their operations. Corporations now have to carry out risk management in response to the worst scenarios of natural disasters such as the Nankai

The Fukushima and Tohoku Disaster. https://doi.org/10.1016/B978-0-12-814078-9.00012-1

Trough Megathrust Earthquake or the Tokyo Metropolitan Epicentric Earthquake, and threats to national security such as terrorism or pandemics.

This chapter reviews how the companies responded to the Great East Japan Earthquake, what problems the earthquake posed for them, and how they readjusted their business continuity planning (BCP). It also examines how they prepared for corporate disaster prevention in the cases of national crises such as the Nankai Trough Megathrust Earthquake or the Tokyo Metropolitan Epicentric Earthquake. It shows how they will maintain their competitive edges and at what cost. This chapter outlines how the companies are prepared from the viewpoint of corporate risk management, and especially studies the case of the Nissan Motor Corporation.

12.2 PROBLEMS IN CORPORATE RISK MANAGEMENT THAT THE GREAT EAST JAPAN EARTHQUAKE CAUSED

Kamei and Takano summarized situations with companies throughout the year that followed the Great East Japan Earthquake [1,2]. The article listed concerns in corporate management: (1) Safety confirmation and stranded employees, (2) Harmful rumors, (3) Unstable electricity supply, (3) Interruption of lifelines, (5) Shortage of fuel and material, (6) Supply chain disruption, (7) Lack of leadership in crisis management, (8) Crisis simulation training of executive officers, and (9) BCP review.

The same concerns apply today after 5 years have passed since the earthquake. We have added a tenth to these nine concerns: (10) Response to overseas risk in forming the supply chain, overseas production and procurement. The political risk has greatly increased with terrorist attacks overseas.

This section focuses on three effects the Great East Japan Earthquake had on corporate risk management: risk assessment, BCP reforming, and response to risk with supply chains.

12.2.1 Conceiving Low-Frequency and High-Risk Events in Risk Assessment

Risk assessment evaluates accidents and disasters in two aspects of their frequency and severity. Prioritizing events with just the two axes of frequency and severity, however, leaves natural disasters occurring once in a thousand years, such as the Great East Japan Earthquake, at a low priority. Nuclear power plants had not had major accidents caused by natural disasters in the past in Japan. From now on, we have to make thorough preparations and plans against natural disasters like the Nankai Trough Megathrust

Earthquake or the Tokyo Metropolitan Epicentric Earthquake, with low frequency but extremely high severity.

12.2.2 **BCP**

12.2.2.1 History and Function of BCP

BCP is the act of building plans that define daily time activities and actions and procedures to follow at times of emergency, such as natural disasters, terrorist attacks, or new strains of influenza, so the business can minimize its asset losses, continue to operate, and quickly recover. Business continuity management (BCM) is the management activity of executing BCP and continuing the activities it defines.

In Japan, the Central Disaster Management Council set a target for the number of companies with BCP as a part of its disaster management, and the announcement of the "Business Continuation Guideline, First Edition" moved companies to start their BCP.

For small to midsized companies, the Ministry of Economy Trade and Industry (METI) published the "Guideline for Setting and Operating BCP for small to mid-sized companies (Feb., 2002)," "Guide for small to mid-sized company BCP (Business Continuity Plan) (Mar. 2008)" and "Tips for setting BCP (Mar. 2009)."

The difference from earlier plans for business disaster management was in setting the "target recovery time" with the assumption that there would be some damage. When we were concerned about an outbreak of a new strain of influenza, H5N1, originally from birds, the Ministry of Health, Labor, and Welfare (MHLW) announced the "Guideline for Businesses and Workplace Actions against New Influenza Strains" in February of 2009. After automobile manufacturers were forced to suspend operations due to parts manufacturers being affected by the 2007 Niigata Chuetsu Offshore Earthquake, the Central Disaster Management Council published the "Business Continuation Guideline, Second Edition."

BCP takes the following five steps [3]:

(1) Specify the primary priority business operations to continue or recover with the operation resources for those businesses.
(2) Set the "time to recovery" for the primary business operations.
(3) Discuss beforehand with the customer the level of supply to maintain at the time of emergency.
(4) Prepare alternate resources and plans for production equipment and procurement at business locations.

(5) Form a common understanding with the employees about guidelines and procedures of business continuation at the time of emergency.

Business impact analysis (BIA) is the prerequisite for steps (1)–(5). BIA sets the primary operations and related resources and evaluates and analyzes the effects on operations and financials when the operations stop due to disasters.

12.2.2.2 Concerns With BCP Identified by the Great East Japan Earthquake

The Great East Japan Earthquake revealed concerns with BCP: (1) The tsunami destroyed the primary businesses, (2) For companies in the evacuation zone of the Fukushima-1 plant accident or those that were completely washed away by the tsunami, target time to recover no longer had meaning, (3) The tsunami destroyed entire areas including customers with a priori discussions, (4) The wide-area disaster damaged alternate resources and plans, and (5) The mega-tsunami and nuclear plant accident were unexpected and had not even been discussed in forming a common understanding.

Takano [2] pointed out the gaps among companies in their BCP and the vulnerabilities in crisis management structures and BCP from the Tokyo Metropolitan Epicentric Earthquake for large corporations. He further reported, to spread BCP wider, the need for clarifying the crisis management structures and BCP evaluation standards, promotional policies to provide incentives for top management of small to mid-sized companies and privately held companies in setting their crisis management structures and BCP, and actions to secure effectiveness in the set crisis management structures and BCP.

12.2.3 Problems With Supply Chains [4–6]

12.2.3.1 Supply Chains and Disasters

Toyota Motor's Just-in-Time production method established in the 1970s is a system for which "necessary parts arrive in the necessary quantity at the necessary time" from parts manufacturers for assembly. It is a method that works without part stocks, to eliminate wasted resources from the manufacturing floor. This type of manufacturing system, called "stock zero," is adopted not only by domestic manufacturers, but by makers over the world. The system is supported by the supply chain of raw material, processed material, parts, and products.

Manufacturing without stock cuts down costs and enhances efficiency and competitiveness; however, it has the fundamental weak point that once a part of the supply chain breaks down, manufacturing of the final product stops. This problem surfaced after the 1995 Great Hanshin-Awaji Earthquake with the damage to the transportation infrastructure. However, the February 1 fire at the Aisin Seiki Kariya Factory 2 in 1997 also turned into a big problem. Toyota then relied on this factory for over 90% of their supply of the proportioning valve to adjust the brake system oil pressure, and were chased into a corner. Toyota, in addition to its affiliated parts suppliers, asked for support from other manufacturers in the area by sending the drawings to produce the part. The efforts ended the factory production suspension in 4 days.

Another incident was the July 16, 2007 Niigata Chuetsu Offshore Earthquake, when the mechanical parts manufacturer Riken suffered damage to its Kashiwazaki Plant and the production of piston rings, a part that was indispensable for the automobile engines, stopped. The supply interruption from Riken, with a 50% share of the domestic market of piston rings, caused all automakers to suspend their production. Each automaker dispatched support workers, peaking at 800 helpers a day, to reconstruct the operation. These support efforts put the operations back in 1 week from the disaster and all the production lines were recovered within 2 weeks [7].

12.2.3.2 Problems With Supply Chains

The Great East Japan Earthquake disrupted the supply chains over a much greater area, beyond comparison, e.g., semiconductor manufacturers for car-mounted microcomputers or material processing plants along the coast. The result was suspension or reduction of production over the nation, affecting industries both in and out of the country.

In the automotive industry, Renesas Electronics, with a 40% share of the world market of car-mounted microcomputers, suffered damage to its Nasu factory and large ripples traveled over the world. The solutions in Aisin Seiki's fire with Toyota quickly recovering its production with alternate sources or Riken's quick recovery from damages with full support from its customer base did not work this time. The reason was that electronic parts such as car-mounted microprocessors have long lead times in production that are heavily customized and alternate manufacturing was difficult. The interruption of the raw material supply of synthetic resin and rubber was another factor in the recovery.

From past experience in disasters, industries had planned their supply chains so the risk in procurement was distributed. The industry had a good grasp of suppliers in the first level of the supplier hierarchy (Tier 1 supplier), but not in the second (Tier 2: supplier to the first level supplier), third (Tier 3), or beyond. The Great East Japan Earthquake revealed that some suppliers were procuring material from the same material supplier and the supposedly pyramid-type structure was actually in the shape of a diamond. The Japanese industry, with strength in its integration between the supplier and maker, again recognized its weakness at the time of this disaster in producing highly customized parts at a high quality: having to keep the competitive edge by differentiating the products and reducing the cost, but at the same time also preparing against the risk of disasters at cost. The never-ending problem in risk management surfaced again.

12.2.3.3 Reviewing Supply Chains

The Great East Japan Earthquake and the half-year later flooding in Thailand in October 2011 forced businesses to accelerate their enhancement of supply chains. They spent efforts in (1) Further clarification of the supply chain = Making the supply chain visible, (2) Modularizing parts = Making specifications and parts common = Reviewing the designs, (3) Diversifying the suppliers including parts that can be imported, (4) Setting and reviewing BCP, and (5) Adjusting the inventory stock.

Research by Mitsubishi Research & Consulting [6] (conducted in December 2011, with 4151 replies from companies) reported that among the companies directly and indirectly affected by the Thailand flooding, 29.0% (large corporations 51.3%, and small to midsized companies 26.3%) answered that the lessons on supply chain from the Great East Japan Earthquake were useful. Among the useful lessons, "Checking the status of procured parts" gained the most points at 57.2%, "Securing alternate suppliers" 35.2%, and "Support preparation for customers and suppliers" 25.5%.

We will next discuss the actual reviews made by each company.

Toyota Motor Corporation requested its 430 Tier 1 suppliers to identify the Tier 2 and beyond suppliers in the supply chain. The supply chain information of 300,000 entities was superimposed on the geographical risk of natural mega-earthquakes like the Nankai Trough Megathrust Earthquake or the Tokyo Metropolitan Epicentric Earthquake to prioritize the risks. This information was stored in a database system so it was visible for Toyota and its suppliers to share. This development led to 70% of the parts confirmed to be ready for alternate production. At the same time, however, the problem of parts with hard-to-copy production or difficulty in generalization due to long lead time or only one technology was also recognized [4].

Daikin Industries has worked on standardization and communalization of its parts since 2008. After experiencing the Great East Japan Earthquake, the company is further developing enhanced initial response for all parts in case of emergency, making long-term contracts for its strategic parts, and developing compatible alternate parts for its strategic and bottleneck parts [4].

SONY has been working on four countermeasures. The first is multiple sourcing including Tier 2 and Tier 3 suppliers. This multiple sourcing involves: (1) Dual sourcing (purchasing from multiple sources), (2) Requesting the supplier to produce at multiple locations in case SONY is relying on a single supplier, and (3) keeping an inventory stock if both solutions are difficult. The second effort is in avoiding geographic risk. The lessons were learned when the multiple Tier 1 suppliers were all in the Tohoku region. The third is in making the total global supply chain visible, and the fourth in knowing the equipment and facilities of partner factories and their alternatives [4].

At Fujitsu, efforts are made to: promote standardization by removing factory-specific special procedures to allow alternate production, planning alternate production from the design stage for multiple sourcing, and by working especially on parts that are single sourced (purchased from a single supplier) [4].

The efforts by these companies are consolidated into the following efforts by Renesas Electronics: (1) Building multiple factories to allow alternate production, and (2) Quality assurance with alternate production. These efforts led to the recognition of keeping a stock of alternately producible parts that lasts for the period of switching to alternate production, and for single sourced parts, keeping a strategically planned stock [4].

The Great Hanshin-Awaji Earthquake (1995), Aisin Seiki factory fire (1997), outbreak of SARS (2003), the Niigata Chuetsu Offshore Earthquake (2007), the Great East Japan Earthquake and flooding in Thailand put the industry through a series of lessons. It is a never-ending problem for the companies to build an effective supply chain for global competition and, at the same time, arrange for corporate disaster prevention and BCM. High-quality custom products with a large share of the global market is the strong point of Japanese manufacturing based on integration of parts suppliers and final product manufacturers. Bracing against mega-disasters, however, calls for high versatility and modularization, with high-quality competitiveness and low generalization on one hand, and high cost competitiveness and high generalization on the other. Corporate risk management in the time of post-Great East Japan Earthquake means making the right decisions between these two directions.

12.3 CASE STUDY: NISSAN MOTOR CORPORATION [8,9]

The author had an interview with Tadashi Sugawara, at Global Internal Audit Office, Corporate Risk Management, Nissan Motor Corporation, the author of references [8,9], on June 26, 2015. This section is based on this interview and his papers.

Nissan faced a critical situation in 1999, having a debt of 2 trillion JPY, 10 years before Hitachi marked the worst-ever loss in 2009. It formed a global alliance with Renault of France and welcomed Carlos Ghosn from Renault as its President. Ghosn quickly revived Nissan in a V-shaped recovery like Hitachi's. President Ghosn excelled in corporate crisis management and Nissan became alert to crises, formed a risk-management structure and prepared against emergency situations with simulation training. The Great East Japan Earthquake severely affected the entire automobile industry, but Nissan was the first to recover and started full production at the end of September 2011. The business world highly respects Nissan for its quick recovery after the Lehman crisis, before the Great East Japan Earthquake and the flooding in Thailand afterwards. We will review Nissan's power in risk management, crisis management, and BCP.

12.3.1 Risk Management Structure at Nissan

Application of the risk management system for the whole corporation, which Nissan had in the background, changed the corporate environment. The Nissan Revival Plan (NRP) that started in the year 2000 revived the company in a V shape from the operations crisis, and then the company shifted its target from an "emergency structure for revival" to "continuous growth."

First, a risk management structure for earthquakes made a serious start. The management viewed a major earthquake as the most critical risk and the top management meeting accepted the "Earthquake Response Plan" in 2003. The plan proceeded with hardware reforms of aseismic reinforcement of buildings, fixing equipment, and relocating the data center to a new building, and for software, established an all Nissan Response Headquarters group and coordinated plans with the suppliers.

In 2005, businesses were bound to report "business risks" in their financial report, and the new Companies Act, enforced in 2006, required establishing "internal control." Nissan recognized the need to build a "corporate system including top management (board members)" and "a total corporate risk management system for corporate continuity." For these purposes, Nissan established a dedicated total corporate administration office and built a total corporate risk management system in 2007. The company, for its opening of risk information, fully revised its business risks in the 2009 Financial Report [10] and 2010 Annual Report [11]. It further broadened open information about risk management activities in its corporate sustainability report (CSR) in 2010 (Table 12.1).

Table 12.1 Risk Management (RM) Systems at Nissan

	Operation Crisis			Lehman Crisis	GEJ Earthquake	Thailand flooding
	2000–01	2002–04	2005–07	2008–12		2011–16
Midterm plans	Nissan Revival Plan	NISSAN 180	Nissan Value-Up	NISSAN GT 2012		Nissan Power 88
	Urgent restructure for revival	Continuous growth with profit	Further development and new challenges	Growth and reliability (suspended in 2009)		Brand and sales force
	Pursuit of cost effectiveness	From a Good Company to a Great Company				Both the market share and profit
Main changes	• Plants closed for efficiency	• Increased sales, higher availability	• New market, local production (BRIC)			
	• Workforce reduction	• Common chassis and parts	• Challenging technology innovation (zero emission, automated driving)			
	• Narrowed down suppliers	• Shorter development time	→ Risk in unknowns, and unexperienced			
	• Global management system	→ added risk of operation and profit risk in quality and recall	• Material cost escalated			
			• Global financial and economical crisis			
			→ Procure and produce locally			
Corporate RM system	• Integrated RM Function in Financial Department	• Enterprise Risk Management (ERM), full corporate risk survey	• ERM function placed with internal auditing			
			• New Companies Act, set internal control committee			
	• Global Headquarters	• Opened risks with businesses	• Strategic information opening			
			• ERM restructured			
			• Started J-SOX compliance	Cooperation with IR/CSR		

Continued

Table 12.1 Risk Management (RM) Systems at Nissan *Continued*

	Operation Crisis			Lehman Crisis	GEJ Earthquake	Thailand flooding
	2000–01	2002–04	2005–07	2008–12		2011–16
Individual RM activities	• Damage prevention activities started (cross department teams)					
	• Compliance committee started	• Supplier reliability risk	→ Enhanced supplier risk management	Internal control, Full corporate RM system		
		• Commodity price risk				
		• IT security and personal information protection				
			• Crisis management committee			
	• Major disaster RM started (earthquake) (fire) (hail) (typhoon) → BCP planning (Niigata Chuetsu Offshore Quake)				Enhanced after 3.11	
			(Pandemic, new flu)			
			(Control Tier2 or lower suppliers)			
Risk finance	• Started Global Insurance Program	• Large self-payment (high exemption)				
		• Apply captives globally	• Expand items covered			

Based on information from Tadashi Sugawara, at Global Internal Audit Office, Corporate Risk Management, Nissan Motor Corporation.

12.3.2 From "Earthquake Response" to "BCP"

Lessons from the 2007 Niigata Chuetsu Offshore Earthquake led Nissan to carry out the following reviews. First of all, Nissan turned the total corporate response headquarters into a more mobile structure. The company assigned its COO to the head of the headquarters and it formed an initial response team in charge of gathering information first. The company arranged a simulation training in which the COO and other primary executives participated. The training compressed the day of the earthquake and the following day into 2 hours to simulate the actions to take during a disaster. The manufacturing department incorporated the experience through supporting the parts manufacturer Riken after the Niigata Chuestu Offshore Earthquake into its action manual. The group identified important processes in the group, executed BCP for them and distributed the plan within the company as a template. For the supply chain, Nissan conducted a supplier survey and implemented improvements to the plans.

During 2008–09, in response to pandemics like the highly pathogenic avian influenza or swine influenza, the company built a global response structure of the headquarters and overseas branches. The overall BCP covered indirect departments like R&D or Headquarters as well. The workforce narrowed down on the priority processes and performed discussions focused on "people." Suppliers were given the almost full BCP of Nissan production factories.

In 2010, Nissan carried out actions for the Tier 2 and beyond suppliers in the supply chain. The actions involved emergency responses for production continuation, enhancing the quality control system for increase in production, identifying parts at risk of supply and checking the factories and supporting their improvements.

12.3.3 **Damage Caused by the Great East Japan Earthquake**

The Great East Japan Earthquake caused simultaneous suffering over a wide area to the entire automobile industry, suspending parts supplies and destroying infrastructures to stop its business activities. In the case of Nissan, the Tochigi and Iwaki factories suffered damage to their facilities on the day, March 11, of the earthquake. Further, on March 15, the subsidiary JATCO Fuji plant, which produced transmissions and continuously variable transmissions (CVTs) was damaged in the earthquake, with its epicenter in Shizuoka Prefecture. All the stock of 2500 new vehicles was destroyed. The 436 Nissan distributors in East Japan suffered damage, including five that were completely washed away by the tsunami.

Iwaki factory, in Iwaki City, Fukushima Prefecture, that produced V-type engines for luxury cars, suffered the greatest damage. The factory floors were damaged by land sinking, the air ducts broke, engines in production fell, lifelines were disrupted and the factories were forced to stop operation. When the earthquake hit, the production for the day had ended and the workers were away from the fabrication areas. Thus, there were no worker injuries. The March 15 instructions to stay indoors forced the workers to standby at home. The reconstruction work restarted on March 23. The aftershocks on April 11 and 12 added some damage; however, on April 22, all machine equipment was recovered. Affiliated companies and parts suppliers sent a maximum of over 200 support workers and the factory was back to its full operation on May 17.

12.3.4 **Initial Response**

When the earthquake hit, it was clear that there would be damage and the Nissan Disaster Response Headquarters immediately started its operation to collect information. Nissan had had their large-scale simulation training just days before on February 21, 2011 and information gathering proceeded as trained. It was during the day on a weekday and the head of the headquarters and other primary members were at work. The safety confirmation system did not work until next morning, but most employees were in the office and Nissan completed most of its safety confirmation within a day using PC mail and intraoffice phones. This event taught the lesson of converting the safety confirmation method to a web-access type and prepare multiple routes for carrying out emergency correspondence.

Nissan had sold its Ginza Headquarters building at the time of the 1999 crisis but with its V-shaped revival had moved to the Yokohama Global Headquarters building. The new headquarters had an aseismic structure with emergency power generators and the company had changed its basic policy at the time of an earthquake to "Headquarters employees to take shelter in the building." Thus, on the day of the Great East Japan Earthquake, policies at the Headquarters were: (1) those in walking distance may go home, (2) those with difficulty returning home were served food, and (3) there were plenty of blankets stocked when the basic policy was changed.

At the same time, Nissan accepted others that had difficulty going home following its policy to "serve the local community." About 150 people stranded in the City of Yokohama entered the building per the city's request and were served drinking water and provided with blankets. Later the building was designated as a shelter building by the City of Yokohama and Nissan made further preparations to accept more people.

12.3.5 **Business Continuation**

Indirect departments shifted to working at home and reconstruction workers commuted to factories. The head of response headquarters, executives and members of the headquarters had meetings almost every day throughout March. The frequency dropped to once a week in April and closed on May 10. Under the directions of the head, the headquarters shared information, guided all units into the same direction, and orchestrated global cooperation by confirming the basic guidelines and priorities and speeding up the processes. A special management meeting was quickly organized to distribute information to the public (users, investors, media, and CSR) and inside (response headquarters reports and messages from the executives).

The actions led by the response headquarters involved: preparation against harmful rumors by radiation measurement and recording (sample checking from March 17, and later all unit checking with a gate-type machine), securing floating funds (adding funds at hand by new loans and issuing corporate bonds), supporting distributors (providing operating capital, providing repair parts and second-hand cars), and lobbying the industry and the government about setting radiation standards.

The data center maintained its function as designed. Damage to the electricity receptor caused 3 days of blackout. However, the data center immediately switched to the emergency generator and continued its operation. Securing fuel, however, required some work. Iwaki recovered its network within 3 days to have access to the intranet and electronic mail.

The recovery support, with experience from the 2007 Niigata Chuetsu Offshore Earthquake, used a human resource map to quickly arrange and dispatch recovery teams. The team recovered the Tochigi plant, JATCO Fuji plant, and Iwaki plant in this order and additionally supported the recovery of affiliated companies and suppliers.

On March 25, President Carlos Ghosn visited the Iwaki plant. He cheered the workers, checked the factory, exchanged opinions and made use of them for reinforcing the ground base. An industrial doctor provided counseling service to the supporters for mental care and carried out lectures on radiation effects in the Iwaki factory. The Iwaki factory also ran an evacuation drill at night in preparation against aftershocks with blackouts. Many workers applied to take part in the drill.

For saving electricity, the plant installed more emergency generators, worked weekends and modified the shift schedule to move the consumption peak, conducted full-time monitoring, and made emergency response plans when the upper limit was threatened.

12.3.6 Regional Support Activities for the Disaster-Struck Areas

In support for disaster-struck areas, Nissan raised funds and donated them, gave 50 four-wheel drive cars, 65 electric cars (Leaf) that ran without gasoline, free use of seven various rental cars, provided emergency supplies, and set special vacations for volunteers (5 days with pay) to carry out volunteer work.

12.3.7 Power for Quick Recovery—Field, Cross-Functional, Cross-Regional

Corporate crisis management by Nissan's President Carlos Ghosn puts a priority on the "field," executes leadership in making "top" decisions by prioritizing actions, and builds and executes "cross-functional" and "cross-regional" problem-solving plans. These policies worked at the time of the

Great East Japan Earthquake as well, and provided the power to recover quickly.

Power in the "field": (1) Damage was minimized with actions learned through evacuation drills (the employees took measures to prevent fire before evacuation); (2) Dedicated efforts were made by loyal employees in the field; (3) The power in the field moved forward through recovery even when they had to rework already finished work forced by serious aftershocks.

Leadership at the frontline: (1) The top executive entered the field, cheered the workers, observed the actual situation and listened to opinions; (2) Total response headquarters shared information, gave the direction for the entire company, and quickly decided on countermeasures.

Cross-functional: (1) For each part, the workers formed teams for each function of development, production and procurement, and coordinated their response activities. The development team made plans and discussed response actions. Production adjusted and made changes to the production schedule. Procurement gathered information about suppliers, and planned and discussed response actions. (2) Formation and dispatch of supplier support teams, planning and execution of response actions, development of alternate parts and material, and selection of alternate suppliers were all carried out simultaneously and progress was shared. (3)

Cross-regional: (1) Supply chain managers in charge at all overseas subsidiaries in the world gathered in Japan and performed their work in a single global Nissan team. They reported the progress to their home countries in their home language to make adjustments. (2) General managers of all overseas factories clarified when they would run out of critical parts. (3) This information was shared on a daily basis and the entire corporate group prioritized acquisition of each part and adjusted the part shipment schedule.

12.3.8 Lessons From the Great East Japan Earthquake and Reviewing BCP

As we saw in the previous section, quick initial response and unexpectedly fast recovery are highly praised, but on the other hand, the Great East Japan Earthquake taught a number of lessons. Nissan Motor reviewed its BCP as follows.

12.3.8.1 Initial Action

The safety confirmation system with a bulk e-mail over the cellular phone system did not function and had to be switched to a web-access system. Training was enhanced to confirm the function of the new safety confirmation system. Also a thorough training in alternate communication methods was conducted. The evacuation plans did not consider tsunami risks and reviewed the evacuation routes and plans based on damage simulation with the up-to-date tsunami prediction. A check was made on whether evacuation would be complete within the set timeframe through training.

12.3.8.2 Supply Stock

The new Global Headquarters Building in Yokohama had an aseismic construction and will serve as a shelter in case of an emergency, and it now stores enough supplies for all employees and visitors. The supplies were stored on one floor (storage level), but the response headquarters realized that if the elevators stop, water would be too heavy to carry to other floors and now part of the water and emergency food is stored on each floor.

12.3.8.3 Total Corporate Headquarters Simulation Training

The Total Corporate Headquarters simulation training is now an annual event. It used to take place during workhours on a weekday, but the new training involves headquarters members to gather on foot or on bicycles during the weekend. In 2013, the first training for weekend gathering was carried out to confirm the coordination of showrooms and the Headquarters. The scenario was made more severe each year, and after the Great East Japan Earthquake, an even more serious scenario was prepared to meet the expected events during a Nankai Trough Megathrust Earthquake or a Tokyo Metropolitan Epicentric Earthquake. Subsidiaries joined the training via video conferencing systems and actual suppliers also participated. Training used to include only reviews afterwards, but preparation is now part of the training for maximum effect.

The actual training proceeds by the Headquarters administration office distributing, at the beginning of each fiscal year (April in Japan), the newest expectations for each job function, and each job function reviews its actions based on the expectations. The results are gathered into a table-top exercise with the related groups to check whether the overall plans are consistent before the training. Although the training is a simulation, it is a test of effectiveness. The findings, problems, and feedback from executives and consulting companies are collected for further review of the action plans.

12.3.8.4 Preconditions and Scope of Responses

After seeing the tsunami devastation during the Great East Japan Earthquake, Nissan reviewed their preparation to save human lives assuming a tsunami from the Nankai Trough Megathrust Earthquake. The earthquake countermeasure plans approved in 2003 prioritized the group and related companies in the Kanto and Tokai areas according to the amount of investment, and aseismic reinforcement was installed in production facilities at the higher end of the value chain. After the Great East Japan Earthquake, however, the same installment was made for those at the low end of the value chain, such as distributors. In addition, for faster production recovery, the ground bases were reinforced as well as secondary structures such as ceiling panels, ducts, piping and electrical wires.

12.3.8.5 Supply Chain

The need to manage the entire supply chain, including Tier-N suppliers, pushed Nissan to identify the bottlenecks and plan countermeasures and execute them. The basic policy of procurement used to prioritize efficiency and the number of suppliers and stock quantity were narrowed down. After the Great East Japan Earthquake, parts produced in only one factory were mandated for elimination and they were distributed over multiple factories. In addition, a database was constructed to quickly monitor the states of suppliers in case of emergency. Global data are collected and the database is expanding. Tier-N suppliers started to actively participate in the move with field diagnosis and making suggestions for improvements. Recently, not only natural disasters but also compliance and CSR activities are now critical for business continuation, and actions to raise awareness and request cooperation from the suppliers are in place.

12.3.8.6 Global Coordination

As overseas operations are becoming more important, requests are now made to overseas subsidiaries to carry out BCP and conduct emergency training. The requests come with self-diagnosis tools and support in the form of introducing the best practice where weaknesses have been identified. The total response headquarters simulation training is carried out during global meetings so overseas officials can watch the exercise.

12.3.8.7 Global Risk Management System Through Lessons From the Great East Japan Earthquake

Nissan's risk management is unique in that "Locations spread over the world fully support Japan at the time of emergency" through its global alliance with Renault. For example, Nissan's Global Vehicle Production Technology Center in Zama City, Kanagawa Prefecture keeps digital drawing data of molds for all Nissan car parts. When the Iwaki factory stopped operation, the data was transferred to the US factories and alternate production of engine parts for the Japanese market started immediately.

Nissan's global risk management system with lessons from the Great East Japan Earthquake now contains the following developments: First, the basic tools of risk management rules, methodology of risk assessment, or risk map format are shared. Second, actual actions are coordinated globally on a job basis. Third, for exchanging information and opinions, global risk management meetings are organized annually, domestic risk management meetings twice annually, and information exchange with Renault and a joint venture in China once or twice a year. Fourth, values are expected to be produced through risk management operations in each region and each company. In other words, risk assessment is made through the viewpoint of each region and each company. "Risks for the entire group" at the global headquarters are not pushed down towards the global companies. Similarly, the risk maps from the companies are not consolidated into a global version risk map. The global risk management system at Nissan is thus a loose coordination with the basic risk management processes shared globally.

12.4 CONCLUSION: LESSONS OF CORPORATE RISK MANAGEMENT

The case of Nissan Motor Corporation shows us the importance of leadership at a time of major disaster, the effect of training against expected risks, and the effectiveness of job-based and region-based cooperation. If we generalize the lessons the Great East Japan Earthquake taught on corporate risk management, they are what Business Policy Forum, Japan showed [4]:

(1) "Building strong factories" with hardware risk control based on thorough expectations of risk
(2) "Thorough emergency training and simulation" as software risk control
(3) "Building cooperative relations with the region and industry groups"
(4) "Enhanced risk communication with the customer" about what risks are expected and how they will be handled
(5) "Building a robust supply chain"
  (a) Complete hierarchical management of suppliers
  (b) Promoting multiple sourcing or means of alternative production for parts and material
  (c) Enhanced risk communication with suppliers
(6) "Building competitiveness in manufacturing"
  (a) Arranging overseas production locations
  (b) Communalization, standardization and strategic application of part generalization and customization (special specification)

The keyword in corporate risk management from now on is "cooperation" with the stakeholders, including the local region. The big lesson in business operation learned from the Great East Japan Earthquake was the fact that it made a strong impression on business people about its importance.

This conclusion matches what Kawata wrote [12]; "The key point in corporate disaster prevention is in 'cooperation' and 'connection' and whether we can overcome national crises depends on whether all units of the society can cooperate or not."

REFERENCES

[1] K. Kamei, K. Takano, The Great East Japan Earthquake and corporate crisis management, in: Review of the Great East Japan Earthquake, School of Societal Safety Sciences, Kansai University, Minerva Shobo, Kyoto, 2012.

[2] K. Takano, in: School of Societal Safety Sciences, Kansai University (Ed.), Disaster Prevention and Responsibilities of the Top Management, School of Societal Safety Sciences for Disaster Prevention and Mitigation, Minerva Shobo, Kyoto, 2014.

[3] METI and Small and Medium Sized Enterprise Agency, BCP Guide for Small and Medium Sized Enterprises, Small and Medium Sized Enterprise Agency, 2008.

[4] Business Policy Forum, Japan, Report on Research of Improving Effectiveness of Corporate Business Continuation after the Great East Japan Earthquake – for the targets of improving power of risk response and securing manufacturing competitiveness under global competition, 2013.

[5] Tokio Marine, Nichido Fire Insurance, Applying lessons from the Great East Japan Earthquake, in: Frontline of Risk Management 2011–16, Tokio Marine and Nichido Fire Insurance, 2011.

[6] Mitsubishi UFJ Research & Consulting, Fiscal year 2011 research on support for small to mid-sized corporations. Report on research of sources of competitiveness of Japan's Manufacturing Industries, 2012.

[7] Cabinet Office, "Business Continuation" a special article in "2009 Disaster Prevention Newsletter." http://www.bousai.go.jp/kohou/kouhoubousai/h21/11/special_03.html (Accessed 21 September 2015).

[8] T. Sugawara, BCM efforts at Nissan Motor and its response to the Great East Japan Earthquake, in: The 146th Symposium, May 17, The Japanese Society for Quality Control, 2013.

[9] T. Sugawara, Corporate risk management at Nissan Motor, in: The 39th National Conference, Sep. 17, Japan Risk Management Society, 2015.

[10] Financial report 2008, Nissan Motor Corporation, March 31, 2009. http://www.nissan-global.com/JP/DOCUMENT/PDF/FR/2008/fr2008.pdf.

[11] Annual report 2010, Nissan Motor Corporation, March 31, 2010. http://www.nissan-global.com/EN/DOCUMENT/PDF/AR/2010/AR2010_E_All.pdf.

[12] Y. Kawata, Corporate Disaster Prevention in the New Age, Chusaibo Shinsho, Tokyo, 2013.

Chapter 13

Problems and Future of Post 3.11 Disaster Journalism

CHAPTER OUTLINE

13.1 DIFFICULTY WITH NEWS COVERAGE OF 3.11—HAVING BOTH QUANTITY AND QUALITY

In this chapter, we will follow the convention of Osawa [1] and refer to the sequence of the Great East Japan Earthquake, Fukushima-1 Nuclear Power Plant accident, and other disasters as "3.11." This 3.11 was absolutely a "super wide-area disaster" [2,3]. Damages were enormous, with the devastation covering a wide area in a variety of forms and the different types of damages affecting one another in complex ways. Recovery from the disaster has been rough and the situation is still quite serious with damage that will persist over years.

How is the media reporting on 3.11? The primary media, which falls into the category of "mass media" like TV, radio channels, newspapers with nationwide coverage, and magazines and books are still sending out a vast amount of information about this unprecedented disaster. The total amount of information that has been passed around is certainly uncommon. However, the loss of interest over time that has been a concern from the beginning is now taking shape (e.g., [4]).

The Fukushima and Tohoku Disaster. https://doi.org/10.1016/B978-0-12-814078-9.00013-3

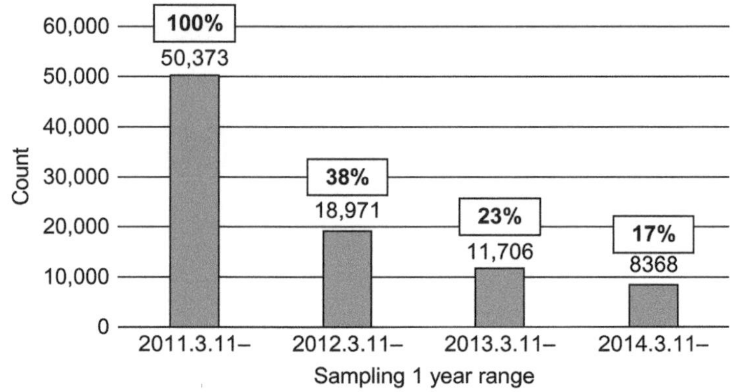

■ **FIG. 13.1** YOMIDAS Rekishikan number of articles with keyword "earthquake" AND "Tohoku OR East Japan."

Newspaper article database services can quickly give a rough idea of media coverage of 3.11. Fig. 13.1 shows the results from YOMIDAS Rekishikan of counting the number of articles in the Yomiuri newspaper, which is said to have the largest subscriber base, using the search condition "(earthquake) AND (Tohoku OR East Japan)." The figure shows the count for each year. If we set the count in the first year of 3.11 at 100, currently (in the fourth year) the count is less than 20. The number of articles has dropped to less than one-fifth. The trend was the same with the search condition "Fukushima AND Nuclear Power," as Fig. 13.2 shows. Macroscopically, the "loss of interest" is certainly taking place.

If we, however, add the key word "loss of interest" and count the number of articles again, the results (Fig. 13.3) are clearly different. As the figure

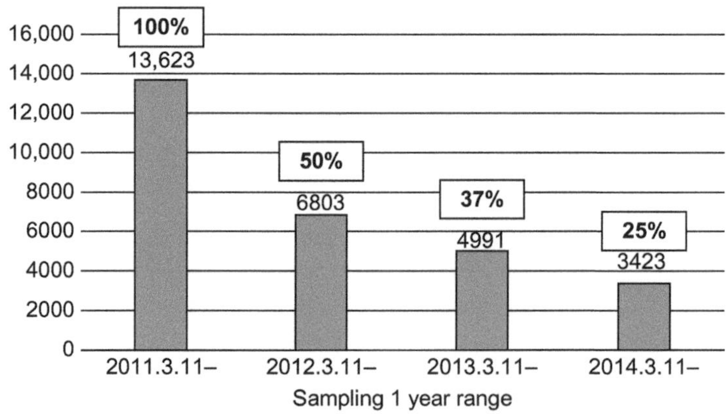

■ **FIG. 13.2** YOMIDAS Rekishikan number of articles with keyword "fukushima AND nuclear power plant."

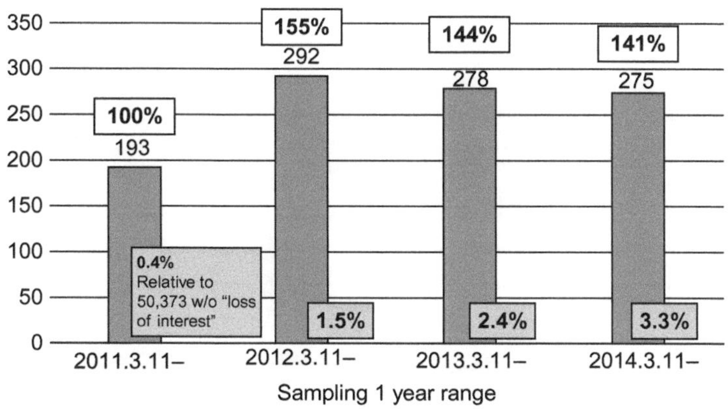

■ **FIG. 13.3** YOMIDAS Rekishikan number of articles with keyword "earthquake" AND "Tohoku OR East Japan" AND "loss of interest."

shows, the second year count was about 1.5 times that in the first year and from then on, was about constant. The percentage of articles containing "loss of interest" among all earthquake-related articles grew from 0.4% to 3.3%, i.e., over eight times. The mass media, in other words, are trying to counteract the loss of interest by repeating that we should not lose interest in the 3.11 lessons but, at the same time, repeating the stereotyped articles telling us that we should not lose interest is giving us the impression that this loss is a reality. It is not just the quantity, but we have to keep our eyes on the quality of the newspaper articles as well.

Another problem with mass media that has been pointed out from the time immediately after 3.11 was the imbalance among affected areas that were highlighted in the news [5]. On one hand, the scenes of greatest devastation caught the eye of the public, but there were other areas that hardly any journalist visited. The imbalance in news coverage led to gaps in support, e.g., in the amount of relief donations [5]. Have these problems been resolved? The answer would have to be "No."

Fig. 13.4 shows the change in media coverage of the 12 municipalities along the coast of Iwate Prefecture that were relieved by the Disaster Relief Act. Some minor changes are seen; however, shifting the coverage from those at the high end to the lower ones did not take place. Municipalities that were not covered well at the beginning had no chance of consolation. And even with higher levels of coverage, region-specific difficulties might not have been shown in the news. We must recognize that the overall quantity and local quality do not necessarily relate to each other.

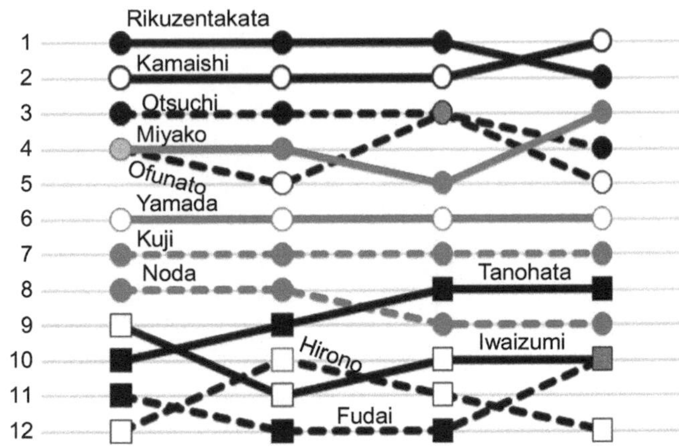

■ **FIG. 13.4** Change in media coverage ranking of 12 municipalities along the coast of Iwate Prefecture.

So far, our discussion has been a very rough outline of the media coverage of 3.11. It has completely neglected "local media" and "media on the net." How the public perceived the coverage has also been ignored. The discussion has merely explained the reality and has simply been objective and external with parameters easy to get a handle on.

The difficulty with media coverage of 3.11 is in achieving both quantity and quality. We will look straight at this problem not only macroscopically, i.e., from a quantitative point of view, but will also make some microscopic observations, i.e., qualitative examinations that may have been difficult in the past. The "unified theoretic framework," which we will discuss later, will open up the current situation and target an improvement in the fundamental viewpoint.

13.2 POINTS IN REVIEWING DISASTER JOURNALISM—WHO OWNS THE INFORMATION?

We have, in the past, suggested "disaster journalism" in place of the conventional "disaster news coverage" [5]. The following three points summarize the functions of "disaster journalism":

1. Save people's lives with "emergency coverage"
2. Support people's lives with "recovery coverage"
3. Protect people's lives with "preventive coverage"

In accomplishing these social missions, it is desirable that more people involved with journalism build relationships that capture the disaster events

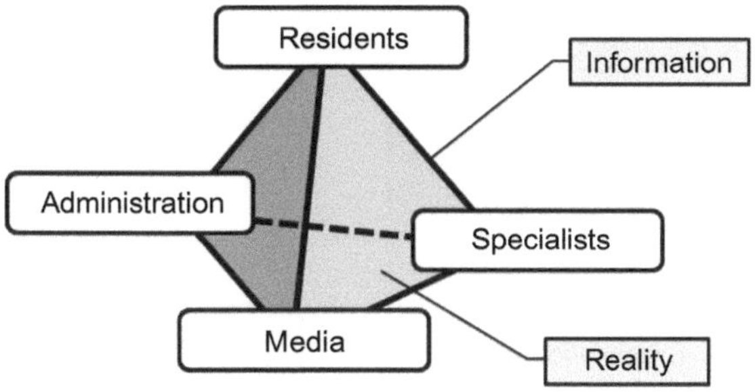

■ **FIG. 13.5** Co-construction of reality model [6].

as their own events, to go beyond conventional "reality reporting." We introduce the "co-construction of reality model" (Fig. 13.5) from the experience of volcanology.

The "co-construction of reality model," in short, is a philosophical model that makes the point that for disaster prevention and mitigation, various people involved with disaster response, e.g., local residents, administrative offices, media, and specialists, should all construct a "3D coordination" that shares urgency and importance in the background of the information, instead of settling on "linear coordination." The word "reality" here means a social structure that is generated by cooperative interaction among the people involved. The phrase includes the meaning of social reality as a theme to discuss. Social reality involves not only whether a piece of information is correct or not, or its speed in dissemination, but it also asks whether people sense "real danger" in the same manner as those under real threat do, and if they share what they value as "important lessons."

The information strategy in the case of a super wide-area disaster is the battle against time and space, and this explanation makes the picture clear. Time has two extremes: one is the "urgency" vector to quickly inform of the danger, and the other is the "tradition" vector, to pass disaster experiences and lessons on to the future generations. Both aspects require realistic views. In terms of urgency, the determining factor is the reality of differentiation that "a danger is right here, now at this moment." And in terms of "tradition," the reality of integration is that "someday the disaster is going to hit me as they have been saying," is effective.

Now, space also has two extremes. One is the "locality" vector, to sense the danger in a small specific area, and the other is the "overview" vector, to look over a wide area in and out of the disaster-struck area. The extreme of "locality" has value in the reality of location, that "this spot is going to get hit by a disaster," and the extreme of "overview" exists in the central

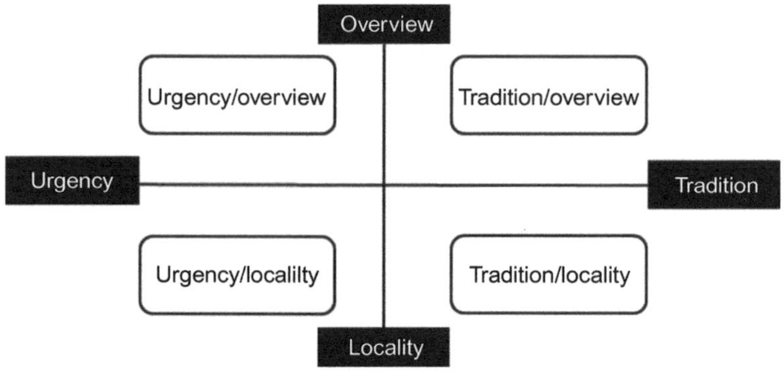

FIG. 13.6 Time and space axes in information strategy.

reality that identifies the distribution of severity and the areas that lack information. Fig. 13.6 shows these relations.

This chapter evaluates the mechanisms placed in the four quadrants in Fig. 13.6 by checking whether they match the co-construction of reality model. For an evaluation made more from a general viewpoint of a citizen, we introduce the key concept of "information ownership." In short, this is an evaluation of whether "the subject owns the information." Once the subject feels that he has nothing to do with the information, the reality of the information is lost and the subject will probably take no action. In contrast, however, information that the subject produced himself will carry vivid reality for him. It will have the power to force the first step in taking an action.

We now live in the age of digital media [6] and the world is making daily advances in transferring information. Who owns this enormous amount of generated information? If critical information about a disaster is of the administration, for the administration, by the administration, then asking for the power to save the lives of local residents is knocking on the wrong door. Similarly, jargons of the specialist, for the specialist, by the specialist may collect some respect from the general public but, in fact, the public places them at a distance and they never see practical value in them.

In the sections that follow, we will review whether post-3.11 disaster journalism is holding the hands of the general public and existing for them; in other words, whether "ownership of information" lies with the general public.

13.3 QUADRANT <URGENT AND WIDE>

We start out from looking at an example from the <Urgent> and <Wide> quadrant in information delivery. During the Great East Japan Earthquake, important alerts such emergency earthquake alerts or megatsunami warnings from the Japan Meteorological Agency were delivered to the whole nation through a variety of media. The disaster-struck areas had hardware problems, e.g., power outages or machine destruction, that interrupted the delivery of information, but moreover many pointed out the information that made its way to their areas lacked the power to persuade the residents [5,7].

The sole public broadcasting company Nippon Hoso Kyokai (NHK, Japan Broadcasting Corporation), for example, after 3.11, implemented the following "improvements" for mega-tsunami warnings [8]: The first point is about an "improvement" of the on-screen messaging on the TV. Instead of showing the calm port before the tsunami attack, it now displays words of action like "Quickly, Run!" on the screen. It also changed the background to gray, used emphasized fonts, and even reviewed all the graphics. The second point is the verbal announcer message saying "Run to a safe elevation, in no time!" in a commanding or definite tone to remove any suspicion or hesitation from the TV viewers and radio listeners. The third is active use of "measured" information instead of "prediction." When an offshore GPS wave recorder catches a tsunami arrival, that information is broadcast in almost real time.

These improvements took the lessons from 3.11 seriously and are high points in the efforts to shift emergency broadcasting to the safer side. Private media are praising the stance of "not fearing a false alarm and not tolerating overlooking." At the same time, however, some negative voices are heard to say, "NHK is stepping in too far"; "The practice will probably not last very long"; or "The false alarms are just the same as crying wolf." How should we look at this confusion? Using the viewpoint of "information ownership" in the co-construction of reality model works here.

The improvement plans by the Japan Meteorological Agency and NHK are technically superb and they live up to the reputation of being a leader in disaster prevention, against other countries. If we, however, review these plans from the people's side calmly and without any prejudice, most disaster information is not "owned" by the people. In terms of information ownership, mega-tsunami warnings (special alerts) belong to "others." The owners are special "experts" like the Japan Meteorological Agency or NHK. Thus, they are sometimes subject to selfish criticism. As long as the people are not the owners, they can blame things on the shortcomings of others and it does not affect them. This situation of ownership suspension is tragic for all and is not constructive.

One way of solving this issue is to use an extremely simple phrase, risk communication. We have to make use of risk communication in order to turn the disaster information ownership back to the "people." For example, asking the people "What kind of screen will urge you to make an immediate evacuation?" or "What words to make you decide to turn the TV off and run for live?" in order to collaborate with the viewers and listeners in the search for "improvements." To arrange a complete broadcasting system before a megadisaster is, in fact, difficult. In the worst case, there will be areas with loss of signal where sharing the reality and preparation with the general public would not be feasible. Risk communication, to begin with, is not just about communicating the risk, but it should involve metadimensional communication to jointly plan the risk communication design.

As long as the mass media struggles alone, the number of people that plan how to make use of disaster information will not grow. In this case, the number of people to praise the success will not grow either, and to make things worse, there is a danger that the number of people to blame failures from the outside will increase. Or else, there will be more people that are not even interested in the issue. Such a bad cycle is nothing but the occluded structure we had before 3.11. This is the core of the problem, which has to change in post-3.11.

One of the points in improving the situation is found in the next section of <Urgent and Local> actions. The entrance to solving this issue is "collaboration" to narrow the gap between the players, so everyone plays and watches to enjoy.

13.4 QUADRANT <URGENT AND LOCAL>

The quadrant <Urgent and Local> is split into two aspects: <Urgent and Local (from Local to Local)> and <Urgent and Local (from Local to Wide)>. We will start this section by looking at the first.

With the <Urgent> aspect, sending out a message with information about a narrow affected area enhances the reality. This has been well known for a long time in information science and psychology (e.g., [9]). We will examine this fact from the viewpoint of "information ownership" in the "co-construction of reality model."

The town of Oarai in Ibaragi Prefecture was attacked by a 4 m tsunami during the Great East Japan Earthquake; however, the death toll was zero. Inoue [10] reported that the head and the firemen of the fire department had the good sense to call for an emergency evacuation order, a nonexistent phrase in the administration, to successfully call for an evacuation to the residents. This was widely introduced after 3.11 as a successful case of direct information transfer by the local media. We, however, can imagine that, although this case taught us a lesson, everyone would not evacuate even if community FM stations and CATV broadcast the same order to the residents.

When the author visited Oarai, Ibaraki Prefecture in March of 2012, many people said they had the "button of their minds" pressed not just because of the phrase "emergency evacuation order" but also because the message came from the trusted mayor side. In other words, the psychological closeness to the mayor turned the "emergency evacuation order" into something "owned by the people." This case indicates that the relationship during normal times largely affects the information ownership at the time of emergency. This is an extremely important point and I would like to emphasize it again: we cannot make an objective evaluation by reading only the words to use. Reality easily changes with the time and place. Looking for a phrase that can always trigger the right action is expecting too much. This is a point that will especially help us in evaluating the "improvement" by NHK discussed in the previous section.

Now with this Oarai, Ibaraki Prefecture case which succeeded in forming "information ownership," we will look at the latter category of <Urgent and Local (from Local to Wide)>. This is an approach to collect a large amount of <local> information and use it for <wide> information from the central position. Many cases exist that used this method before 3.11. For example, the joint business by Weathernews and the Asahi Newspaper of weather forecasts with viewer participation is a good example [11]. Viewers of the TV news can post the weather conditions around them, such as the sun is shining, hot, cold, heavy rain with thunder, can hear cicadas, or so on, and the information is consolidated and shared on a web map. Each viewer owns the piece of information, and the wide-area total view is built by information from the viewers. Thus the weather forecast turns into something closer to vivid reality. The person in charge of this business (at the time), Takayuki Kido says some of the posted information may be wrong, but consolidating the large amount of information can cancel the contamination.

This weather forecast business is rather for ordinary times, and how it can function in extreme times of <emergency> is unknown. Weathernews's website reports, however, that it is challenging to build a guerrilla rainstorm prediction system that can "predict at a rate of 90%" on their "Guerrilla rainstorm defense team" page with a number of aggressive participants. We will have to keep our eyes on the media to see how they can make maximum usage of anonymous local information.

As a midlevel measure, a way to improve the accuracy and reliability of the information is needed. NHK, Osaka station Radio-1, solicits "volunteers" from its listeners to send information about tsunami attacks, or isolation by landslides not anonymously but with their names to build a "network with faces." Also a private broadcasting company is working on a system to collect wireless information from taxi drivers. These moves show us a revolutionary change in the ownership concept by getting the viewers' hands not only on information but also on the media themselves [12].

13.5 QUADRANT <TRADITION AND WIDE>

To review a disaster from a macroscopic viewpoint and pass the lessons on is a strong area for mass media. The daily activities of collecting news naturally build a database. What the media has been doing for a long time is to reedit the stored information into special broadcasts or news, or into compressed versions, or onto DVD. The rapid advancements of digital technology have made the information archiving and opening up the archives easy.

Again, if we take NHK, for example, the webpage "NHK Great East Japan Earthquake Archive" collected a large number of testimonies of the disaster victims and has been showing them to the world. The contents are in short video and text for easier viewing [13]. The Japanese version has a map interface to identify each testimony with location. One of the pages is dedicated to studying disaster prevention, with cases categorized into the lessons learned, e.g., "Should we drive cars to evacuate?" "What to do when you are stranded" or "How the handicapped should survive."

The development of web archives is a feature of post-3.11. Many have expectations for "earthquake disaster big data" (e.g., [14]) and visualization technology. The accomplishments are highly praised, with the primary purpose being to take an overview from the central position. If, however, we want to pass on the pain and sorrow of the disaster as fellow humans standing by the victims, we need to secure another viewpoint.

We have to bear in mind here that much of the contents on "NHK Great East Japan Earthquake Archive" was collected, edited, and arranged by NHK. The people in the clips are, of course, the main characters of them. However, for each clip, there was another main character whose name is lost, i.e., the interviewer on the spot. The information has a uniform texture created by the "ownership" of the huge organization. We will explain what this means with another case with a clear contrast.

The Japan News Network (JNN) stations under Tokyo Broadcasting System Television (TBS) created the series "3.11 Mega-Disaster, Views of the Journalists." The series won the Grand Prize in the Documentary Series Section of Asian Television Awards and the Award of Excellence in the 66th Agency for Cultural Affairs Art Festival, Documentary Section. The

DVD book has 18 programs of this series [15]. In each program, the journalist talked as "the first person" throughout the series.

For example, one episode followed a harsh experience when a journalist participated in the search for corpses in the Ishinomaki Municipal Okawa Primary School. He, by accident, found one and interviewed the parents, who identified the body. Just before the interview, his narration says, "I do not have children. I was not sure how to speak to the parents who confirmed the death of their child with their own eyes. (silence.) I blanked out my mind and pointed the microphone to them." The interview that followed captured the high-toned voice of the strained journalist followed by the father's voice in a completely opposite dull tone; the father probably could only speak that way then. Interviewee and interviewer, one with a great loss and one without: a big gap that cannot be crossed. This program, however, did not cut out the interviewer who ended up revealing the big gap. Such an artifact, i.e., editing, was suppressed and the names of both the interviewer and interviewee were left in the program. The interviewer kept contact with the family later and built a relationship with them to make this choice possible. Information does not just exist. Someone's intention records and creates it. By having information ownership, we can stand at the "edge of imagination" and think that maybe we could have been the "somebody," or think what we would have done if we were there.

The series producer said, "I had the journalists value their egos and report what exactly they felt and suffered in the field. We, the media are at a turning point to rethink our stance. The series was a small step. In short, the journalists had a chance to realize that they themselves are also human beings" [16]. No matter how big the data you have, or how many agents you hire, people cannot touch the "deep layer" of disasters. Disaster journalism, in order to protect people's lives and keep support going on to the very last person who needs it, cannot bear its true value unless it resonates in people's hearts. Didn't we repeat over and over since the last century the lessons of "war journalism": that playing with pieces of information in a clean room can cause great damage? In post-3.11, we shall not let our eyes follow the renderings of glamorous technology and throw the heart of journalism back into the old days [17].

13.6 QUADRANT <TRADITION AND LOCAL>

To overcome the dead-end in the <Tradition and Wide> quadrant in the previous section, we need stronger activation in the <Tradition and Local> quadrant. Although, there have already been sincere and down-to-earth activities in the field, and my article has nothing new to add to those.

When we view an information series that closely followed the local events through local media that experienced 3.11, their passionate presentations sometimes brings tears to our eyes. On the other hand, the long-run local series by the mass media (e.g., in the case of NHK, "Future Academy from Tohoku," or "TV that keeps the eyes on Fukushima") have their unique attractions. We can say the media, instead of a dry detachment that merely passes for information, has a strong commitment to "solidarity," to dive into the situations in order to share the reality [7]. These are the mind processes of the players that lead to "attachment."

In the quadrant of <Tradition and Local>, therefore, information ownership is often discussed. In the disaster area the following types of conversations were often heard: "Ah, this article is something that he wrote. I kind of like him because he visits us so often"; "That announcer has now grown strong. She was always in tears in the evacuation camp. So I told her that she couldn't carry on the interview like that."

Hideya Terashima of the newspaper company Kahoku Shinpo, based in the City of Sendai, started his own blog after 3.11 while working as a newspaper editor. In his blog, he has been writing articles in the first person point of view. He takes responsibility for what he produces without any limit on the number of words, to record what he saw and felt. The title of his blog series is "Producing newspaper in the aftershocks," still unchanged 5 years after the disaster. The aftershocks are still felt in the area. His work is still shaking the hearts of the people. A newspaper editor is also a human being, and when one human meets another, an article is born. If such a natural process can only be expressed in a blog, the mass media has to rethink the meaning of articles. "Information ownership" for disaster journalism does not belong just to a company or a boss. It lies in the tight relationship between the writer, with a trembling heart, to squarely face the disaster victims' sufferings, and the victims living in the middle of the disaster-struck area.

13.7 FUTURE OF INTERWOUND JOURNALISM—BEYOND REGIONS AND BEYOND GENERATIONS

We quickly reviewed post-3.11 disaster journalism with the viewpoint of "information ownership." To simplify our discussion, we split the explanation into four quadrants; however, each quadrant is deeply and dynamically interrelated with the others. Let me emphasize this point at the end.

First, let's look at the horizontal axis of <Urgent> and <Tradition>. These two concepts supplement each other. The <Urgent> reality, differentiated into "when needed," shall be passed on without losing its sense of urgency. However, the on-the-ground action of passing on the information leads to a dead end of facing mannerism. To keep the reality of integrated <Tradition>, we have to inject power from the differentiated reality. What happened "at that time" and why were the precious lives lost "at that time." When the generation with the experience and those unaware of "that time" try to confirm the events with open hearts is when the reality of "that time" comes to life and the passage to revival opens up. We have to connect the generations beyond generations with continuous efforts of "intergenerationality" [18].

Next is the vertical axis of <Wide> and <Local>. As we saw in Section 13.4, the quadrants function independently at times, but in principle, they have the functions of supplementing one another. The <Wide> information of the total area helps the <Local> and when information from the <Local> fields are gathered, <Wide> information is confirmed. Now here we have another viewpoint of an approach to make use of the relation of <Local> to <Local>. Here is an example.

In the mountain area of Kyoto Prefecture is the Town of Kyotamba. In 2004, a cable television service started. Today a clerk from the town administration office works the roles of cameraman, reporter, and editor to broadcast news for this town. This <Local> to <Local> media started its own service after 3.11. The town is a sister community with the Town of Futaba in Fukushima Prefecture. Residents of the two towns had been visiting each other since before 3.11. The nuclear plant accident did not break this tie. The Kyotamba CATV crew went around the temporary housing for the people of Futaba and covered the scenes of receiving donations from Kyotamba, like the Tamba's famous black beans, and broadcast them in the local news program of "Closeup Kyotamba." Koki Nishimura, in the Planning and Policy Department of the Kyotamba administration office, carried out interviews with the Futaba Mayor and covered scenes in the hard-to-return areas. He has so far produced over 30 reports.

This is a new form of "reconstruction journalism" based on a strong interlocal connection of the people that love their towns. The efforts do not put together a <Wide> package from the "Tokyo View," often the case of mass media, but it is also not enclosed within the <Local> alone. The media, with its foundation in the people that love the hills, the ocean, and their farmland keeping company with each other have already established a sense of ownership in the viewers that "they are mine." This type of interlocal activity had already caught some attention with the communication among disaster areas of the Great Hanshin-Awaji Earthquake and the Chuetsu Offshore Earthquake [19]. If the victims and nonvictims start to hold each other's hands like this, the moves will allow a close and vivid flow of information to start forming new realities, unlike taking the grand stance of "reconstruction" or "disaster prevention." This interlocality has the possibility of driving "intergenerationality."

Post-3.11 journalism should not be busy with short-term criticisms, one-time heartwarming stories, or waving flags for only a short-term occasion based on a success story. Rather, it should stand on the firm base of a "unified theoretic framework" and heighten its potential to contribute to society. How much 3.11 journalism has accomplished in the theme of having both quality and quantity is for the victims and nonvictims to decide, not for the journalists or the specialists. We do not have to voice it, but disaster journalism is for the people.

REFERENCES

[1] M. Osawa, Into an Awakening Deeper than a Dream—The Philosophy After 3.11, Iwanami, Tokyo, 2012.

[2] Y. Kawata, Learning From an Urban Mega-Disaster The Great Hanshin Earthquate, Kinmiraisha, Nagoya, 1995.

[3] Y. Kawata, Surviving a Super Urban Disaster, Shinchosha, Tokyo, 2006.

[4] Fading Lights and Shadows—Interim Report of 'Earthquake Disaster Disappearing from the Media', E-lock.planning, 2013.

[5] S. Kondo, K. Yamori, O. Yoshihiro, F.H. Lee, The social reality of the Great East Japan Earthquake as a media event—the content analysis on the NHK-TV reporting, JASDIS 10 (2012) 77–90.

[6] C. Hiratsuka, New Edition, Disaster Information and the Media, Liberta Publishing, Tokyo, 2012.

[7] S. Kondo, The Role of Disaster Journalism in Post-3.11, in: School of Societal Safety Sciences (Ed.), Societal Safety Sciences for Risk Management, Response and Action against Natural and Social Disasters, Minerva Shobo, Tokyo, 2015.

[8] NHK website, Special news: our tsunami warning has changed, 2015. http://www3.nhk.or.jp/news/0307newkeihou/ (Accessed 29 September 2015).

[9] H. Cantril, The Invasion From Mars, Princeton Legacy Library, Princeton, 2014.

[10] H. Inoue, Why did the town of Oarai Call for evacuation—case report of call expression in disaster prevention administration broadcast during the Great East Japan Earthquake. Broadcast Study and Research, Sept. 2011.

[11] Asahi Newspaper, Weather reporting by all, popular participation type program site, May 13, 2010.

[12] Weathernews, Protecting the people of Japan from guerrilla rainstorm, Guerrilla Rainstorm Defense Team. http://weathernews.jp/door/html/guerrilla2009_join/ (Accessed 29 September 2015).

[13] NHK, Great East Japan Earthquake Archive. http://www9.nhk.or.jp/311shogen/en/ (Accessed 23 January 2017), Japanese version: http://www9.nhk.or.jp/311shogen/.

[14] H. Abe, Earthquake Disaster Big Data, <the reality of 3.11>, <key to recovery>, <disaster prevention for the next generation> visualized, NHK Publishing, Tokyo, 2014.

[15] JNN, Omnibus Documentary, 3.11 Mega-Earthquake, Views of the Journalists, TBS Service, Tokyo, 2012.
[16] H. Akiyama, "Why the 'Views of the Journalists'," JNN, "Omnibus Documentary, 3.11 Mega-Earthquake, Views of the Journalists", TBS Service, Tokyo, 2012.
[17] P. Virilio, original French version "L'accident original" 2005, Éditions Galilée, English version "The Original Accident", translated by Julie Rose, 2007, Polity Press, Japanese version, translated by Masami Kobayashi, Seidosha, 2006.
[18] T. Atsumi, Locality and inter-locality, in: K. Yamori, T. Atsumi (Eds.), Word Map, Human Science of Disaster Prevention and Mitigation, Supporting Lives, Standing in the Field, Shinyosha, Tokyo, 2011.
[19] T. Atsumi, Disaster Relief Volunteer, Group Dynamics for the New Society, Kobundo, Tokyo, 2014.

Chapter 14

Nuclear Plant Disaster and Safety

CHAPTER OUTLINE

14.1 INTRODUCTION

The Fukushima Daiichi (Fukushima-1) nuclear power plant disaster that started on March 11, 2011 surprised not just the Japanese; people all over the world were shocked and it triggered all of them to rethink what safety meant for them. Everyone saw the myth of absolute safety with nuclear power plants (NPPs) fall apart before their eyes. However, thinking about the next form is not easy. A number of accident investigation reports have pointed out that "those in the field believed in their safety and only made superficial preparations against severe accidents." After the Chernobyl accident in 1986, the International Atomic Energy Agency (IAEA) worked on defense in depth with additional layer 4 and 5 to the conventional triple layers, but Japan alone was late in joining the move and the actions taken by Japan were not serious ones. It has been pointed out that this was one of the reasons that the Fukushima-1 accident was so severe. When the IAEA investigated the Chernobyl accident, it developed a new concept of "safety culture" and, this time it is likely reaching that same conclusion again. Engineers in advanced countries, especially those that are familiar with the technological level of Japan in normal times, must have been surprised and had

The Fukushima and Tohoku Disaster. https://doi.org/10.1016/B978-0-12-814078-9.00014-5

to question what the safety culture in Japan actually was. The reaction was the same with regular Japanese people. Many Japanese were surprised at the collapse of the safety myth, but at the same time puzzled at the difference in the actions of other advanced countries. Why had it been so disrespected in a society that frequently voices the word "safety"?

Most of the people learned about the constructed safety from discussions that surrounded the number of published investigation reports. The government and the Diet each formed its own accident investigation team, and other private sectors established similar organizations as well. Such reactions from all directions were unthinkable before this event, and these movements made us all think about the depth of the problem. Along this line of thought, we can say that these past 5 years have been well spent for us to learn and review how Japan was different in its concept of safety from other advanced countries with their nuclear power plants. On the other hand, nuclear power is an actual business activity and there are also strong movements pushing for its restart. Japanese society is approaching a deadline for making its choice. For the society of Japan in the middle of its transition, this chapter will take on the topic of nuclear power–related litigation and discuss the concept of safety, what actions to take, and the concerns, i.e., the reason behind the difficulty in the concept of "safety culture" and the possibility of overcoming that difficulty.

Nuclear power–related litigation involves issues concerning safety of the plants and the influence on people and the environment, and it often takes the form of administrative or civil lawsuits. The former starts by a suit against the Minister of Economy, Trade and Industry (the nation), who approved the installation, to cancel their approval, and the latter starts with residents filing a case against the business owner, such as the electricity company, requesting them to stop the construction or operation of the facilities based on their human and environmental rights.

In Section 14.2, we will discuss what nuclear-related litigation is, its history, and what problems are associated with it. Then in Section 14.3, we will review what Japan needs for its safety culture and the possibility of establishing it.

14.2 NUCLEAR PLANT LITIGATION AND SAFETY REVIEW

14.2.1 How Nuclear Power–Related Lawsuits Are Different

As discussed previously, the starting point for forming a sense of safety and how it was different in Japan were the issues. Here, I will focus on nuclear-related litigation as an arena for readers to squarely face the problem of safety, to organize the problems associated with it and expand the discussion to question what is right and what is wrong. When we talk about nuclear power generation, you might have the idea that the difference in the knowledge level between a specialist and an amateur would be a major obstacle. However, the court is where regular residents receive support from legal and technical experts to challenge the nation or business units on highly technical themes. Compared to mere "risk communication" type procedures, these processes involve serious knowledge about actual problems, their characteristics, and their limitations in order for the judges to make their decisions.

The mass media often only talks about who won and who lost. Those attitudes are insufficient because the knowledge gained through the proceedings of the arguments are what those involved should share with all people, no matter which side you are on. The accumulation and spread of sophisticated knowledge develops into societal growth, spreading the roots of safety culture so it can potentially grow in the right directions. I must admit that legal actions have their limitations. However, they offer the only arena for serious discussion and sophisticated processes in Japan. In modern Japanese society, the court stands out as a precious opportunity for such growth.

Nuclear power plants are highly advanced, complex technical systems. Thus, even if we concentrate only on their safety review, administrative decisions based only on empirical data are insufficient and the administration must prepare a group of people with special knowledge and expertise, must value their opinions, and must make administrative decisions in order for the nuclear power business to proceed safely. Nuclear-related litigation concerning these administrative decisions takes the form of legal disputes, but it leads to scientific judgments on the practical possibilities of modern technological science, and at the same time, it is also said to be cultural rulings that determine the direction of our civilization.

The judicial judgment that requires such special technological discussions always incorporates a concern as to how far the court should step in and whether it can really understand the disputes. The past litigation can be categorized into two types. One type starts from a blank page and has thorough discussions, as if the court were the administrative office itself, to reach a final conclusion. The other is to review whether the judging processes undergone by the administration office, supported by technical decisions of the experts, have made any nonnegligible errors or oversights.

The theory has been commonly supported that administrative decisions that require specialized technical judgment by administrative offices should have wide ranges of discretion.

The Act on Control of Nuclear Reactors itself has the abstract and comprehensive expression of "(Article 24) ... will not hinder the preventions of disasters..." and it has been pointed out that legislators expect broad discretion from the administrative office. The technical discretion of the administrative office specified in the Act on Control of Nuclear Reactors has two types: (1) Specific technical judgment about standards or judgment standards of the actual safety review, and (2) Specific technical judgment about the applicability of requirements. The first (1) comes from continuously changing technology and knowledge. If the detailed standards were spelled out in the approval requirements, then the judgments would be inflexible. The Act on Control of Nuclear Reactors, thus, allows the actual judgment standards to be set within lower-level regulations or internal rules. The understanding is that the judgment is left to the administration office. For (2), nuclear reactor facilities are complex and constructed with the involvement of all the highly technical scientific fields, and judging their safety requires specific technical knowledge, experience, and academic knowledge in all related fields. Moreover, safety judgment also requires estimation about the future, which cannot be accurately predicted, and thus specific technical discretion has been given a wide range. This was the situation before the Fukushima accident, where the administration was given a wide range of discretion over judging safety.

However, as the causes of the Fukushima accident have been revealed, the question has surfaced that this kind of structure, which gives the administration a degree of flexibility, might be actually hiding the problem in its assumptions.

14.2.2 Lawsuits Before the Fukushima Accidents

The lawsuit against Ikata NPP was the first NPP lawsuit in Japan. Among all nuclear plant–related lawsuits, the plaintiffs, such as the local residents, have never won any cases so far. There have, however, been a couple of plaintiff victories in the lower courts. One is the Appeal Trial of the Fast Breeder Reactor (FBR) Monjyu case and the other is the District Court case of Shika NPP Unit-2 operation suspension.

14.2.2.1 The District Court Ruling (1978) and the Supreme Court Ruling (1992) of the Ikata NPP Case

The first NPP-related lawsuit in Japan was filed in August 1973 against Ikata NPP Unit-1 of Shikoku Electric Power Company. The plant was built in response to a strong enticement by the residents of the Ikata community, which was suffering from depopulation at the time. However, 35 of the residents demanded cancelation of the nuclear reactor licensing, and the town was split into two sides. The District trial, which lasted almost 5 years, uncovered the poor safety reviews with a number of absences of the committee members and no records of the meeting minutes. Furthermore, the plaintiff side was supported by the nuclear chemist Sanshiro Kume of Osaka University and Hiroaki Koide of Kyoto University Research Reactor Institute in nuclear power safety and radioactivity measurement. The two specialists beat the government's witnesses in scientific debates and the media reported that the government side was in bad shape. In April of 1977, when the court finished laying down the facts, the Chief Judge who had been on the case from the beginning left for a job rotation. The newly assigned judge postponed the court proceeding twice for health reasons and, without sitting through a trial even once, passed on the ruling to the Matsuyama District Court Head Judge Kenkichi Kashiwagi. This uncommon change of judge attracted a number of complaints towards the Supreme Court. Moreover, towards the end of this lawsuit in September of 1977, the government claimed the plaintiff lacked the standing to sue because the plaintiff was not the addressee (electricity company) of the installation permit. The claim denied the plaintiff's qualifications for filing an administrative lawsuit; this was called the starting problem of administrative lawsuits.

In April of 1978, the Matsuyama District Court ruled a dismissal with prejudice on the merits. The ruling acknowledged the plaintiff's qualifications, but denied all the dangers the plaintiff pointed out about nuclear power and said the decision whether to license a nuclear power plant installation or not is with the government.

The Appeal Trial at Takamatsu High Court ruled the same dismissal with prejudice on the merits on December 14, 1984, and the Supreme Court also ruled against the plaintiff on October 29, 1992. Yuichi Kaido explained the Supreme Court ruling as follows [1]: (a) The purpose of safety review is "to prevent a disaster in one out of ten thousand" and to conduct a safety review in the broad range to prevent occurrence of accidents. (b) Respect for scientific and expert opinions: The adequacy of standards should be left to the rational judgment by the Minister (Prime Minister at the time, now it is the Minister of METI) who respects scientific and expert opinions from the Atomic Energy Commission. (c) The court's judgment standards of illegality are based not on the time of the event, but on current times, that is when the proceedings are in process. A case is judged illegal if "the review standards were irrational" or "the process of safety review had serious mistakes or flaws." (d) The responsibility of proving stays with the government side and if the administrative institute fails to insist or to prove, then its judgment is conjectured to be irrational. (e) The objective of ruling is limited to

Table 14.1 Safety Review of Nuclear Facilities

Administration Stance	Items to Check for Nuclear Reactor Licensing	
Make overall judgment of the reactor's conformance to licensing standards from various angles (and pay respect to scientific expertise and advice at the time)	Nuclear reactor facilities	Engineering safety
	Radiation effect	[During normal operation] effects on workers, local residents, and surrounding environment
		[Upon accident] effects on surrounding regions
	Location A	Natural conditions like topography, geology, and weather conditions at the proposed construction site
	Location B	Social conditions like population distribution around the proposed construction site
	Operation A	Future forecast
	Operation B	Technical capacity of the licensee

safety-related matters of the basic design. Table 14.1 shows a summary of items listed in an explanation of the Supreme Court ruling of the Ikata NPP lawsuit case [2].

14.2.2.2 Appeal Trial Ruling (2003) and the Supreme Court Ruling (2005) of FBR Monjyu

The lawsuit against FBR Monjyu was filed in September 1985. First the standing of the plaintiffs was questioned and the Fukui District Court, on December 25, 1987, ruled all the plaintiffs lacked the standing. However, the Kanazawa Branch of the Nagoya High Court, on July 19, 1989, approved the standing for residents within the 20 km range, and the Supreme Court, on September 22, 1992, ruled all members of the plaintiff had the standing to sue and sent the case back to the District Court to carry out the proceeding. On March 22, 2000, the Fukui District Court ruled dismissal with prejudice on the merits. However, prior to the ruling on December 8, 1995, Monjyu had had a sodium leakage and fire accident in the secondary cooling system and had been stopped for years. The owner, Japan Nuclear Cycle Development Institute (JNC) (formerly Power Reactor and Nuclear Fuel Development Corporation (PNC), now Japan Atomic Energy Agency (JAEA)), filed a Design Modification Permit Application in June of 2001 which was approved in December 2002 before the Appeal Trial ruling. The Kanazawa Branch of the Nagoya High Court ruled in the Appeal Trial, on January 27, 2003, for the plaintiffs to cancel the nuclear reactor installation license. The court explained the following three reasons [1]:

a. The plant had neglected corrosion caused by sodium. The fact that steel reacts with sodium compound at a temperature of 600°C, lower than the steel melting point of 1400°C, was well known in steel refining. However, this point was overlooked for the nuclear reactor safety review. The safety review assumed steel would not melt unless the temperature reached its melting point.

b. Possible rupture of steam generator. JNC claimed the rupture of steam generator tubes would only propagate to four tubes, but PNC had experimental records of simultaneous rupture of 25 tubes in 1981. PNC had failed to report this serious results to the Science and Technology Agency. Also in 1987, there was an accident in the United Kingdom where a rupture of a single tube propagated to a total of 39 tubes within 8 seconds.

c. Error in evaluating core collapse accident. The biggest possible accident with an FBR is a "core disruptive accident." PNC had performed a simulation of this event, but also had been covering up the results. Core disruptive accidents include not only a prompt criticality followed by a nuclear explosion, but also a fuel pin disruption going through the "transition phase" of propagating to fuel assemblies, causing core melt out to pool formation and a core explosion caused by prompt criticality of a large volume of melted fuel. The study on this transition process started in the United States and when licensing of an FBR in Germany was debated, the state government against the licensing supported this hypothesis.

Concern (a) was based on facts revealed when Monjyu had a sodium leakage accident in 1995. Those in the field knew about (b) and (c), but covered them up without reporting them. The High Court presented data that the administration had overlooked, and pointed to the irrationality that could not be ignored, to rule for the plaintiffs.

The Supreme Court, however, on May 30, 2005 overruled the High Court decision that canceled the installation license and ruled that the license was legal.

Kaido explained [1] this ruling as the Supreme Court adding facts the High Court had not approved as "facts confirmed legal with the original ruling" and completely ignored High Court approved facts that conflicted with the newly added evidence. Kaido stated [1] the Supreme Court ruling directly saved the administration with this process, whereas it could have ordered the High Court to redo its proceedings. The ruling viewed the guideline by the Nuclear Safety Commission (NSC) as something higher than the laws and performed no evaluation of possible irrationality of the guideline. Ito writes

[3]: "The Supreme Court supported the safety review that followed the NSC guideline without questioning whether the guideline was rational or not." The High Court ruling, in contrast, objectively evaluated the safety review from a third party stand point to see if sufficient safety was assured or not. The fact is FBR Monjyu has been stopped since this accident without a schedule for restart. The technical difficulties of this type of nuclear reactor have caused the United States, Germany, the United Kingdom, and France to terminate their development plans. Japan is, in fact, the only country among those that started it that carries on the development.

14.2.2.3 Shika NPP Unit-2 Ruling of Suspension of Operation (2006)

The plaintiff of this civil case lawsuit was 135 residents from 17 prefectures to stop the construction (later when the construction was complete, it was changed to stop the operation) of the reactor unit. It was filed in August 1999 in Kanazawa District Court. The residents won the District Court case on March 24, 2006, but the High Court ruled, on March 18, 2009, against the residents' appeal for stopping the construction and operation and the Supreme Court dismissed the final appeal by the residents on October 28, 2010.

Though it was overturned, the original District Court ruling was suggestive of nuclear power-related lawsuits in the future. The court requested the plaintiff to prove the high possibility of damage and then had the defendant rebut their arguments. The court then made the judicial ruling. This method was aligned with the Supreme Court ruling of the Ikata case (d) that "the responsibility of proving stays with the government side and if the administrative institute fails to insist or to prove, then its judgement is conjectured to be irrational."

The Kanazawa District Court ruled that the plaintiff did not provide sufficient evidence to prove technical danger during "normal" operation and the resulting serious accident and release of radioactivity. The ruling, however, acknowledged that the claims by Hokuriku Electric Power Company (Rikuden) and the Guideline for Aseismic Design Review by the central government were questionable in seismic motion and earthquake resistance and safety of the nuclear reactor. The court requested Rikuden to show counterevidence or to make countermeasures, but the defendant Rikuden would rely on proving safety only by showing that it had already obtained the license from the administration and it had no interest in the situation around them at the time. For example, the government Earthquake Research Committee had pointed out in March 2005 that the Ouchigata fault zone near the nuclear reactor had four to five short faults in close vicinity and if they started moving in sync, they would invoke an M7.6 or so earthquake. However, the stubborn Rikuden did not change its stance in relying on the old Safety Review Guideline.

This ruling acknowledged that the plaintiff proved to a certain extent that the residents in the area have a real possibility of suffering from exposure to radioactivity over the tolerable limit, and criticized that Rikuden's rebuttal was unsuccessful. The ruling also determined that the review results by NSC did not provide additional evaluation with later events of the 2000 Tottori Prefecture West Earthquake and the 2005 Miyagi Prefecture Offshore Earthquake, thus the NSC results did not affect the judgment.

The point of argument with this lawsuit, "Guideline for evaluation of earthquake motion and earthquake resistance and safety of nuclear reactors," had the history of development shown in Table 14.2. The 1995 Great Hanshin-Awaji Earthquake triggered a study to revise this guideline. Knowledge about seismology also advanced at that time and the revision work was intended to incorporate the new knowledge. During the discussion, "Osaki's method (developed in 1979 by Prof. Yorihiko Osaki at the School of Engineering, University of Tokyo)" that Rikuden had been using was

Table 14.2 History of Guidelines for Aseismic Design Review

1978–09	AEC set guideline for aseismic design	
10	NSC established	
1981–07	NSC, partial revision of aseismic guideline	
1995–01	17th: Great Hanshin-Awaji Earthquake (M7.3)	
01–09	NSC started to evaluate adequacy of aseismic guideline	
07	Headquarters for Earthquake Research Promotion established	
1996–2000	Research on aseismic safety of nuclear power facilities	
2000–10	6th: Tottori Prefecture West Earthquake (M7.3)	No record of active fault
2001–06	Aseismic guideline review committee started review	
2005–08	16th: Miyagi Prefecture Offshore Earthquake (M7.2)	S1/S2 seismic motion exceeded guideline values
2006–03	Plaintiff wins Shika NPP case in District Court	Problem in guideline pointed out
04	Aseismic guideline review committee drafts revision	
05–09	Solicited opinions and discussed them	
09	NSC set new guideline for aseismic design	
2007–03	25th: Noto Peninsula Offshore Earthquake (M6.9)	S2 exceeded guideline
2007–07	16th: Niigata Chuetsu Offshore Earthquake (M6.8)	Maximum seismic acceleration twice the standard

questioned because the 2000 Tottori Prefecture West Earthquake and 2005 Miyagi Prefecture Offshore Earthquake had motions that exceeded the conventional predictions. The revision work took a long time and the March 2006 Kanazawa District Court ruling was made. The ruling pointed out that: (1) The unknown direct hit earthquake magnitude prediction is too small, (2) Evaluation of active faults was insufficient, and (3) Calculation of seismic motion was inadequate. The guideline revision work was pushed by this ruling (March 24, 2006) and the new guideline was set on September 19 of the same year.

The ruling by the Kanazawa District Court that suspended the operation was overruled on March 18, 2009 by the Kanazawa Branch of the Nagoya High Court. By that time Rikuden had taken measures for the new aseismic design review guideline and the High Court reevaluated the aseismic safety based on the new guideline and acknowledged Rikuden's claim of assuming an M6.8 earthquake without specifying the earthquake origin and that active faults in the vicinity would not make simultaneous movements. The Supreme Court, on October 28, 2009, dismissed the final appeal.

One journalist unveiled records that suggested why revising the aseismic design review guideline took so long [4]. During his investigation as a member of the Fukushima Nuclear Accident Independent Investigation Commission of the National Diet of Japan (Diet Commission), Takashi Soeda identified the relationship between the Federation of Electric Power Companies of Japan (FEPC) and the Japan Society of Civil Engineers (JSCE) from the meeting minutes the FEPC had filed for a duration of more than 10 years. The electric power companies would gather specialists that sided with them to discuss tsunami heights and countermeasures. The Nuclear and Industrial Safety Agency (NISA) accepted their reports and without checking the contents took the word of the electric power companies that "safety is assured." Soeda pointed out the repetition of this practice gave no opportunities for bringing in up-to-date knowledge of the seismologists or mechanisms for justice. The reports and meeting minutes were never opened, to avoid outside verification, and the "meetings behind doors" buried the nuclear plant weaknesses in the face of tsunamis, discovered with effort, and responsibilities were obscured.

Table 14.3 shows how FEPC evaluated the impact of tsunamis on nuclear power plants in its February 2000 report [4]. The evaluation was made for three levels of 1.2 times, 1.5 times, and twice the water levels for uncertainty in measuring parameters. Fukushima-1 had a numerical analysis result of about a 5 m water level. When it faced a tsunami of height 1.2 times this estimate, the 5.9–6.2 m of water stopped the seawater pump and it was

Table 14.3 Summary Evaluation of Tsunami Impacts on Plants

	Water Level Rise			Scheduled or Already Decommissioned (Years in Service)
	×1.2	×1.5	×2.0	
Tomari 1, 2	o	o	o	
Higashidori 1	o	o	×	
Onagawa 1–3	o	×	×	
Shika 1, 2	o	o	o	
Fukushima-1, 1–6	×	×	×	In process: 1(41), 2(37), 3(36), 4(33), Waiting: 5(35), 6(34)
Fukushima-2,1–4	o	o	o	
Kashiwazaki-Kariwa,1–7	o	o	×: 1–4, o: 5–7	
Hamaoka 1–5	o	×	×	In process: 1(32), 2(30)
Mihama 1–3	o	o	×	
Takahama 1–4	o	o	o	
Ooi 1–4	o	o	o	
Shimane 1, 2	×	×	×	Waiting: 1(44)
Ikata 1–3	o	×	×	
Sendai 1, 2	o(o)	o(o)	o(×)	
Genkai 1–4	o	o	×: 1, o: 2–4	Waiting: 1(39)
Tokai 2	o	×	×	In process: Tokai 1 (35)
Tsuruga 1, 2	o	o	o	Waiting: 1(45)
Oma	o	o	o	
Monjyu	o	o	o	

o, no effect; ×, affected.
Based on T. Soeda, The People Who Buried the Warnings Against Nuclear Power and Great Tsunami, Iwanami, 2014 (in Japanese).

clear that the reactor cooling function had troubles. The only other plant that would face trouble with 1.2 times water levels was Shimane NPP (Chugoku Electric Power Company). Almost half of the 28 plants could survive even the worst scenario of twice the water levels. So at least in 2000, Tokyo Electric Power Company (TEPCO) must have known that Fukushima-1 was the weakest plant in the country against tsunami events.

14.2.3 **Movement After the Fukushima-1 NPP Accident**

The government, in May 2011, requested Hamaoka NPP to shut down the operation and, in press releases, hinted at a direction of moving away from nuclear power generation. The problem of restarting nuclear plants that were in periodic maintenance, however, surfaced shortly and fell under the new regulation system established in September 2012 by the Nuclear Regulation

Authority (NRA). NRA put a new set of standards in effect in July of 2013 and as of June 16, 2015, 15 plants and 25 units have requested reviews for restart. Rulings that suspended the operations have, so far, been the following two cases, and since these two rulings were made by the same judge, where we will head in the future is, as we have seen so far, unclear.

14.2.3.1 Ohi NPP Operation Suspension Ruling by Fukui District Court (2014)

On May 21, 2014 Chief Judge Hideaki Higuchi of Fukui District Court ruled for the plaintiff that requested suspension of operating units 3 and 4 of Ohi NPP in Fukui Prefecture. In his ruling, he stated "Operation of this plant has the actual danger that can cause damage to the human rights and thus the plaintiff's request is approved." The lawsuit had 166 residents as plaintiffs residing within the 250 km range from Ohi NPP.

He had started out saying "Organizations that conduct business that can cause serious damage to many people's lives, bodies, and their foundation of life, once a serious accident takes place, should be required to provide safety and high level of reliability according to the size of damage." He acknowledged that as many as 150,000 people were forced to evacuate with the Fukushima NPP accident and that at least 60 hospitalized patients lost their lives in the process of evacuation. He also referred to damage from the Chernobyl accident and stated, "In terms of protecting human rights based on people's survival from the danger of radioactivity, safety technology and facilities of this nuclear power plant are not only suspected that they are insufficient, but I have to admit that they are so weak that they can only stand upon an optimistic vision without solid roots." He also mentioned a concern, not directly related to the ruling, about the problems of electricity cost and CO_2 gas, saying that the Fukushima NPP accident was the worst-ever pollution and environmental contamination in our history and citing that using the environment as a reason for continuing operation of nuclear power plants was a great misunderstanding.

14.2.3.2 Takahama NPP Provisional Ruling to Ban Operation (2014)

On December 5, 2014, residents filed a request for provisional ruling to stop the restart of units 3 and 4 of Takahama NPP operated by Kansai Electric Power Company (KEPCO). NRA informed KEPCO, on February, 2015, that the two units passed the safety standards and KEPCO was preparing to form an agreement with the locals.

On April 14, 2015, Fukui District Court pointed out that the new regulation standards by NRA were "too lax and irrational" and even if the units were conformant to the standards, restarting them was not approved. A provisional ruling to ban nuclear plant operation was the first in the country and, differently from a ruling, the decision took effect immediately. The judgment was made by the same Chief Judge that wrote the ruling to suspend operation of Ohi NPP on May 21, 2014. The newspapers had split opinions: one side praised the judgment saying, "Justice made a heavy warning," (Asahi, Mainichi, and Tokyo) and the other side raised concerns that the judge went overboard by denying the regulation standards (Yomiuri, Sankei, Nikkei).

As I write this manuscript, the Chief Judge that made this ruling has moved to Nagoya Domestic Court, and Fukui District Court, with a new Chief Judge Jun Hayashi) has conducted court hearings to complete by November 2015. Two such hearings have taken place so far. NRA started the inspection before startup of Unit 3 in August of 2015. It is the final procedure before restarting the unit.

14.3 THE CHANGES JAPAN NEEDS IN ITS LITIGATION SYSTEM FOR NUCLEAR POWER

14.3.1 Problems With Nuclear Power–Related Lawsuits From the Viewpoint of Safety Science

How to take a nuclear disaster differs greatly with the interest angle. This section discusses the problems in changing systems or structures in supporting the administration side, which needs technical expertise. This is the condition for the nuclear power–related lawsuits.

Modern Japanese society is supported by an advanced civilization of scientific technology. Administrative dispositions that require highly technical judgment will continue to increase, as well as judicial decisions. Nuclear power litigation is one type of such decisions. However, "nuclear power technology," not limited to power generation, is vastly different from other technologies in terms of control, with some doubting if it can ever be really controlled by humans. There is a large possibility of it being a singular technology. But, no matter the process, Pandora's box is now already open and whether we utilize it or put it back in the box, we need to develop our power to control it. Otherwise, we will not be able to master either route. Unless we irresponsibly choose to desert the technology, we will need the best of our scientific and technical knowledge to gain the power to control it and at the same time maintain a high sense of ethics. Thus, this chapter is trying to find some possibility of self-cultivation by discussing litigation cases, which Japanese tend to shy away from, as arenas for serious social discussion.

From this standpoint, as we viewed the progressing of nuclear power–related litigation, I would like to direct your attention to the two rulings for the plaintiffs and the higher court rulings that then denied them. There is a lingering question whether the denials by the higher courts were really appropriate. If the two original rulings had prevailed, they could have contributed to quality improvement of nuclear power plant safety. Strict logic, development of consistent discussions, and their evaluation are the spirit that justice is after. This would form a spiritual backbone for the people that develop the scientific technology, while safely controlling the development, as well as for the people that watch the development. That is one reason we believe the original rulings for the plaintiffs could have contributed to higher safety. Another reason is using the judicial method itself in safety checking.

An important point with the Ikata NPP Supreme Court ruling in 1992 was "Nuclear power, from its nature, have a great danger. The nation's safety review, therefore, serves the purpose of avoiding disasters by any chance, thus, the judgement is based not on 'whether it is safe or dangerous' but on whether 'dangers are prevented by all means' is accomplished or not. Thus, sufficient review has to be made." Now, if this were true, what was the real intention in ruling for the administration in the Appeal Trial of FBR, by denying the real problems of "gap" and "overlook"? FBR Monju was a "prototype reactor" before building a commercial FBR that uses MOX (mixed oxide of plutonium and uranium) to produce more fuel than the amount consumed. Developing an FBR is one of the national projects as part of the nuclear fuel-cycle program. The circumstances of not allowing the judgement of canceling the installation license to take effect were understandable. Forcing development, however, shall not be separated from consideration for using the technology for safe control. The two concepts should be integrated. Of course, without a priority being placed on safety concerns, scientific lawsuits will collapse from their feet. Scientific technology, as long as people are interested, continues to grow and develop, and without the will and technology to keep them under control, they can run away from us. Furthermore, nuclear waste, whether we are interested in it or not, requires years and years of extended firm control. The case of FBR Monjyu is special, with it being a prototype, and to simplify our discussion, we will review the case of "Shika NPP Unit-2 District Court ruling."

This was a case that NSC should have reviewed and they should have revised the guidelines much earlier. However, the plaintiff, in search of problems, caught onto the weak point by chance and found irrationality.

Pointing out the problem in this manner was so effective that it moved NSC. However, that was only because NSC was in charge of the guideline. The internal rules of NSC had a principle of making information public, but when it came to maintenance, how long they took and the actual processes undergone were often unclear. The District Court case of Shika NPP Unit-2 proceeded by narrowing down the concern and forming a rebuttal against it. It was a safety check within a fairly narrow topic. The strategy was probably effective in a field with such large differences in specialized knowledge. The method has its limitations so it does not apply to the general safety check, which has priority in making sure that all safety concerns are reviewed. The plaintiff and the defendant in a lawsuit basically have a tension between them. The tension between NISA and NSC, for example, the two organizations for double-checking reviews, had between them their different types of tension, and also they had a cooperative relationship for providing knowledge on nuclear power. Since there are such differences, maybe we should accept a variety of methods for safety checking. In other words, let there be a number of ways for safety checking and combine the methods as needed for the particular location. When this idea is adopted for double- and triple-checking, an effective safety-checking mechanism may be established.

We wonder if there was really a need to save a defendant who could not make sufficient rebuttal against a narrowed-down safety topic. There will be lawsuits about lack of manuals for safety approval. This case may be offering a chance to demonstrate the defendant's ability in providing an answer to such challenges. The time and expense spent by the side on trial are not wasted. We view the Shika NPP District Court case as an emergency response training for the defendant.

In either case, we see consideration of "national projects" in the ruling. When it comes to a highly advanced scientific technology, there should be more thought given to the fact that a national project does not necessarily mean a wise development of scientific technology (being developed in a safe way). While national projects need protection, a ruling that is illogical, unethical, or technically mistaken cannot be made. Such a ruling will develop a weak scientific technology that is only a burden when it comes to practical application. A litigation, differently from a discussion that is an arena for pure competition, always determines a winner and loser. If the loser reasonably understands the winner's logic, there can be a bright pathway to a solution. However, if they lose face, the loser is left with bitterness. A nuclear power–related safety litigation comes with action to take after the ruling. Concerns with safety are not just oral promises but the claims, actions and results are subject to later total evaluation. This pressure enforces a well-balanced decision making for a complex topic. These circumstances lead to a repeated learning experience from the total evaluations after the judiciary makes their decisions. Their courage and commitment to justice in bravely pursuing their role places major guideposts for all the other industries.

14.3.2 A Society That Does Not Learn "Safety Culture"

When IAEA first brought up the concept of "safety culture" in 1986, Japan followed the move and the number of reports and textbooks containing the phrase "safety culture" rose. People frequently used the word safety and it appeared that we had reached a level of safety culture. Investigations after the disaster, however, revealed that many of those in the industry believed the plants were safe and ended up treating severe accident safety measures as just "a formality." There was evidence that "safety" and "safety culture" were only empty words. Nevertheless, the NRA keeps "safety culture" as a major theme. This is an ironic reality.

There were a few people before the Fukushima-1 accident who pointed to obscurity and bias in the "nuclear village" of the nuclear administration system and organizations. Katsuhiko Ishibashi has sent out warnings since 1997 [5] with his concept of a "quake and nuke disaster complex" about catastrophic multiple disasters of earthquakes and quake-induced nuclear accidents. The message, however, did not reach the nuclear village. Under such circumstances, we would have expected journalists and the mass media to respond. Unfortunately, however, the situation had gotten to the point of it being quite difficult to point out and discuss the obscurity and bias in order to correct the ill state, until the disaster gave its devastating blow.

There is no easy way to force changes in people's attitudes once they have built up a belief. They rarely try to change by themselves unless there is some external pressure.

We need to discuss whether we can find a possibility of escaping the bind on our own, through the movements of nuclear power litigation. This does not mean making changes to the judicial system; rather, it means to change its social meaning. This is one approach of safety science.

14.4 THE MAGIC WORD "SAFETY" MEANS WE HAVE TO STAND ON OUR OWN

When we line up the various problems with safety and take a good look at them, we find the fact that people and the organizations they build are living organisms. What we need is perhaps to recover the power of creation and bravely make changes where and when needed. Adding an addendum to an existing manual is not very difficult, unless there is a special strong intention. We should, however, focus our lessons on cultivating people and building organizations to have the strength to produce new manuals.

The problem we have is clearly extracted from the nuclear power–related litigation cases we have seen so far. National projects in which the administrative organizations may have made mistakes are the basic sources of administrative lawsuits. Japanese have high trust in administration, and many people think that "There is no mistake in administrative judgment". The administrative side, in fact, has hardly lost any cases in administrative lawsuits. As we have seen in the nuclear power plant cases, the administrative side followed the winning scenarios even when the side opposing nuclear power had strong evidences or clearly superior arguments. The over-protectionism can easily lead to transformation and spoiling within related organizations. We Japanese need to change the idea of the over-protectionism.

Modern society starts the discussion from "accidents happen" and administrative infallibility is just a myth. The stability of the legal system and national projects both can only show healthy growth when the judicial system makes strict evaluations. Unless such rules are applied, at least for concerns of safety evaluation in scientific litigations, it will be difficult for us to properly develop scientific technology, or to end it. There may be pathological reactions, however; a living organism under healthy development will generally resist hardship by seriously improving its logic and searching for high-quality evidence. That is the natural reaction of life and the basic solicitude for cultivation. In other words, from the broad viewpoint about safety concerns, we Japanese need to break the concept of "something to cultivate" away from its weakness. The concept of "something to cultivate" means to make advancements as the related organizations keep healthy activities without going bad.

REFERENCES

[1] Y. Kaido, Nuclear Power Plant Lawsuits, Iwanami Shinsho, Tokyo, 2011.
[2] T. Takahashi, Supreme Court Ruling of the Ikata and Fukushima-2 NPP Cases, Yuhikaku, Tokyo, 1993. Jurist No.1017.
[3] Y. Ito, Ahh, the Supreme Court ruling of Munjyu, People's Lawyer Yoshinori Ito's Site. http://www.shomin-law.com/essaymonjuhanketsu.html.
[4] T. Soeda, The People Who Buried the Warnings Against Nuclear Power and Great Tsunami, Iwanami, Tokyo, 2014.
[5] K. Ishibashi, Quake and nuke disaster complex, Sci. J. 67 (10) (1997). Kagaku, Tokyo.

FURTHER READING

[1] M. Osawa, Into an Awakening Deeper Than a Dream—The Philosophy After 3.11, Iwanami, Tokyo, 2012.

[2] Y. Abe, Hanrei Times, vol. 362.

[3] Research Organization for Information Science and Technology, Revision of design review guideline for nuclear power plants. http://www.rist.or.jp/atomica/data/dat_detail.php?Title_Key=11-03-01-30.

Chapter 15

Socioeconomic Issues, Reluctance in Buying Produce and Dealing With Highly Radioactive Waste

CHAPTER OUTLINE

15.1 EFFECTS THAT THE FUKUSHIMA-1 NPP ACCIDENT PRODUCED ON CONSUMERS AND CITIZENS

What changes did the Fukushima Daiichi Nuclear Power Plant (Fukushima-1 NPP) accident cause in opinions and actions of the consumers and citizens concerning socioeconomic issues of nuclear power generation? A variety of effects on consumers and citizens are caused by nuclear plant accidents. In this chapter, we will discuss the problems of harmful rumors about Fukushima produce as a consumer's problem and deep geological disposal of high-level radioactive waste as a problem in politics and the economy. The problem of harmful rumors is a direct response from consumers not buying from regions hit by the nuclear accident and the problem of high-level radioactive waste disposal is the indirect effect of citizens' reviews of the government's nuclear power policies since having to face the nuclear power accident.

The Fukushima and Tohoku Disaster. https://doi.org/10.1016/B978-0-12-814078-9.00015-7

The reason for looking particularly at these two problems is because they still continue to be affected by the nuclear plant accident today and neither has been solved yet. One factor is the environmental risk of radioactive contamination still around today in 2015, 4 years past the accident, and the concern is that it will probably continue into the future. Managing contaminated water and blocking underground water from flowing into the reactor building are works still underway by Tokyo Electric Power Company (TEPCO) and the decommissioning of the multiple reactors with meltdown cores is said to take 30–40 years.

The consumers and citizens know, through mass media, that the accident has still not ended and that is probably affecting their attitudes in shopping and their views against nuclear power–related policies. We need to clarify what the effects of the nuclear power accident are and how they are persisting in peoples' minds, in order to seek solutions to these problems. Solving the issue of harmful rumors is absolutely necessary for the recovery of the regions affected by the Great East Japan Earthquake and Fukushima-1 NPP, and forming a nationwide agreement on whether deep geological disposal of high-level radioactive waste is allowed or not is a problem we cannot avoid in how Japan will direct its energy and nuclear power policies for the future.

For the problem of harmful rumors, we will see how the consumers changed their shopping habits during the year immediately after the accident, from the results of panel survey research conducted three times during the year to see what psychological factors affected their judgment of whether to hesitate in buying Fukushima produce, or not.

For the problem of deep geological disposal of high-level radioactive waste, we will review the results of panel survey research before and after the nuclear accident to find out how the citizens' perceptions changed about the government policies and what psychological factors affect their decisions in their opinions about the policy of deep geological disposal.

15.2 LOSS OF CREDIBILITY IN NUCLEAR POWER TECHNOLOGY AND THE PLANT OPERATORS

The Fukushima-1 NPP accident and the resulting contamination of the environment greatly affected peoples' perception of nuclear power generation and related scientific technologies. For example, WIN-Gallup international reported [1] that 62% of Japanese answered favorably and 28% unfavorably about nuclear power generation in 2010 before the accident, and after the accident the rate of favorably dropped to 39% and unfavorably raised to 47%. The majority of opinions against nuclear power generation flipped over in Japan and many that thought nuclear power generation was safe changed their perception and now think that it is dangerous and they dramatically switched their attitude to antinuclear.

The Fukushima-1 NPP accident not only changed citizens' perceptions of nuclear power generation, but it also lowered the trust in scientists and the administration working with nuclear technology. Fig. 15.1 shows the change in citizens' trust in the government about deep geological disposal of high-level radioactive waste [2]. The figure shows a big drop in people's trust in the organizations and technology of deep geological disposal. Another report showed that the trust in scientists and engineers in general dropped after the nuclear power accident [3].

The loss of trust in the science and technology of nuclear power and the related organizations after the nuclear power accident propagated to a socio-economic problem that needed solutions and measures. Citizens' understanding and cooperation are necessary for countermeasures for the harmful rumors, wide-area disposition of earthquake and tsunami debris, and plans for setting disposition land for specified waste contaminated with

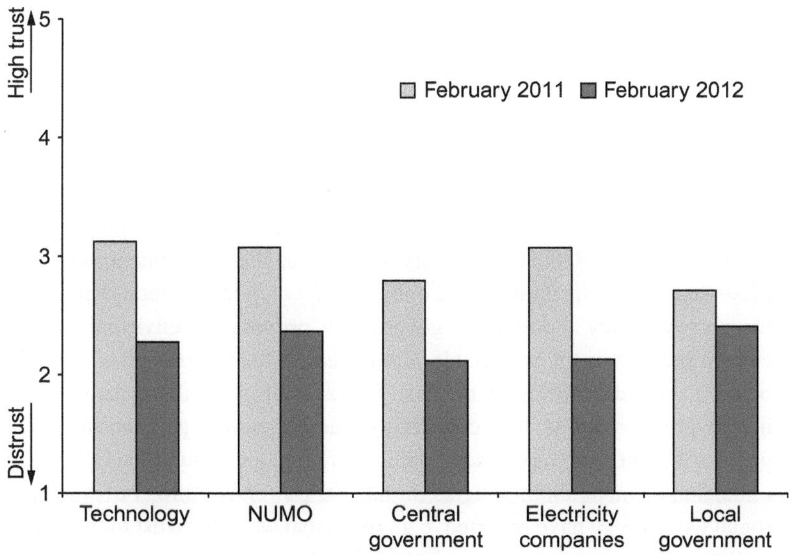

■ **FIG. 15.1** Trust towards nuclear power–related organizations before and after the Great East Japan Earthquake. *(Based on S. Ohtomo, Y. Hirose, Consumer's psychological process of hoarding and avoidant purchasing after the Tohoku earthquake, Jpn. J. Psychol. 84 (6) (2014) 557–565.)*

radioactive material. Further trust is needed in specialists and administrative organizations that plan and execute reviews of national policies of the environment and energy and deep geological disposal of high-level radioactive waste, in order to form a nationwide agreement.

Under the public opinion that the majority of the people of Japan are against restarting nuclear power plants, forming an agreement about national plans and policies concerning nuclear power will be difficult without strong trust in the specialists, administration, and corporations in nuclear-related scientific technologies. Citizens' evaluations of many nuclear-related policies are becoming more severe, e.g., against the national vision of how to position nuclear power generation in the energy strategy for the future of Japan.

In 2012 after the Fukushima NPP accident, the government had difficulty in deciding how to incorporate the people's opinions in positioning nuclear power generation in the review of the environment and energy policies. The difficulty came from the majority opinion that nuclear power should be discarded. The government then decided to apply, instead of an opinion poll that reflects the emotional reaction to the nuclear power accident, a deliberative poll developed in the United States that shows the well-thought-out, calm opinion from the people of Japan, so it could reference the results in the policy review. A deliberative poll is a participation-type survey that invites randomly selected citizens from the whole nation to have thorough discussions on social issues. This was the first time the method was said to be applied in forming a national policy. Nevertheless, after discussions, the majority of the selected citizens said "nuclear power generation should be abandoned" [4].

We conducted social survey research of the entire nation using the internet to hear the evaluation by general citizens about this deliberative poll [5]. Our results also showed that the majority did not trust the government; however, those that judged the deliberative opinion poll was a fair method in planning the energy policy thought the government proposal of environment and energy policies were reasonable, compared to those that had a negative opinion about the fairness of the poll. These results show that when the trust in the government is lost concerning nuclear power generation, a fair process that reflects the citizens' opinion is of major importance in forming a social agreement about policies for nuclear power. A number of studies in social psychology have reported, e.g., [6,7], that when the trust in the authority of the government is low, citizens decide whether to accept the directions of the policies or not by judging whether the procedure to advance the policies is fair or not.

15.3 RUMORS AND CONSUMERS' RELUCTANCE IN PURCHASING FROM FUKUSHIMA AND SURROUNDING AREAS

The influence of the nuclear power plant accident still persists today after 5 years in Fukushima. Those affected are not just the residents evacuated from the area that surrounds the nuclear power plant, but the people in agriculture, forestry, and fishery living in Fukushima and surrounding prefectures are also facing troubles.

For example, harmful rumors are circulating about products of the agriculture, forestry, and fishery industries from Fukushima. People have different opinions about the definition of the phrase "harmful rumor," whether it means a case when there is no objective risk, or when the radioactive contamination is below the announced standards but nonzero. Here, for this chapter, we will define it to mean the case when there is no clear evidence of risk to health, while the consumers, from their anxiety against risk, are reluctant to purchase the products they purchased before, causing economic damage to the producers. For the case of the Fukushima-1 accident, the radioactive cesium measurement from products such as vegetables, rice, mushrooms, milk, fish, and shellfish exceeded the preliminary standards and their shipment was banned [8]. Products that were not banned did not exceed the preliminary standards and were declared safe; however, consumers were reluctant to buy them.

Even 4 years after the nuclear accident, prices of farm, forest and fishery products from Fukushima are lower than the level before the accident. The 2014 Tokyo Metropolitan Central Wholesale Market average price of Fukushima peaches was 20% less than the national average. Beef was also close to 20% less. Peaches and vegetables are sampled in the production area for radioactivity inspection. Radioactivity inspection is performed for all bags of Fukushima rice, and while safety is assured, some of them are more than 20% less than the nation average [9].

Ujiie conducted an internet survey twice: in March, 2011 and in February, 2012. He surveyed among consumers in the Tokyo area and Osaka area about risk to the health from radioactive contamination of vegetables from Fukushima Prefecture and Ibaraki Prefecture [10]. The survey showed that consumers' reasons for their behavior of not buying the produce was fear of a health risk caused by contamination. The survey confirmed the reluctance in buying produce from Fukushima. Because it is the prefecture where the nuclear accident took place, a negative image was present with Fukushima food products and the consumers were reluctant to make purchases. This effect still persists.

The consumers' anxiety about radioactive material from the nuclear accident is a factor of the harmful rumors, but how many of them are actually reluctant to make purchases? The Consumer Affairs Agency of the Japanese government conducted five surveys from February, 2013 to February, 2015 about consumers' consciousness about harmful rumors. The results showed that about 20% of the consumers answered, "I am conscious about the origin because I want to buy food products free of radioactivity." Those that listed Fukushima Prefecture as "regions that I avoid because I want to buy food products free of radioactivity" were also a little less than 20%, and the percentage remained almost the same. Consumers that hesitate to buy from their anxiety about health risks from radioactivity are not in the majority; they make up a small portion of the entire consumer base.

However, even if the percentage of consumers that do not purchase specific products is small, the effect is not. Sekiya explained that the bullwhip effect worsens the effects of harmful rumors as we move the supply chain back up from retail towards wholesale [11]. Even when the number of consumers that hesitate to buy certain products is small, the demand in retail drops. The retailer that foresees the reluctance in the consumer would switch the products to the same from regions other than Tohoku that includes Fukushima Prefecture or stops or reduces purchasing Fukushima and Tohoku originated products because of the fear of dead-stock or drop in sales. The wholesalers would now strongly ask the Tohoku producers for a discount. The consumers, retailers, and wholesalers all responding to the harmful rumors magnify the effect and the producers suffer major damages. Kasperson et al. called this phenomenon of a chain reaction by social classes causing the risk to spread to the society and magnifying its effect "social amplification of risk" [12]. The theory explains the magnification of harmful rumors effect. Although the percentage is small, what judgments are the consumers making when they hesitate to make purchases? Did they panic immediately after the accident without calmly thinking about their choices? Are they still hesitant to make the purchases from anxiety or fear even now after years have passed since the accident? Or are the consumers more consciously avoiding the purchase after making a calm judgment about health risks from radioactivity contamination? Clarifying the consumers' mental process of their hesitation in making purchases will give important clues in planning effective risk communication to disperse the harmful rumors.

We carried out internet surveys three times in April, June, and November of 2011 about consumers' hesitation in making purchases of Tohoku food products in the Tokyo Metropolitan area [13].

Fig. 15.2 shows the percentage of the consumers who hesitated in buying Tohoku food products from immediately after the quake to the end of June, and from the end of June to the end of November. The ratio of the consumers that hesitated in all products increased from June to November. At the end of November, the hesitation rate is large with vegetables, fishery products, rice and fruits. The rate of consumers that hesitated in purchasing at least one product to the end of June was 29%, whereas the rate increased to 37% towards the end of November.

We analyzed whether the consumers continued their hesitation in buying through the two research periods. Among those that hesitated to the end of June, 81% continued to hesitate to the end of November; 19% of them ceased their hesitation. Among those that did not hesitate to the end of June, 82% still did not hesitate to the end of November, whereas the remaining 18% started to hesitate by the end of November. Most of those that were hesitant to make purchases up to the end of June continued to be hesitant after that.

So the trend in being hesitant to buy Tohoku food products continued, but did the consumers make the same judgment throughout that timeframe? The mass media continued to send out information about the spread of radioactive contamination immediately after the nuclear power plant accident. Were the consumers affected by the news, causing them to turn hesitant and fearful of radioactive contamination of food products, and then stopped

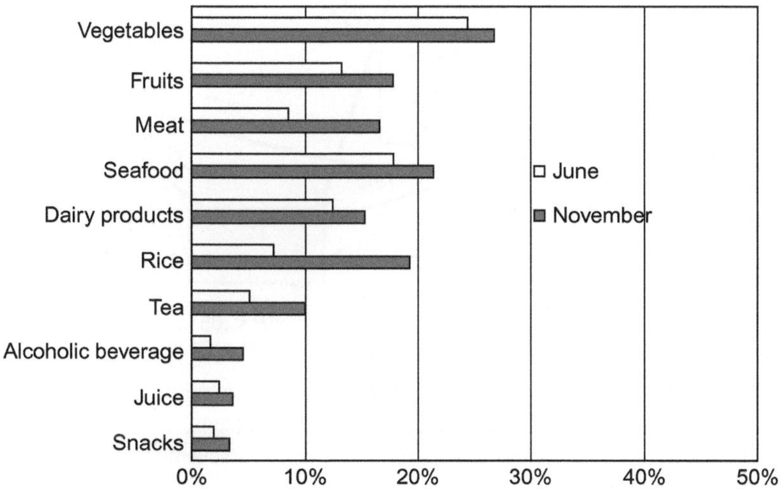

■ **FIG. 15.2** Tokyo area consumers' hesitation in buying food products from Tohoku. *(Based on S. Ohtomo, Y. Hirose, Consumer's psychological process of hoarding and avoidant purchasing after the Tohoku earthquake, Jpn. J. Psychol. 84 (6) (2014) 557–565.)*

reaching for Tohoku food products on the shopping floor? Or did they make their own evaluation of the health risk by radioactive contamination of Tohoku food products that did not have the shipping ban?

In general, consumers choose a product by conscious judgment of benefit gained, risk evaluation, and expectations from family members, but also in addition they are affected by guided images of what comes to mind with how they are displayed and other consumers' behaviors [14,15]. To clarify which factor was strongest in the hesitation of purchase, we sought the consciously intended judgment of hesitation in buying by asking about the attitude of hesitation of "cannot help because will feel anxiety if I do not hesitate," and the family member expectation of "family wants me to hesitate"; further we asked about action intention of "will continue to hesitate buying Tohoku food products." For the unconscious and guided hesitation, we asked if "many other people are hesitating in making the purchase" and "will probably continue to hesitate if the situation prolongs." Fig. 15.3

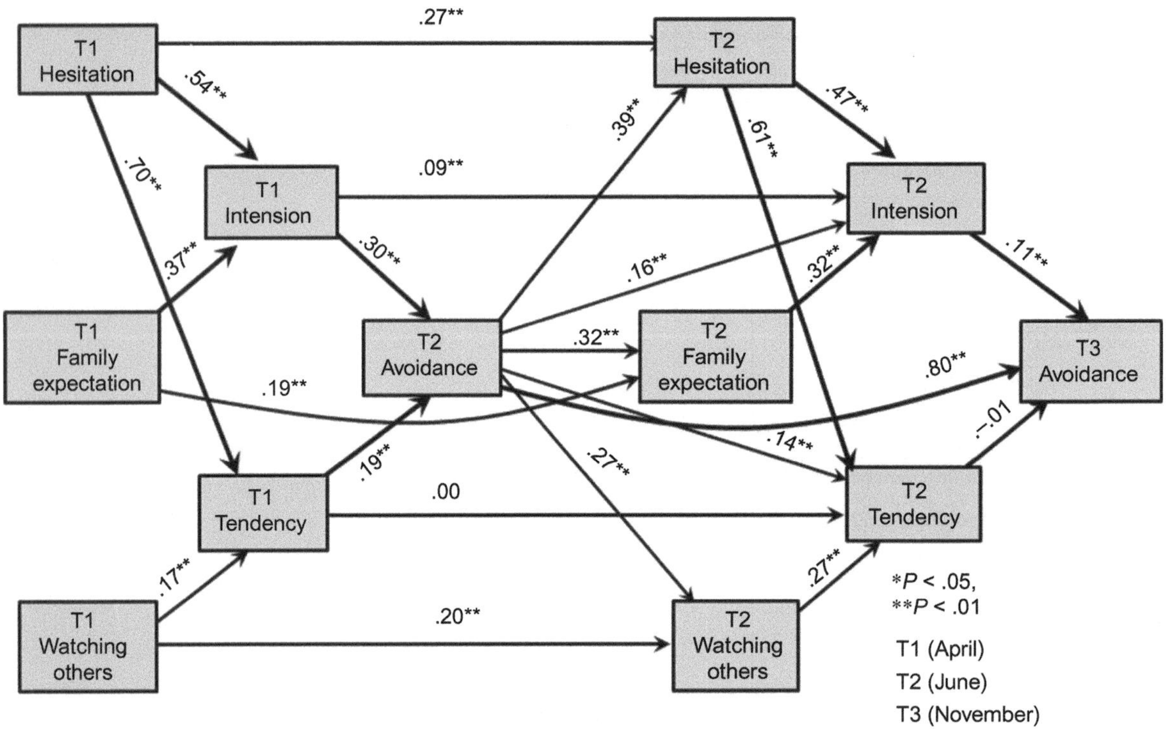

■ **FIG. 15.3** Relation of factors in hesitation in buying. *(Based on S. Ohtomo, Y. Hirose, Consumer's psychological process of hoarding and avoidant purchasing after the Tohoku earthquake, Jpn. J. Psychol. 84 (6) (2014) 557–565.)*

shows the results of our analysis on how these factors affected the hesitation in buying in June and November.

The hesitation in buying from the accident to June was affected by both conscious intention and guided behavior. The conscious intention was influenced by both the anxiety and family member expectations. The guided behavior was due to hesitation in buying and actions by others. In other words, up to 3 months after the accident, the hesitation in the consumers to make purchases had both effects from conscious intention and unconscious reactive behavior from the shopping floor. In contrast, for the hesitation from June to November, the guided hesitation from outside disappeared and the hesitation was only affected by conscious behavior. What mostly affected the hesitation up to November was the hesitant action to June. That is, after 8 months from the accidents, the consumers hesitate to make purchases from conscious intention formed by anxiety of radioactivity contamination and family member expectations. Also, once they hesitate in the past, the behavior continues.

The subject of our research this time was not just food products from Fukushima Prefecture but it covered food products from the Tohoku area to study the trend in consumers in the Tokyo area about their hesitation in buying those products. The problem we identified was that a certain percentage of consumers continues to hesitate in buying Tohoku products once they did so immediately after the accident. The aforementioned report by the Consumer Affairs Agency also pointed out the trend of hesitating to buy Fukushima food products for the conscious judgment of health risk by radioactivity contamination even after 4 years from the accident. Ambiguity avoidance [16] and status quo bias [17] are probably the psychological reasons for the consumers to continue their behavior of hesitation in buying [13]. The current situation, with scientists having different opinions about the health risk of the radioactive contamination, allows consumers to avoid the uncertainty and ambiguity by choosing not to purchase Tohoku and Fukushima food products. Furthermore, carrying on with the current choice of not buying is more attractive for the consumers than taking a new option of abandoning their hesitation in buying such products, with the small chance of a health risk by radioactive contamination. The consumers' behavior can be explained using this reasoning.

This explanation tells us that it is not easy to disperse the harmful rumor this time; however, we must come up with a countermeasure given these conditions.

15.4 CHANGE IN CITIZENS' ACCEPTANCE OF HIGH-LEVEL RADIOACTIVE WASTE TREATMENT

Japan, instead of disposing of them as they are, collects nuclear fuel consumed through nuclear power generation and treats the radioactive material for recycling. The high-level radioactive solution that remains after this treatment is mixed and melted together with glass material and placed in stainless steel containers as high-level vitrified waste. This nuclear waste has an extremely high level of radioactivity and it takes several tens of thousands of years for the radioactivity to drop to a safe level [18]. The government, in 2000, established the "Act on the Final Disposal of Designated Radioactive Waste" and decided to bury the nuclear waste in deep-geologic stable layers at least 300 m deep. The Nuclear Waste Management Organization of Japan (NUMO) was established for implementing safe geological disposal of the radioactive waste. NUMO has publicly sought municipalities from the nation, since 2002, to start the evaluation of land for building deep geological disposal facilities; however, so far, there has been no municipality that has responded to the announcement.

Discussions on where to position nuclear power energy in the national environment and energy strategies have included concerns about deep geological disposal of high-level radioactive waste. The suggestion by the former Prime Minister Koizumi that Japan should eliminate nuclear power generation because of its difficulty in deep geological disposal triggered people's concern and the problem of radioactive waste was a hot topic in the previous Governor of Tokyo election. The current government is steering towards restarting the nuclear power plants; a Cabinet decision revised the "Basic Policy for Final Disposal of Designated Radioactive Waste" and the government is taking a proactive role in establishing a location for the disposition. The new basic guideline states that the government will list the candidate locations from a scientific viewpoint and will ask the local governments for cooperation in starting the evaluation. It also states that the government will assure that the wider population will demonstrate gratitude and respect to the areas contributing to the projects [19]. In the background of this basic policy is the fact that it is a typical "Not In My Backyard (NIMBY)" type of deep geological disposal project and there is no sufficient consensus among the people of Japan. Any region will have difficulty in accepting such a disposition.

We can assume from this discussion that the Fukushima-1 NPP accident affected the citizens' evaluation of deep geological disposal of high-level radioactive waste related to nuclear power generation. We had carried out social research using an internet survey for finding the psychological factors among the citizens, in forming their attitudes about accepting deep geological disposal just before the accident. A year after the accident, we then conducted the same social research with the same people and same questionnaire to see if there were changes in the psychological factors in forming their attitude about the subject. The study reviewed whether people not directly affected by the Fukushima-1 NPP accident had changed their attitudes towards the deep geological disposal project [2]. The following explains our method of research and the results.

The survey consisted of about 2000 samples from around the nation and had them assume the project would carry out geologic investigations for deep geological disposal where they lived; it asked about psychological factors that would affect whether they would accept geological disposal or not.

The research assumed that factors that affect people's judgment of accepting the project or not included: evaluation of the benefit of geological disposal and risk, trust towards the project leader, stigma concerns, sense of responsibility towards past and future generations, fairness of procedures in accepting the geologic investigation, and emotional feelings towards geological disposal. The questions were: What benefit and convenience would deep geological disposal of high-level radioactive waste bring to the entire society and the local community? And how large were the environmental and health risks with geological disposal? We also conjectured that the trust towards the project leader NUMO and the central government would affect the judgments about benefits and risks. Further, the community that accepted geological disposal would dislike the negative image that comes with it (stigma concerns). High-level radioactive waste is a social risk beyond the individual and the land investigation would require acceptance by the regional residents, so the process of gaining agreement would have to be fair. The project stands out in its extremely long risk of tens of thousands of years until the high-level radioactive waste becomes safe, so the effect is not just on the current generation but it affects the past and future generations as well. The residents would have to think about their responsibilities to their forefathers and parents who had protected the land and also to the children and future generations that would continue their lives there. And lastly, the emotional feeling of deep geological disposal of high-level radioactive waste would also affect the judgments towards all these factors. Our two surveys, one made a month before the nuclear power accident and the other a year later, showed a negative change in attitudes towards accepting land investigation for building the facilities. Also, all factors, except evaluation of individual benefits, turned negative (Fig. 15.4). Emotions and risk evaluation of geological disposal also made turns in negative directions. The drop in trust towards the project leader in general was especially drastic.

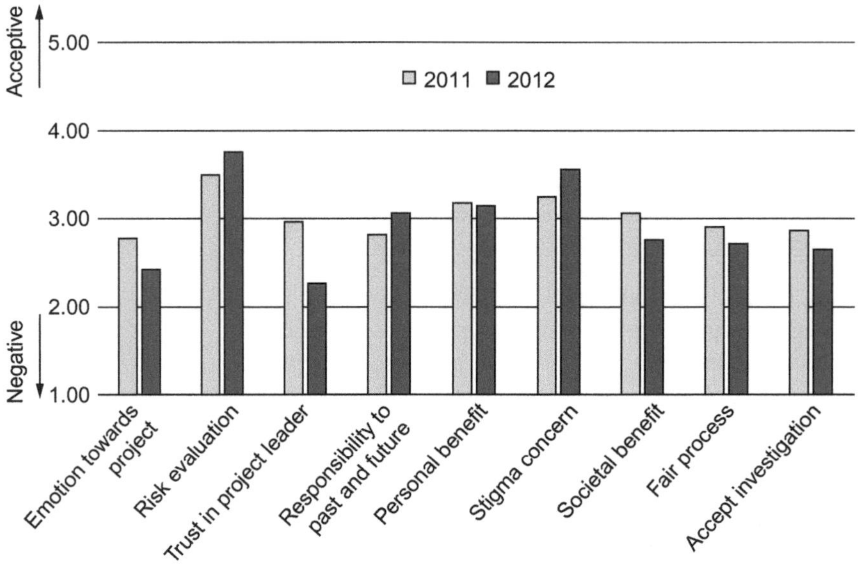

■ **FIG. 15.4** Evaluation of factors in relation to geologic investigation before and after the nuclear accident. *(Based on S. Ohtomo, Y. Hirose, Consumer's psychological process of hoarding and avoidant purchasing after the Tohoku earthquake, Jpn. J. Psychol. 84 (6) (2014) 557–565.)*

In other words, we confirmed that the Fukushima-1 NPP accident largely influenced almost all of the citizens' attitudes and their psychological factors concerning the deep geological disposal of high-level radioactive waste project to turn negative.

When we analyzed our research results on the psychological factors against accepting land investigation before and after the accident, factors that directly affected the attitude against acceptance did not change. From the largest in the order of descending size, they were: fairness in the process, risk evaluation, sense of responsibility to the past and future generations, evaluation of social benefit of deep geological disposal, benefit to the community and self (Fig. 15.5). The relative sizes of these five psychological factors did not show significant changes before and after the accident. In other words, the nuclear plant accident did not affect what people value in deciding their attitudes about deep geological disposal. The citizens would have kept their eyes on whether the project proceeded in a fair manner or not, rather than worrying about their own benefits or risk evaluations before the nuclear plant accident, and the accident did not change their sense of values.

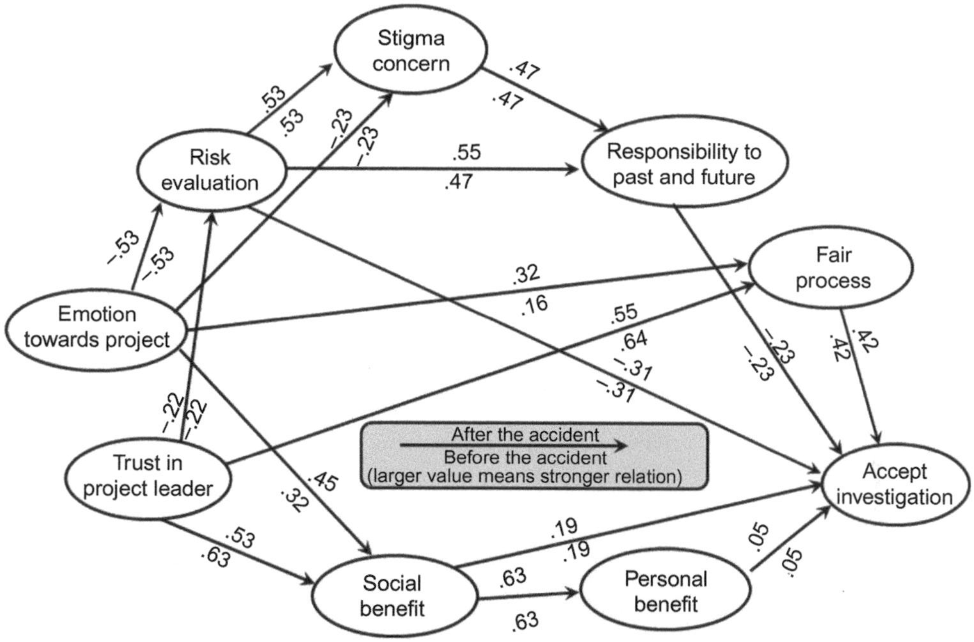

■ **FIG. 15.5** Relations among factors in accepting geologic investigation before and after the nuclear accident. *(Based on S. Ohtomo, Y. Hirose, Consumer's psychological process of hoarding and avoidant purchasing after the Tohoku earthquake, Jpn. J. Psychol. 84 (6) (2014) 557–565.)*

Fig. 15.5 shows the relation of the acceptance before the accident with the eight psychological factors. Emotion towards deep geological disposal of high-level radioactive waste, trust towards the project leader, and stigma concern do not directly affect the acceptance; however, among the aforementioned five factors, four, excepting only individual benefit, influence the acceptance indirectly. Among them, emotion towards geological disposal and trust towards the project leader affected procedural fairness and social benefit; however, the effect changed after the accident. The influence of trust towards the project leader was smaller and the emotion towards geological disposal became larger. People became more conscious about radiation contamination after the Fukushima NPP accident and rather than judging if the geological disposal process is fair or not with the trust towards the project leader, they judged the process as being unfair with their negative emotion towards deep geological disposal of high-level radioactive waste. However, the emotion towards geological disposal did not directly determine the acceptance of the process before and after the accident, so the citizens did not set their attitudes because of emotional repulsion, but they made calm judgments by thinking over the related factors.

15.5 SOLVING SOCIOECONOMIC ISSUES FOR OUR FUTURE

So far, we have looked at the issues of reaction to the nuclear accident by the consumers and citizens through consumers' hesitation in buying food products from the nuclear disaster–hit areas and neighboring regions and their attitudes towards the policy of deep geological disposal of high-level radioactive waste. We explained that the nuclear accident left a large impact on the two issues through our social research results. This section will examine what are important, based on our discussion so far, for phasing out the harmful rumors and promoting the formation of an agreement about deep geological disposal of high-level radioactive waste.

First, about the harmful rumors. Our research results confirmed that a certain percentage of consumers still hesitate to make purchases even after time has passed since the nuclear accident and that it is a conscious choice due to anxiety about health risks from radioactive contamination. After the nuclear accident, Fukushima Prefecture has been performing radiation inspections for all its agriculture, forestry, and fishery products and Fukushima products in distribution are safer than those from other prefectures; however, just being produced in Fukushima causes some consumers to shy away from them.

Koyama claims the problem lies in the anxiety towards the system of radiation safety inspection [20]. He listed the reasons that the effect of internal exposure through food intake has not been clarified, split evaluations and opinions about the safety standards, and the inspection is only made on random samples for products other than rice. Koyama further suggests eliminating harmful rumors will require checking of the areal distribution of contamination of the farmland, and developing ways of controlling radioactive material from entering produce. These measures will assure safety of the products and it is important to open the results.

Agriculture, forestry, and fishery products from Fukushima have a negative image that they, immediately after the accident, exceeded the safety standards. For eliminating the harmful rumors, the decommissioning of Fukushima-1 NPP has to proceed smoothly, for one thing, and further, to disperse the negative image, the information that the products do not exceed the standards and the measurements are relatively the same as products from other prefectures has to be announced repeatedly. This information being attached to the products on the shelf would catch the attention of the consumers to recover their trust.

Next, about deep geological disposal of high-level radioactive waste. Our research results of the citizens' attitudes towards the issue gave us a suggestion of how to proceed in forming an agreement.

The Science Council of Japan (SCJ) announced in 2012 its suggestion of the process of forming an agreement on geological disposal [21]. In 2010, the Atomic Energy Commission of the government of Japan requested SCJ to deliberate on the topic of "Process of high level radioactive waste disposal." SCJ's response stated that one of the reasons for the difficulty comes from social disagreement on the nuclear power policy of the nation. SCJ suggested to set a moratorium period to temporarily store the high-level radioactive waste and offer a place for the people of Japan to discuss the matter, so the agreement will be formed gradually in a multistage manner. The suggestion says that the evaluation standards for forming an agreement should include safety, value to life and health, economy, risk, and benefits, and furthermore, the ethical topics of fairness of the burden, responsibility to the future generations, and fairness of the process.

Our survey results also showed that the attitude towards geological disposal was set by fairness of the process, responsibility to the past and future generations, risk evaluation, and social benefits. The results confirmed that the citizens valued the same evaluation standards SCJ suggested for determining their attitude towards deep geological disposal of high-level radioactive waste.

The citizens placed the highest priority on fairness of the process. Studies in and out of the country on forming an agreement about plans that affect the environment [22,23] list as factors for citizens' decisions on fairness of process: whether sufficient information about the plan is opened, if there were sufficient opportunities for the citizens to discuss and ask questions to the stakeholders, and whether the citizens' opinions and discussion results are incorporated into the policies. Forming agreement on geological disposal should also meet these standards for fairness of the process. Now is a time in which trust of policies and their operations are lost, with the Fukushima NPP accident. Unless a variety of citizens with different attitudes and interests can all accept the fairness of the process, a social agreement will never be formed. Past studies [22,23] also point out that advancing the formation of public plans with citizen participation fosters trust towards the administration.

The government is trying to proceed to installment by revising the basic policy. However, unless the citizens judge the process is fair, there will be no widely accepted understanding nor an agreement from the people of Japan. The government and NUMO should be well aware of these facts.

REFERENCES

[1] WIN-Gallup International. Japan earthquake jolts global views on nuclear energy, 2011. http://www.nrc.co.jp/report/pdf/110420_2.pdf (Accessed 29 January 2014).

[2] S. Ohtomo, H. Osawa, Y. Hirose, S. Ohnuma, The impacts of Fukushima nuclear accident on public acceptance of geological disposal of high level radioactive waste, Jpn. J. Risk Anal. 24 (1) (2014) 49–59.

[3] National Institute of Science and Technology Policy, About the change in perception against scientific technology, 2012. http://www.nistep.go.jp/wp/wp-content/uploads/12gatu-made_shinraido.pdf.

[4] Executive Committee for Deliberative Poll About Selection of Energy and Environment, Report on deliberative poll about selection of energy and environment, 2012.

[5] H. Maeda, Y. Hirose, S. Ohnuma, K. Sato, H. Nonami, S. Ohtomo, Procedural fairness and social acceptance of the deliberative poll on future energy and environmental policy, in: 10th Biennial Conference on Environmental Psychology, 2013.

[6] K. Van den Bos, H.A.M. Wilke, E.A. Lind, When do we need procedural-fairness?: The role of trust in authority, J. Pers. Soc. Psychol. 75 (1998) 1449–1458.

[7] Y. Hirose, S. Ohtomo, When does procedural fairness influence on acceptance of environmental plan?: Moderating effect of trust in authority, in: XXIX International Congress of Psychology, 2008.

[8] Asahi Newspaper, "Vague criteria", anxiety about produce, September 8, 2011 morning issue.

[9] Yomiuri Newspaper, "Even though it is 'safe'", March 12, 2015, morning issue.

[10] K. Ujiie, Consumer's evaluation on radioactive contamination of agricultural products in Japan: decomposition of WTA into a part due to radioactive contamination and a part due to area of origin, J. Food Syst. Res. 19 (2) (2012) 142–155.

[11] N. Sekiya, Harmful Rumors and Their Mechanism, Kobunsha, Tokyo, 2011.

[12] R.E. Kasperson, O. Renn, P. Slovic, H.S. Brown, J. Emel, R. Goble, J. X. Kasperson, S. Ratick, The social amplification of risk: a conceptual framework, Risk Anal. 8 (1988) 177–187.

[13] S. Ohtomo, Y. Hirose, Consumer's psychological process of hoarding and avoidant purchasing after the Tohoku earthquake, Jpn. J. Psychol. 84 (6) (2014) 557–565.

[14] F.X. Gibbons, M. Gerrard, H. Blanton, D.W. Russel, Reasoned action and social relation: willingness and intention as independent predictors of health risk, J. Pers. Soc. Psychol. 74 (1998) 1164–1180.

[15] S. Ohtomo, Y. Hirose, C.J.H. Midden, Cultural differences of a dual-motivation model on health risk behavior, J. Risk Res. 14 (2011) 85–96.

[16] H.J. Einhorn, R.M. Hogarth, Decision making under ambiguity, J. Bus. 59 (1986) 225–250.

[17] D. Kahneman, J.L. Knetsch, R.H. Thaler, Anomalies: the endowment effect, loss aversion, and status quo bias, J. Econ. Perspect. 5 (1991) 193–206.

[18] Agency for Natural Resources and Energy, "Nuclear Water", The First Lessons of Energy, 2015. http://www.enecho.meti.go.jp/category/others/tyousakouhou/kyouikuhukyu/modeling/pdf_ss/27.pdf (Accessed 15 February 2017).

[19] METI, Basic policy on the final disposal of designated radioactive wastes was revised, 2015. http://www.meti.go.jp/english/press/2015/0522_01.html (Accessed 18 June 2015).

[20] R. Koyama, Reviving the food and agriculture through regional leadership, in: T. -Hamada, R. Koyama, M. Hayajiri (Eds.), Recovering Agriculture, Forestry, and Fishery in Fukushima, vol. 2, Misuzu Shobo, Tokyo, 2015, pp. 67–126.

[21] Science Council of Japan, High Level Radioactive Waste Disposal, Science Council of Japan, Tokyo, 2012.

[22] Y. Hirose, S. Ohtomo, Determinants of social acceptance of a waste management plan by citizen participation and citizen's trust for administration, Soc. Saf. Rev. 4 (2014) 43–50.

[23] S. Ohnuma, Y. Hirose, H. Nonami, J. Sugiura, Procedural fairness as a determinant of policy support via a citizen participation project: a case study of planning-cells in Lengerich, in: 10th European Congress Psychology, 2007, p. 154.

Chapter 16

The Government and TEPCO Problems in Communicating Information With the Public During the Fukushima-1 NPP Accident

CHAPTER OUTLINE

16.1 RADIATION DAMAGE FROM THE FUKUSHIMA-1 NPP ACCIDENT

The Great East Japan Earthquake caused a huge tsunami that hit the prefectures of Iwate, Miyagi and Fukushima; 11.5–17 m tsunami waves attacked Tokyo Electric Power Company (TEPCO) owned Fukushima Daiichi

The Fukushima and Tohoku Disaster. https://doi.org/10.1016/B978-0-12-814078-9.00016-9

Nuclear Power Plant (Fukushima-1 NPP). Fukushima-1 lost all external electrical power due to the earthquake, and the tsunami flooded almost all the main building areas with the important facilities of switchboards and emergency generators and the plant experienced a station blackout (SBO). The nuclear reactors had emergency shutdowns immediately after the quake, but the loss of electrical power disabled the cooling of the reactors and the fuel pools. Vents, hydrogen explosions and damage to the containment vessel released a large amount of radiation [1].

The radioactivity released from Fukushima-1 fell with rain in Fukushima Prefecture and the Northern Kanto area, on the land and houses. Aircraft monitoring revealed that from October 24 to November 5, in 2011, Fukushima Prefecture had a wide area that spread into the northeast from Fukushima-1 with a dose of 3.8 μSv/hour or higher at 1 m above the ground. The area of 0.2 μSv/hour or higher was much more widespread, from the southern part of Miyagi Prefecture, northern part of Tochigi Prefecture, Gunma Prefecture, the southern part of Ibaraki Prefecture, and the northern part of Chiba Prefecture [2]. Fukushima Prefecture distributed a questionnaire by mail (details in Section 16.4.2.2) to estimate the residents' detailed exposure. The estimation showed that the percentages of residents with exposure of 3 mSv/year or over were 2.7% in Soso, and 0.4% in the middle and northern sections of the prefecture. The estimated largest exposure in Fukushima Prefecture was 25 mSv/year in the southern section of the prefecture, and 99.9% of the residents were estimated at 2 mSv/year or lower. In Aizu and Minamiaizu, 99.3%, and in Iwaki, 99.1% of the residents were estimated to have been exposed to 1 mSv/year or less [3].

16.2 RISK COMMUNICATION REQUIRED AFTER THE FUKUSHIMA-1 NPP ACCIDENT

Risk communication in a broad sense can be defined to mean information exchange about events related to the risk. Today, however, it is often viewed as an activity to form agreement about social arrangements among stakeholders (in most cases, administration vs people, or people vs people) with conflict of interest. This chapter calls this narrowly defined risk communication "consensus communication." Risk communication, in the broad sense, in addition to consensus communication, involves care communication for providing help and intervention concerning proper procedures for people at risk. It also covers risk communication of information exchange when critical events, like severe accidents, have occurred.

During normal times when there is no disaster, consensus communication often takes the place of risk communication. Once a disaster breaks out, care communication for those with anxiety, who are suffering, takes place.

Crisis communication was needed when the Fukushima-1 NPP accident broke out and there was a state of emergency. As the emergency state extended, several tens of thousands of people were forced into evacuation and many had fears of exposure to radiation, mainly in Fukushima Prefecture. Care communication for these people was strongly needed.

The use of nuclear power especially calls for approval by society. For power companies, nuclear power generation is just a commercial activity, but its operation legally requires the blessings of the government, e.g., the nuclear regulation body, and of the community where plants are built. Therefore, restarting nuclear power generation after this Fukushima-1 NPP accident has to include consensus communication in order to gain agreement from the society.

16.3 CRISIS COMMUNICATION IMMEDIATELY FOLLOWING THE ACCIDENT (END OF MAY 2011)

The Fukushima-1 NPP accident that started on March 11, 2011 vented Unit-1 past 9 a.m. on March 12 and released radiation. Unit-2 followed at about sunset on March 12, and Unit-3 before 9 a.m. on March 13; these ventings were also responsible for radiation release. Furthermore, Unit-1 at about 15:36 on March 12, Unit-3 at about 11:01 on March 14, and Unit-4 around 6 a.m. on March 15 suffered a hydrogen explosion, releasing large amounts of radiation into the atmosphere. Later, on March 29, finally concrete pumping trucks succeeded in pouring large amounts of water to cool the nuclear reactors and the state of emergency started to converge [1].

16.3.1 Crisis Communication Performed

The Ministry of Economy, Trade and Industry (METI) [4] and the Ministry of Education, Culture, Sports, Science and Technology (MEXT) [5] summarized their risk communication during the roughly 3 months following the outbreak of the Fukushima-1 NPP accident.

16.3.1.1 Providing Information to the People and Media

The Nuclear Emergency Preparedness Act assigns the Chief Cabinet Secretary to the General Manager of disaster management. The Chief Cabinet Secretary, Yukio Edano (at the time), actively made announcements with press conferences and TV broadcasting to directly open information to the public. He directly communicated evacuation orders to the people in the area through the media [4].

The Nuclear Industrial Safety Agency (NISA) was the government organization at the time with the responsibility for directly handling nuclear accidents. After the Kashiwazaki-Kariwa NPP accident caused by the 2007 Chuetsu Offshore Earthquake, NISA had built "Mobile NISA," which was designed to quickly give accident information to the people. The system was designed to send out accident information to preregistered mail addresses (28,484 as of May 10, 2011) at times of emergencies. NISA sent out its first mailing about Fukushima-1 NPP from Mobile NISA at 15:16, i.e., 20 minutes after the earthquake. Later, information from NISA primarily took the forms of press releases and press conferences. It released 155 press releases and gave 182 press conferences by May 30 [4].

MEXT, in charge of supervising research and development of nuclear power technology, conducted an environmental radiation level survey over the prefectures of the nation. The ministry also joined forces with Fukushima Prefecture, Japan Atomic Energy Agency (JAEA), and TEPCO to monitor the air dose rate, dust, and soil near Fukushima-1 NPP [5]. In addition to MEXT, the police, self-defense force, and TEPCO started their own monitoring. However, this was without proper coordination. On and after March 16, the monitoring results were all gathered to MEXT to make public announcements about the results from that point on [6].

The Nuclear Safety Commission (NSC) conducted, on its own, daily press conferences from March 25 to April 24, and from April 25, a total of eight press conferences after NSC had meetings.

16.3.1.2 Evacuation Orders to the Residents

The risk communication needed immediately after the Fukushima-1 NPP accident was crisis communication and care communication. Above all, the most urgently needed was evacuation instructions to the residents in the area and surrounding areas. As mentioned previously, the General Manager of nuclear disaster management was the Chief Cabinet Secretary and the task of announcing such evacuation instructions was the duty of this Secretary.

At 21:23 on March 11, evacuation instructions were given to the residents within 3 km from Fukushima-1 NPP, with instructions to stay indoors to those within the 3–10 km area. Then at 5:44 on the following day, March 12, the range for evacuation instructions was extended to residents within 10 km, and at 18:25 of the same day, further to within 20 km [1,6].

At 11:00 on March 15, the government instructed residents in the 20–30 km range from Fukushima-1 NPP to stay indoors, but the City of Minamisoma, within 30 km from the plant, carried out guided evacuations starting on that day to those that requested evacuation. Without monitoring data and radiation spread forecasts by SPEEDI, many residents of Minamisoma ended up evacuating towards Iitate and Kawamata, later found to have had high levels of contamination. A similar situation took place in Namie [1,6].

Evacuation instructions to the residents in the area were to be given via phone calls to the municipalities by the Nuclear Emergency Response Headquarters. The earthquake and tsunami, however, had damaged the telephone network and a large number of municipalities had no telephone connections, and the Nuclear Emergency Response Headquarters called the local response headquarters and the Prefecture of Fukushima to relay the instructions to the municipalities. The Headquarters also used police wireless communicators so the instructions could be transferred to the residents. Such evacuation instructions rarely reached the municipal residents and the Chief Cabinet Secretary appeared in press conferences on TV and radio in efforts to inform the residents about the evacuation instructions [4]. The cabinet office distributed a questionnaire by mail to the 59,373 householders in the 22 municipalities under evacuation instruction and the 10 municipalities surrounding them from February to May 2014, and they had 19,535 (32.9%) valid responses [7]. The results of it showed that 93.1% of the respondents evacuated on March 11 and 12, 2011, but 26.0% of them could not get any information about the evacuation. Souses of evacuation information were TV/radio (49.9%), Neighborhood association (43.3%), and family/neighbors (28.8%), and little information was given by police (3.0%) and TEPCO (2.3%). As the most difficulties they experienced during March 11 to April 30, the respondents listed "no information where to evacuate" (57.7%), "disabled for using mobile phones" (51.8%), and "administrative officers provided no information about evacuation" (49.7%) as many as "gasoline shortage" (74.3%), "shortage of foods, water, daily commodities" (57.7%), and "traffic jam" (42.3%) [7].

16.3.1.3 Transferring Information to Overseas

Japan was obligated to quickly inform the International Atomic Energy Agency (IAEA) about nuclear accidents under the treaty of Convention on Early Notification of a Nuclear Accident. International Health Regulations (IHR) had assigned the same responsibility to Japan to report to the World Health Organization (WHO).

NISA made its first communication to IAEA at 16:45 on March 11. It then informed IAEA in a total of hundreds of transmissions of press releases, plant parameters, and monitoring results. The Japanese government also, via its diplomatic channel of Permanent Mission of Japan to the International

Organizations in Vienna, followed the Convention on Early Notification of a Nuclear Accident and provided information to IAEA. IAEA relayed this information to member countries, mass media, and the general public. The Japanese government also provided information to WHO following IHR.

Furthermore, the Japanese government explained and answered questions, during various occasions at international conferences, about the status of the accident and how they were managed. To counter the number of import restrictions on Japanese products by other countries, the Japanese government explained the postaccident situation through foreign diplomats in Tokyo and diplomatic bodies abroad to governments of foreign countries and requested scientific explanations for such restrictions.

Ministries and Agencies of Japan held joint press conferences for foreign media from March 13 on. Simultaneous translators previously arranged for press conferences by the Prime Minister were also made available for conferences by the Chief Cabinet Secretary.

Starting on March 12, ministries and agencies of Japan posted accident-related information on their websites in English, Chinese, and Korean. In addition, the Prime Minister and his cabinet, on its Twitter and Facebook accounts, posted summaries of the press conferences held by the Chief Cabinet Secretary. Information was relayed to foreign governments via local diplomatic offices as needed and the offices posted accident-related information on their websites. The efforts resulted in website information available in 29 languages.

16.3.2 Discussion on Crisis Communication

The communication of "information" immediately after the Fukushima-1 NPP accident by the government and TEPCO revealed a number of problems in information gathering, making political decisions, making announcements, communicating the information, and so on. In the background was, as written in a number of accident investigation reports, a basic lack of a framework for emergency management and poor crisis awareness on the part of the responsible people. This unpreparedness led to a number of incidents of inadequate recognition of the basics and poor action in the emergency state after the accident.

16.3.2.1 Definition of Crisis Communication

Leonard and Howitt, political scientists who study organizational actions against crises, set the four states as [8]: (1) normal operations, (2) minor operating problems, (3) routine emergencies, and (4) crises.

Leonard and Howitt defined a crisis to be within an unexperienced state that can result in unforeseeable dangers. According to their definition, accidents also classified as: (1) no accident, (2) minor problems that are not quite accidents, (3) expected accidents, and (4) unexpected accidents.

The Fukushima-1 NPP accident was a "crisis" as defined here. Crisis communication is the exchange of information with the public in the case of a crisis (severe accident). One of the biggest difficulties at the time of crisis is to accomplish accurate recognition of the state based on a proper understanding of the situation. The difficulty comes from inaccurate measurement of the state with machines in the field and psychological factors that distort the recognition of reality. These situations raise the importance, in addition to the field information from those responsible to the public, of transferring information from the public to those in charge so the responsible can check the objectivity of their recognition and judgment for crisis communication.

16.3.2.2 The Public During Crisis Communication

In crisis communication, the relations between those in charge of managing the situation and the public on the other side of information transfer are as follows [9]:

(1) Residents/general public
(2) Mass media: Today with its highly advanced level, the mass media opens the window to the residents/general public for the response team.
(3) Administrative organizations: Administrative organizations in different fields and at various levels execute their authorities for the residents/general public. Close communication is needed between the responsible administrative organization and those that carry out the responses in the field, e.g., the fire department, police, or self-defense force.
Administrative organizations also serve, like mass media, as a window of communication with the residents/general public. The residents/general public expect administrative organizations to execute their power through their council members, mayors and governors.
Communications among administrative organizations are important in fulfilling these functions. The basic municipalities gather to form prefectures and prefectures to form the nation. All these players, including the central government, have tight relations among them; however, they are different organizations for those that carry out actual emergency responses, so tight communication among them is important.
(4) Related companies and others in the same business: Nuclear power generation in Japan is not performed by single electricity companies but

they all have affiliated, secondary and tertiary contractors. Communications among these companies are important at the time of crisis. In addition, during a crisis, human and other resources available during normal times are lacking and to compensate for lack of sufficient resources, information exchange with affiliated companies and others in the same business is important, even if there is not much of it during normal times. Businesses need to prepare themselves beforehand for ways of communication with partners and others in the same business at the time of crisis.

(5) Other industries and organizations in research and development: A crisis is a state that faces abnormal situations not encountered during normal times. Thus, businesses in other industries or unrelated research and development may become useful in responding to crises. That is because an abnormal situation for a business may be a normal state of an expected event for others. Therefore, when the situation is highly irregular, the possibility of information in other fields or research and development turning effective is higher.
The communication channels with other industries and unrelated research and development organizations are, in general, not ready at the time when needed. Such communication, therefore, will probably have to take the form of calling for cooperation to the unspecified public as in the case of public offering.

(6) Foreign countries: For the globalized international society, crisis communication at the time of a nuclear accident is required to be made in accordance to those made domestically.

16.3.2.3 Need for Crisis Communication

The reason for requiring crisis communication from those in charge of responding to a crisis at such times is summarized as follows [9]:

(1) Ethical responsibility: The organization responsible for causing the accident ethically (to compensate for the trouble to the society it caused) has responsibility to explain the situation to the public. The reason for demanding crisis communication is purely ethical only when the crisis is contained within the responsible organization. Otherwise, when the crisis affects not only the responsible organization but causes damage to others, the following reasons turn important.

(2) To gain public understanding: For public opinions to form, based on grounds that can stand rumors or criticism, the responsible organization must have the public understand the crisis with sufficient and accurate information. For this reason, the responsible party has to provide the information necessary for the public to form an understanding, and it needs to respond to the feedback from the public about the information provided.

(3) To help the public: If damage from the crisis is reaching the public, information about the danger and evacuation must be opened swiftly to the public. The responsible party also needs to respond quickly to information from the public about the possible danger associated with the crisis.

(4) To seek help from the public: As the crisis becomes more serious, the responsible organization alone cannot handle the whole situation. In such cases, it has to seek support from the public (the whole nation or maybe the entire human race). What is important in asking for help from the public is to clearly state what support is needed. Giving information that is easy for the receiver to understand is also necessary for crisis communication that seeks help.

16.3.2.4 Discussion About Crisis Communication at Fukushima-1 NPP

In about the second week of the Fukushima-1 NPP accident, when water injection to the reactors and fuel pools had difficulties and hydrogen explosions had taken place, causing anxiety in the people, what turned things around to drive the accident towards control was a citizen's information to the government that a German-made concrete pumping truck was sitting in the port of Yokohama being transported [10].

If help-seeking crisis communication had taken place earlier, the existence of the concrete pumping truck would have been known earlier and the accident would probably have been contained under control quicker.

During the Fukushima-1 NPP accident, government organizations and TEPCO had troubles in information transfer with the municipalities because the off-site center was not functional, many municipalities suffered devastating damage from the earthquake and tsunami, and the telephone network was not functioning [1,4]. Most municipalities and residents, thus, had to respond to the accident using their own judgment. In the case of a crisis, distributed responses with decisions made against local situations are more effective than a central response [8]. Therefore, information to support the municipalities in making their decisions, for example, the radiation spread forecast by SPEEDI would have helped if given by the mass media or other means.

The primary responsible party, TEPCO, actively carried out crisis communication with the administrative organization in charge [1,6]. It also performed crisis communication immediately after the accident for the mass media as well. The announcements, however, made in TEPCO's press conferences and in those by NISA had some discrepancies, and the government instructed "to enhance accuracy and clarity, provide a single source matching information over a wide range of the accident and in detail" and from April 25, joint press conferences were carried out by Nuclear Emergency Response Headquarters. Then Assistant Chief Cabinet Secretary, Goshi Hosono, NISA, TEPCO, Nuclear Safety Commission, and MEXT were present at these press conferences [4].

METI explained that the reason for having joint press conferences was that information transfer from specialists to the disaster victims shall be made in "one voice (without giving different instructions and information)" [4]. Information transfer in one voice is desirable at times of emergency when people are fleeing towards safety. Instructions based on simple information are far more effective than complex information for people in the middle of evacuation, when their information processing power is weakened. It is, however, questionable if it was necessary to consolidate information transfer into one voice after one and a half months after the accident and when further drastic damages were unlikely. The people would have appreciated a variety of information about the accident. The joint press conferences by the Nuclear Emergency Response Headquarters deprived TEPCO of its role as the primary responsible party and thus it is likely that real information from the accident field were hard to expose.

TEPCO, during the Fukushima-1 NPP accident, performed "help asking" crisis communication only with the legally higher decision maker, the government [1,6]. At a time of crisis, operations unexpected during normal time are called for. Those that are responding to the situation have to think that others may have better technologies and knowledge about such unexpected operations. Carrying out "help asking" crisis communication with detailed specifications about what is in need, e.g., "remotely controlled heavy machinery with electronics that do not malfunction in this highly radioactive environment" to smaller companies, research and development organizations, or foreign countries would have been effective. Furthermore, those in the nuclear industry shall have systems to provide mutual personnel assistance and shared preparatory storage of machines and consumables for use at times of accidents.

16.4 CARE COMMUNICATION AFTER THE FUKUSHIMA-1 NPP ACCIDENT

16.4.1 Care Communication Immediately After the Fukushima-1 NPP Accident

The Fukushima-1 NPP accident released a large amount of radiation to the outside of the plant. As we mentioned previously, MEXT conducted radiation monitoring of the atmosphere and ground near the plant starting immediately after the accident. The monitoring probably led to reasonable estimates of the radiation dose of the residents in the area. The government issued evacuation instructions and evacuation zones were set. At the same time, plenty of information was given out by the government and mass media, but care communication for the residents who were likely to have suffered from radiation exposure was needed.

The National Institute of Radiological Sciences (NIRS) took the central role of medical responses to radiation exposure. Experts on medical treatment for radiation exposure were dispatched to the off-site center/J village ever since March 12, 2011. The emergency radiation exposure medical structure was established through coordination with the local Nuclear Emergency Response Headquarters with NIRS at the center, together with 49 universities and 66 university hospitals to provide the aid [11].

16.4.1.1 Exposure Screening and Residents

The government surveyed the field to find the residents' exposure to radiation and at the same time explained the situation to the residents. MEXT set its bases at Fukushima Medical University. NIRS and JAEA took the central role in conducting exposure screening with whole body counts and special examination of internal exposure. At the same time the residents were given an explanation of the screening results [5]. JAEA reported that by January 31, 2012, 15,408 residents of Fukushima Prefecture had a special examination of internal exposure based on whole body counts and the results showed that 15,383 (99.8%) had 1 mSv/year or less exposure [12].

JAEA, from July 11, 2011 to January 31, 2012, conducted special examinations of internal exposure with whole body count to 9927 residents of the municipalities of Iitate, Kawamata, Namie, Minamisoma, Katsurao, Futaba, Okuma, Tomioka, Kawauchi, Naraha, and Hirono all in Fukushima Prefecture. The examination was first given to expecting mothers and infants. Of those that received the examination, the 17 years old or under made up 68.5%. The examination proceeded with registration, preexplanation, checking for surface contamination, whole body count, postexplanation, and consulting. Many came to the examination worrying about their children and themselves, but it turned out that hardly anyone had suffered from serious radiation exposure and the examiners provided consultations until the people were thoroughly satisfied and most returned home with their minds eased, JAEA reported [12].

16.4.1.2 Phone-Based Consultation for the Residents

A number of organizations set up phones for the residents who wanted to hear explanations and receive information about radiation exposure.

NISA, with support from the Japan Nuclear Energy Safety Organization (JNES), opened 24-hour lines to respond to inquiries from the general public. The center received about 15,000 calls from March 17 to May 31 [4].

MEXT, with support from JAEA, set up a health consultation hotline on March 17, 2011 to respond to health concerns of the general public and to spread accurate understanding. About 25,000 calls were made during March 17 to May 31 [5].

NIRS also set up its health consulting hotline. It received about 7800 calls by May 31 and answered health questions and provided information about radiation exposure treatment [4].

The Prefecture of Fukushima set up a consultation window about radiation within the prefectural government building and received over 14,000 calls and visits by May 31.

16.4.1.3 Web Replies From Academic Societies

Academic societies such as the Atomic Energy Society of Japan or The Society for Risk Analysis-Japan set up pages in their homepage to reply to questions from the general public about radiation exposure and the Fukushima-1 NPP accident.

16.4.2 Care Communication After the Emergency Situation was Lifted

16.4.2.1 Government Actions chaired by Ministry of the Environment

The government established a "Nuclear Disaster Victims' Support Team" about a year after the Fukushima-1 NPP accident. The team had its first meeting on April 20, 2012.

The Minister of the Environment chaired this committee with vice-chairs, the State Minister of the Environment, the State Minister of Education, Culture, Sports, Science and Technology, the State Minister of Health, Labor and Welfare, and the State Minister of Economy, Trade and Industry (Executive Director of Nuclear Disaster Victims' Support Team). Over 10 other key officials in the central government were members of this team. It was a team for which the whole government unified its strength together [13].

During the first meeting, a document titled "System of Policies for Resolving Anxiety against Health Effects from Radiation [14]" was distributed. This document revealed what the government thought was necessary about a year after the accident in care communication to the residents, and how it evaluated its actions. Table 16.1 summarizes how the government explained its policies.

First, the document set the targets (receivers of information) to "parents and children" and "others" showing their priority was on children and their parents. The communicators (senders of information) were those related to health and medical welfare, educators, and local government staff. The document also discussed support for public relations of safety of food and drinking water by the farmers and distributors.

The document listed for care communication (source of anxiety) (1) Checking health status of the subject (radiation exposure, health effect from radiation, and health effect from stress or sickness), (2) knowledge about radiation effect on health and consultation and treatment about health and stress, and (3) living environment.

The shaded areas in Table 16.1 are areas that the document pointed out as "insufficient." This means that the government expected those related to health and medical welfare, educators, and local government staff to take the roles of what NIRS and JAEA had offered.

Systematic Policies for Resolving Anxiety about Radiation Effects on Health had a total of 3 meetings and on May 31, 2012, it reached a conclusion of "Action Plans for Resolving Anxiety about Radiation Effects on Health of Nuclear Accident Victims." The decision set the following four actions as important plans [15]:

(a) Establish coordination and common understanding among those involved (conduct meetings among the Committee for Coordinating Action Plans for Anxiety about Effects on Health, central government and local public organizations, and have a portal site for announcing consistent information).
(b) Develop personnel for radiation effects and offer communication with the people (government to compose unified basic documents about effects on health, develop personnel for health and medical welfare and education, and design active-type programs).
(c) Improve central location for concerns about radiation effects and enhance coordination (set Fukushima Medical University, Health Management Center to be the central location for Fukushima Prefecture and enhance its coordination with other related organizations).

Table 16.1 Policies for Removing Anxiety About Health Risk From Radiation (Conference of Coordinating Countermeasures for Health Anxiety of Nuclear Accident Victims)

Anxiety				Corresponding Policies: Policies for Information Recipient		Policies for Information Source			Support for PR of Product Safety
				Parent/Child	Others	Medical Staff	Educator	Other Administrative Staff	Producers and Distributors
(1) Want to know health condition	Want to know own exposure	Initial exposure	External	MEXT		MEXT			
			Internal	Support Team, SNRA					
		Additional exposure	External	Support Team	Support Team				
			Internal	Support Team					
	Want to have health effects by exposure checked			Fukushima Health Management Survey, Support Team, MOE					
	Want to have health effects by stress and lack of exercise checked			Fukushima Health Management Survey, Support Team					
(2) Want to know or have consultation on radiation effect on health	Want detail knowledge of exposure risk (health effect from radiation)			MEXT, MHLW, Enecho, Consumer Affair Agency, SNRA, Food Safety Commission, Special Child Care Grant		MEXT, MHLW	MEXT	Consumer Affair Agency, Food Safety Commission	
	Want detail knowledge of various standards (e.g., their interrelationship)			Consumer Affair Agency, Food Safety Commission				Consumer Affair Agency, Food Safety Commission	METI
	Want consultation for health effects from radiation			MEXT, SNRA		MEXT			
	Want consultation of health effect from stress. Want medical treatment for PTSD			MEXT, MHLW, Fukushima Pref. Grant, Special Child Care Grant		MHLW, Special Child Care Grant	MEXT		
(3) Want to protect living environment	Want exposure reduced	External	Want areas with high dosage clarified	MEXT, MOE					MAFF
			Want high dosage areas decontaminated	MEXT, MOE					
		Internal	Want food checked	MHLW, MEXT			MEXT, Support Team	MHLW, MAFF, Consumer Affair Agency, Support Team, Decontamination Grant	MAFF, Support Team, Emergency Response Funds
			Want drinking water checked	MOE, MHLW					
	Want to reduce radiation ourselves			MEXT, SNRA, Consumer Affair Agency, Support Team, Decontamination Grant		MEXT			
	Want research and development for proper treatment of cancer								
	Want stress and lack of exercise resolved			MEXT, MHLW, Support Team			MEXT		

Light gray boxes are where support is insufficient

Dark gray boxes are where support is needed

METI: Ministry of Economy Trade and Industry
Enecho: METI, Agency for Natural Resources and Energy
MAFF: Ministry of Agriculture, Forestry and Fisheries
MEXT: Ministry of Education, Culture, Sports, Science and Technology
MOE: Ministry of the Environment
MHLW: Ministry of Health Labor and Welfare

NRA: Nuclear Regulatory Authorization
SNRA: The Secretariat of the NRA

Support Team: Team for supporting lives of nuclear accident victims
Special Child Care Grant: Temporary Special Provisions Grants for Child Care Support
Fukushima Health Management Survey

Based on Addendum 1 "System of Policies for Resolving Anxiety against Health Effects from Radiation" during The 1st Conference on Health Anxiety of Nuclear Accident Victims" (partly modified by the author).

(d) International cooperation (Establish cooperative relations with victim countries of the Chernobyl accident on information exchange and joint research. Build cooperative relations with international organizations like IAEA).

16.4.2.2 Survey by Fukushima Prefecture of Its Residents' Health [16]

The Prefecture of Fukushima decided to conduct a prefectural resident health survey to clarify the amount of radiation exposure caused by the Fukushima-1 NPP accident and at the same time, provide care communication to the residents. This survey was scheduled as an annual event starting from 2011.

The prefectural resident survey consisted of a basic survey and a health exam.

The first basic survey was mailed out to those who lived in Fukushima Prefecture during the period between March 11 and July 1, 2011. The residents filled out the questionnaire and returned them in the mail. The form asked the residents about where they lived and worked and what food and beverages they took to estimate the radiation exposure of the prefectural residents.

For the 2011 basic survey, for the 2,057,053 applicable residents, 2,042,058 surveys were mailed out and 371,039 were filled and returned. The response rate was 18.0%; however, it was much higher with the prior survey for those in the evacuated areas (Yamakiya area of Kawamata, Namie, and Iitate) with 29,103 applicable residents, 29,048 surveys mailed, and 14,324 recovered, making a response rate of 49.2%, as of Nov. 30, 2011 [17].

The Fukushima Prefecture committee of this survey reached a conclusion, as mentioned first in this chapter, that "Radiation caused health effects is believed to be nothing" and opened the results to the residents of the prefecture [3].

The health exam involved a thyroid gland test, health checkup, questions about mental health and living habits, and questions for expecting mothers. The thyroid gland test had 367,685 Fukushima residents of age 0–18 (including fetuses) at the time of the earthquake. The 2011 survey was sent to 13 municipalities in the evacuation zone: Date, Minamisoma, Tamura, Kawamata, Namie, Futaba, Okuma, Tomioka, Naraha, Hirono, Iitate, Katsurao, and Kawauchi. In 2012, the 12 municipalities of Fukushima City, Nihonmatsu, Motomiya, Kooriyama, Shirakawa, Kunimi, Koori, Miharu, Otama, Tenei, Izumizaki, and Nishigo received the survey, and in 2013, the remaining 34 municipalities. Among the 367,685 that were eligible to have the exam, as of June 30, 2015, 300,476, including 9510 that had the exam outside of the prefecture, received the exam, i.e., a rate of 81.7%. After the first, second, and cytological diagnosis, those diagnosed with malignity were 15 in the 2011 test, 56 in 2012, and 42 in 2013, making a total of 113 [18].

16.4.3 Discussions on Care Communication

Care communications concerning the Fukushima-1 NPP accident have been made, not only by the central or Fukushima government, but also by various organizations including other local communities, schools, medical welfare organizations, NPOs, voluntary organizations, the media, and even private companies. The type of communication and activities spread over a wide variety as well. An example was the activity by JAEA that continually organized meetings in Fukushima Prefecture and Northern Kanto area in response to requests by schools and concerned parent groups to explain the effect of radiation exposure to the health [19]. This chapter discussed only the primary activities by the central government and the Prefecture of Fukushima.

At the time of the Fukushima-1 NPP accident, the central government instructed evacuation on the day of the accident when radiation was released from the nuclear reactor. This means there were people in need of care communication from the very beginning of the accident. It was difficult to conduct care communication as an emergency action when situations of damage from the accident and how it would spread were unclear.

Furthermore, most people made evacuations in groups; however, there were some that fled to as far as Okinawa or Osaka. The fact that people, in order to receive the care communication, spread all over the country made this action difficult.

16.4.3.1 Difficulty in Understanding Effects of Radiation Exposure to Health

Many Japanese, with knowledge of the devastation after the atomic bomb attacks on Hiroshima and Nagasaki, have fear and anxiety about exposure to radiation. Unlike the acute symptoms from relatively strong radiation that exceeds 1 Sv/year (=1000 mSv/year), the effect on health from low-level radiation exposure of 100 mSv/year or less, which is below the detection limit of diseases (cancer), is hard to explain and experts cannot show any firm scientific threshold. The difficulty causes a number of experts to state different opinions about health effects from low-level radiation exposure. In addition, regulations about protecting people from low-level radiation exposure do not necessarily match scientific viewpoints. Many of the general public, thus, think that health effects from radiation exposure are hard to understand and they developed distrust towards the experts.

If one wants to avoid the danger of radiation exposure, it is natural to stick to the safest (lowest) threshold. Risk analysis, however, makes judgments with the balance between risk and profit/cost, and it leads to the conclusion that the safest threshold in most cases will result in cost that is too large compared to the profit of avoiding the danger; thus it is not reasonable.

Care communication made after the Fukushima-1 NPP accident explained the risk analysis way of thinking while clear standards were unavailable about health effects. It also communicated that even without the Fukushima-1 NPP accident, our natural environment has about 2 mSv/year of natural radiation (=background radiation) and that we have been exposed to it throughout the history of mankind. The communication also explained that our food has radioactive material in it from the beginning and that our bodies also have radiation inside. This care communication intended to have those that suffered lead peaceful lives and to put a stop to rumors about farm and marine products from Fukushima Prefecture.

To understand the explanation is one thing, but how the people feel about living there is another. It must have been difficult to live there peacefully even with the reasonable explanations given.

16.4.3.2 Government Actions and the Current State of Fukushima Prefecture

The government policy, as shown in "System of Policies for Resolving Anxiety about Health Effects from Radiation," took a two-stage strategy of having people in medical welfare, educators, and local administrative staff assume the roles of opinion leaders to carry out risk communication to the general public. It carried out its plans by organizing seminars for the people to form an opinion leader group about health effects from radiation exposure and risk communication [20].

We are not quite at the stage to firmly discuss what the government has achieved through its plans; however, I would like to open the reaction by the people of Fukushima Prefecture about the health survey and exams.

The prefectural resident health survey gave the examinees answers to whether they had suffered health effects from the exposure to radiation and, if there were effects, they could expect to receive proper support. Thus, reasoning tells us that those who were more skeptical about health effects would have been willing to answer the survey. The percentage of the residents who answered the survey, however, was only 49.2%, even in the

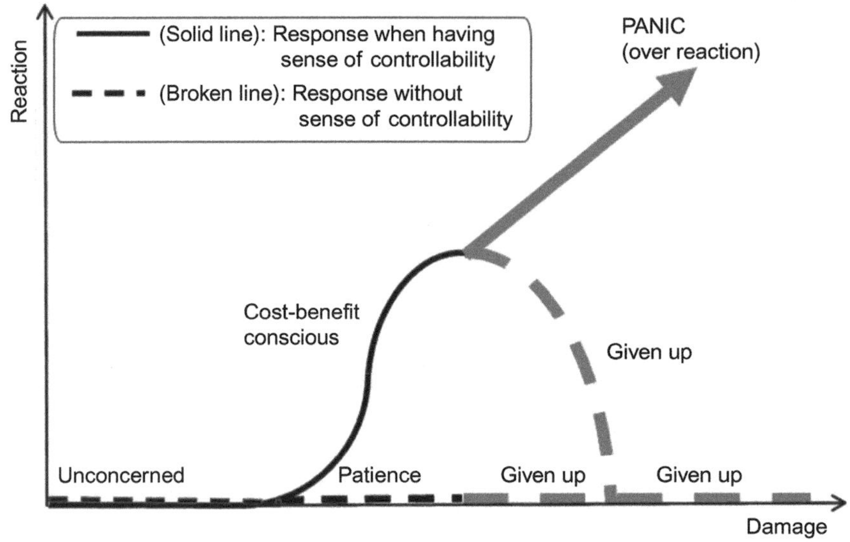

■ **FIG. 16.1** Psychological reaction at the time of disaster.

evacuation zone, and the number dropped to 18.0% for the entire prefecture. One explanation is that most of the residents of Fukushima Prefecture were not worried about health effects from radiation exposure.

Here is a hypothesis by Hirotada Hirose. He wrote, "Whether one can make a change to the damage determines whether one has sense of controllability over the disaster" [21]. In other words, people take actions to escape from damage when they believe they have the ability to do so. When the damage is so large and if the victim thinks he/she cannot run away from the damage, she/he gives up and accepts his/her destiny. Fig. 16.1 shows this idea by Hirose [22]. If this hypothesis is right, the fact that most prefectural residents did not answer the prefectural resident health survey (basic survey) means they had given up and were ready to accept their destinies because they could not go back to the way things had been before the disaster, instead of not being worried about health effects from the Fukushima-1 NPP accident.

16.5 SUMMARY

Among risk communications made in relation to the Fukushima-1 NPP accident, this chapter primarily talked about the government crisis communication immediately after the accident, in emergency response and care communication about radiation exposure.

16.5.1 Crisis Communication

The Great East Japan Earthquake devastated the areas attacked by tsunami, and many municipalities lost their main administration facilities and the numbers of available staff were small. Many municipalities could not perform their administrative functions. The phone networks were destroyed as well, with no estimates of when they would be fixed. Fukushima-1 NPP was also attacked by tsunami. In the middle of this chaos, the government instructed (=ordered) evacuation and many residents of the municipalities had to decide on their routes and destinations on their own. Information that would have helped the residents, e.g., forecasts on spread of radiation by SPEEDI, were rarely given to them. In fact, the residents of Minamisoma and Namie ended up relocating themselves to areas with higher levels of contamination than where they started from. Some point out that in a crisis local judgment in the field would have led to better results than centralized decisions. It would have been better if TVs and radios broadcasted information that would have helped municipality residents in the field to make their own judgment.

The accident started to reach an end because a concrete pumping truck happened to be sitting in the port of Yokohama. The idea to use this concrete pumping truck was raised by a citizen to government officials, and it went into service 2 weeks after the start of the accident. It took so long because neither TEPCO nor the government made "help-seeking communication" to the public. The more serious the emergency is, the more situations are beyond the capacity of those responsible; therefore they should have made help-seeking communication to the public and any organizations giving them precise information about the situation.

After the accident broke out, the primary responsible party, TEPCO, gave information in the form of press releases. Separate releases by TEPCO and NISA had discrepancies and TEPCO was banned to give out its own press releases on April 25. This caused TEPCO to lose its position as the primary responsible party and the real voices from the accident location became difficult to find.

16.5.2 Care Communication

When a nuclear accident with radiation release takes place, proper care communication first estimates the exposure dose based on radiation monitoring, conducts screening by checking surface contamination, and measures internal exposure with whole body counts to provide proper information to people that suffered and those that suspect radiation exposure.

Many people, in order to receive this care communication in response to the Fukushima-1 NPP accident, stayed in Fukushima and nearby prefectures; however, a number of people fled to remote locations in Japan, like Tokyo, Osaka, Okinawa, and so on. The spread of the subjects of care communication to remote locations was a factor that made the situation more difficult.

Care communication has to start immediately when the accident is triggered. However, the Fukushima-1 NPP accident was prolonged for days, making the care communication difficult as well.

The Prefecture of Fukushima is carrying out a care communication to have all residents of the prefecture take health exams. The prefectural resident health survey estimated the radiation dose with a questionnaire asking about their actions after the accident. The prefecture reached a conclusion that there were none in Fukushima Prefecture that suffered serious exposure and the results were opened publicly.

Care communication secondly has to give scientific explanations to the subjects so they can resist excessive anxiety. Such understanding is different from the mindset that exposure to radiation is terrifying and this communication often faces difficulty. NIRS and JAEA have been active in this care communication.

The government set its first target subjects of this care communication as children and their parents. It set up a two-stage communication by first assigning opinion leaders of medical welfare, education, and local government administration workers. These leaders, at the beginning, had hardly any knowledge of radiation exposure, so the action plans called for educating these people first.

This type of care communication was made not only by the central government, but also local governments, organizations like coops, and many others have been carrying out similar efforts. We can expect the results that would control rumors about vegetable and fish products from Fukushima.

16.5.3 Consensus Communication

At the end, I would like to discuss consensus communication about nuclear power generation after the Fukushima-1 NPP accident. So far, we cannot find any consensus communication by the government aimed at forming a social agreement on the future of nuclear power generation. The outcome could be one to abandon nuclear power generation, to restart it, or without settling on either way, to continue discussion and form a mutual reliance. There are some research activities made for this purpose. Sendai NPP by Kyushu Electric Power Company was approved by the Nuclear Regulatory

Authority and with agreements from the local municipal and Kagoshima Prefecture governments, it has restarted. The local and Ehime Prefecture governments passed resolutions to request restarting Ikata NPP owned by Shikoku Electric Power Company. And it also has restarted. These facts mean that the local communities where these plants are located have formed some social consensus to restart the nuclear plants. On the other hand, there are definite strong public opinions to abandon nuclear power generation. After the Fukushima-1 NPP accident, the government is facing a situation in which it, at least formally, should not try to conduct consensus communication in Japan now.

REFERENCES

[1] Investigation Committee on the Accident at the Fukushima Nuclear Power Stations of Tokyo Electric Power company, Interim Report, 2011. http://www.cas.go.jp/jp/seisaku/icanps/post-1.html (Accessed 30 September 2015).

[2] T. Torii, Y. Sanada, T. Sugita, K. Tanaka, Distribution of dose-rates and deposition of radioactive cesium by the airborne monitoring surveys, J. At. Energy Soc. Jpn. 54 (3) (2012) 160–165.

[3] Prefectural Oversight Committee for Fukushima Health Management Survey, Fukushima Prefecture, Basic survey (radiation dose estimates) (reported on 31 August 2015), 2015. http://fmu-global.jp/?wpdmdl=1157 (Accessed 26 October 2015).

[4] Ministry of Economy, Trade and Industry (METI), Japan, Communications after the accident, 2011. http://www.meti.go.jp/earthquake/nuclear/backdrop/pdf/09-communication.pdf (Accessed 30 September 2015).

[5] Ministry of Education, Culture, Sports, Science and Technology (MEXT), Japan, Document 2 "The Accidents of Fukushima Dai-ichi Nuclear Power Station and Countermeasures against them", Meeting (Mar. 11, 2012) of Committee on Atomic Science and Technology, 2012. http://www.mext.go.jp/b_menu/shingi/gijyutu/gijyutu2/055/shiryo/__icsFiles/afieldfile/2012/05/21/1306760_2.pdf (Accessed 30 September 2015).

[6] Investigation Committee on the Accident at the Fukushima Nuclear Power Stations of Tokyo Electric Power company, Final Report, 2012. http://www.cas.go.jp/jp/seisaku/icanps/post-2.html (Accessed 30 September 2015).

[7] Cabinet Office, Japan, Report of Field Survey on the Evacuation from the Nuclear Power Plant Accident at Great East Japan Earthquake (reported on 18 December 2015), (2015). http://www.bousai.go.jp/jishin/sonota/hinan-chosa/ (Accessed 20 December 2015).

[8] H.B.D. Leonard, A.M. Howitt, Routine' or 'Crisis': the search for excellence, Crisis/Response J. 4 (3) (2012) 32–35.

[9] S. Tsuchida, Crisis communication at the Fukushima accident and the concept of crisis management, J. At. Energy Soc. Jpn. 54 (3) (2012) 181–183.

[10] Y. Funabashi, Countdown to Meltdown, Bungeishunju, Tokyo, 2013.

[11] National Institute of Radiotogical Sciences, Outline of the countermeasures against the nuclease disaster by TEPCO Fukushima Dai-ichi Nuclear Power Stations, 2012. http://www.nirs.go.jp/information/info.php?i16 (Accessed 30 September 2015).

[12] T. Momose, C. Takada, T. Nakagawa, K. Kanai, O. Kurihara, N. Tsujimura, Y. Ohi, T. Nurayama, T. Suzuki, Y. Uezu, S. Furata, Whole-body counting of Fukushima residents after the TEPCO Fukushima Daiichi nuclear power station accident, in: Proceedings of the 1st NIRS Symposium on Reconstruction of Early Internal Dose in the TEPCO Fukushima Daiichi Nuclear Station Accident, National Institute of Radiological Sciences, Chiba, 2012, pp. 67–82.

[13] The 1st Coordinating Meeting (Apr. 20, 2012), Health anxiety of nuclear accident victims, Document 1 about the coordinating meeting for structure of measures to reduce the nuclear disaster victims' health anxiety, 2012. https://www.env.go.jp/jishin/rmp/conf-health/01-mat02.pdf (Accessed 30 September 2015).

[14] The 1st Coordinating Meeting (Apr. 20, 2012), Health anxiety of nuclear accident victims, sub-document 1 "System of policies for resolving anxiety against health effects from radiation," 2012. https://www.env.go.jp/jishin/rmp/conf-health/01-mat08.pdf (Accessed 30 September 2015).

[15] The 3rd Coordinating Meeting (May 31, 2012), Measures for nuclear disaster victims' health anxiety, action-plans for resolving anxiety against health effects from radiation. http://www.env.go.jp/press/files/jp/20079.pdf (Accessed 30 September 2015).

[16] Fukushima Prefecture, Fukushima health management survey, 2014. http://www.pref.fukushima.lg.jp/sec/21045b/kenkocyosa-gaiyo.html (Accessed 30 September 2015).

[17] Prefectural Oversight Committee for Fukushima Health Management Survey, Fukushima Prefecture, Overview of Fukushima Health Management Survey [Basic Survey (Estimates of external radiation doses), thyroid ultrasound examination] (reported on 13 December 2011), 2011. http://www.pref.fukushima.lg.jp/uploaded/attachment/6492.pdf (Accessed 30 September 2015).

[18] Prefectural Oversight Committee for Fukushima Health Management Survey, Fukushima Prefecture, Final report of thyroid ultrasound examination (preliminary baseline screening) (reported on 31 August), 2015. http://fmu-global.jp/?wpdmdl=1222 (Accessed 26 October 2015).

[19] Sector of Fukushima Research and Development, Japan Atomic Energy Agency, Communication activities, 2014. http://fukushima.jaea.go.jp/initiatives/cat02/index.html (Accessed 30 September 2015).

[20] Ministry of Environment, Japan, Measures of risk communication concerning health fear of radiation, 2013. http://www.env.go.jp/chemi/rhm/workshop.html (Accessed 30 September 2015).

[21] H. Hirose, Why do people fail to escape? Psychology of Disaster, Shueisha, Tokyo, 2004.

[22] S. Tsuchida, Social psychology of accidents, in: Faculty of Safety Science, Kansai University (Ed.), Safety Science for Accident Prevention and Mitigation, Minerva Shobo, Kyoto, 2013, pp. 182–198.

Chapter 17

The Great East Japan Earthquake and Nuclear Power Safety Regulation

CHAPTER OUTLINE

17.1 OVERVIEW OF CHANGES IN NUCLEAR POWER REGULATION AFTER FUKUSHIMA-1 NPP ACCIDENT

The Great East Japan Earthquake struck at about 14:46 on March 11, 2011, and the huge tsunami that followed the earthquake about 1 hour later caused loss of all external power supply and almost all the internal power supply that had been prepared at the Tokyo Electric Power Company (TEPCO) owned Fukushima Daiichi Nuclear Power Plant (Fukushima-1). The nuclear reactor and spent fuel pool lost their cooling functions, causing radiation leakage to the outside of the plant. The accident was level 7 on the International Nuclear Event Scale (INES), i.e., a major accident, which is the worst on this scale. A level 3 incident also took place at TEPCO's Fukushima Daini Nuclear Power Plant (Fukushima-2) about 11 km away from Fukushima-1. Many residents near Fukushima-1 were forced to evacuate.

The Fukushima and Tohoku Disaster. https://doi.org/10.1016/B978-0-12-814078-9.00017-0

The earthquake, tsunami and the nuclear accident altogether forced 150,000 people to be evacuated at the peak.

As we know, throughout the history of nuclear power plant accidents over the world, the 1986 Chernobyl accident in Ukraine in the at-that-time Soviet Union was the only accident previously ranked at Level 7. During this accident, one of the four nuclear reactors in the plant had a core meltdown and exploded. In the case of Fukushima-1, three of the six nuclear reactors had core meltdowns and core damage. These facts complicated not only the emergency response immediately after the accident, but also the contaminated water disposition, still underway, and the decommissioning work to go into full swing later.

During the Chernobyl accident, 33 workers and firemen died during their emergency response immediately after the accident, and a large number of residents were exposed to radiation. How this accident affected the health of those exposed is, after about 30 years from the accident, still being researched by international organizations [1,2]. There have been no direct victims or radiation exposure with the Fukushima-1 accident. We, however, have to note that a large number of victims were marked in the evacuation process immediately after the accident, at hospitals due to delayed initial treatment caused by loss of medical functions, and in the totally changed lifestyle at evacuation camps.

We can see this problem through data of "earthquake-related deaths." People that lost their lives, not because of a direct hit by the earthquake or tsunami, but due to injuries or stress caused by them are called "earthquake-related deaths." This is a category introduced after the 1995 Great Hanshin-Awaji Earthquake and, if approved as such, the "Act on payment of disaster condolence grant (1973 Law #82)" provides a condolence payment. The newest Reconstruction Agency data counted 3331 such deaths from the Great East Japan Earthquake. Fukushima Prefecture had the most with 1914, followed by Miyagi at 910, Iwate at 452, and Ibaraki at 41 [3]. As Table 17.1 shows, the number of disaster-related deaths in Fukushima stands out over Miyagi and Iwate when compared with the total number of direct deaths and missing. Especially in the towns of Namie and Tomioka, within 20 km from Fukushima-1, the direct death toll was 150 and 18, whereas the related death count was 374 and 320, respectively. Furthermore, in Kawauchi, the related death toll count was 85, while no one died from a direct cause (as of October 9, 2015) [3].

The direct cause of the Fukushima-1 accident was the natural phenomena of earthquake and tsunami. However, what led to the serious nuclear disaster of INES Level 7 were the lack of preparedness and weaknesses in *a priori*

Table 17.1 Number of Victims in the Great East Japan Earthquake

Prefecture	Dead	Missing	Earthquake-Related Deaths
Fukushima	1612	199	1914
Miyagi	9541	1237	910
Iwate	4673	1125	452
Ibaraki	24	1	41
Others	73	3	14
Total	15,893	2565	3331

Note: "Dead" and "Missing" are as of December 10, 2015, and Earthquake-related deaths as of March 31, 2015.
Data from Reconstruction Agency, Current status of the number of the Great East Japan Earthquake Related Death, Report on 31 Mar. 2015. http://www.reconstruction.go.jp/topics/main-cat2/sub-cat2-6/20150630_kanrenshi.pdf (Accessed 20 September 2015) (in Japanese); National Police Agency of Japan, Damage and police responses in the Tohoku District–off the Pacific Ocean earthquake, 2013. Available from http://www/npa.go.jp/archive/keibi/biki/higaijokyo.pdf (Accessed 10 October 2015) (in Japanese).

safety regulation and accident prevention measures, TEPCO actions after the accident, and government countermeasures against nuclear disasters. These points, especially the major flaws and problems in the government nuclear safety regulation, were pointed out by the Investigation Committee on the Accident at the Fukushima Nuclear Power Stations of Tokyo Electric Power Co. (Governmental Accident Investigation Committee) [4,5].

The government, therefore, has revised the laws concerning nuclear safety and has made major reconfigurations in the regulation system since the accident: the Nuclear and Industrial Safety Agency (NISA) and Nuclear Safety Commission (NSC) were closed and the Nuclear Regulation Authority (NRA) was newly started. Responsibilities for nuclear safety administration were also transferred to the Ministry of the Environment from the Ministry of Economy, Trade and Industry (METI).

This chapter discusses the changes in the nuclear safety regulatory system in Japan before and after the Fukushima-1 accident and recommends a preferred ideal nuclear safety system.

17.2 REGULATORY SYSTEM AT THE TIME OF FUKUSHIMA-1 NPP ACCIDENT

As the International Atomic Energy Agency (IAEA) states in its Fundamental Safety Principles, Principle 1: "The person or organization responsible for any facility or activity that gives rise to radiation risks or for carrying out a program of actions to reduce radiation exposure has the prime responsibility for safety" [6]. In the case of commercial power generation, this is the electricity company. Under this principle, the electricity companies in

Japan have taken on the prime responsibilities for safety. The business of nuclear power generation, however, as an electricity supplying business, is subject to regulations for public benefit, and especially from the need to secure safety, must follow safety regulations enforced by the government.

When the Fukushima-1 accident took place, the regulatory body for safety of nuclear power generation businesses was METI, and the entity in charge of regulation was the now-decommissioned NISA. NISA, based on the Act on Control of Nuclear Reactors and the Electricity Business Act, carried out the following three regulatory operations for nuclear power facilities:

(1) Safety regulation of design and construction
(2) Safety regulation during operation
(3) Safety regulation for decommissioning.

Separately from NISA and in addition, the now-closed NSC was another nuclear regulatory organization. The NSC separated out from the Atomic Energy Commission (AEC) in 1978 and was given strong authority to make recommendations about nuclear power safety through the Prime Minister to related administrative organizations. The primary operations of this commission were:

(1) To carry out the second review of the results of the first by administrative agencies for the approval of permission to start construction, i.e., to conduct double checking
(2) To define the Guideline for Safety Review for the second review
(3) To check the regulatory actions by administrative agencies
(4) To take actions upon nuclear emergencies

In addition, there was an independent administrative agency, the Japan Nuclear Regulation Authority (JNES). JNES, as a technical support organization for NISA, carried out nuclear plant inspection with NISA, and provided technical support in nuclear plant safety reviews and defining safety regulation standards. As of April 2012, JNES had 423 employees, a size equivalent to NISA at the time of its shutdown. JNES, in March of 2014, merged into NRA, a year and a half from NRA's establishment.

This was the nuclear safety regulation system in Japan at the time of the accident; however, there was hardly any tension among the electricity companies and the regulatory bodies, and the regulators rarely had the necessary technical expertise to carry out their functions. The National Diet of Japan Fukushima Nuclear Accident Independent Investigation Commission (National Diet Accident Investigation Committee) described the situation as follows: "The lack of expertise resulted in 'regulatory capture,' and postponed implementing relevant regulations. They avoided their direct responsibilities by letting operators apply regulations on a voluntary basis" [7].

The knowledge and problems to solve concerning the safety of nuclear power are in the plant fields; thus it may be difficult for the regulating organizations to maintain high-level expertise in securing safety. For the regulatory organizations to carry out their tasks, they need high technical skills in review and operations, in addition to practical expertise in safety and technology at levels equivalent to engineering in electrical companies. As Ref. [4] pointed out, NISA was extremely weak in this aspect.

In their response to the Fukushima nuclear accident, both NISA and NSC failed to meet their duties, revealing their organizational limitations. At the same time, problems in their past regulatory activities and performance were uncovered. Those were the reasons both organizations were decommissioned in September 2012 and reorganized into the NRA.

17.3 DEVELOPMENT AND SAFETY REGULATION OF NUCLEAR POWER GENERATION IN THE WORLD

The 1953 speech by the 34th US President Eisenhower at the United Nations (UN) triggered a major advancement of nuclear power generation technology. The IAEA was established in 1957 within the UN for international cooperation in nuclear technology, due to its large across-the-border social effects. This IAEA now serves as the organization for international coordination of nuclear technology, including nuclear power generation.

Fig. 17.1 shows [8] nuclear power plants (NPPs) built since the 1950s up till now, with their years of startup and power. Data for the figure were extracted from the World Nuclear Association (WNA) database [8]. There have been 323 pressurized water reactors (PWR) (among which 44 have been decommissioned), 114 boiling water reactors (BWR) (34 decommissioned), 83 gas cooled reactors (53 decommissioned), 60 pressurized heavy-water reactors (10 decommissioned), and 9 fast breeder reactors (1 decommissioned), making the total count 589 reactors (142 decommissioned).

The leader in atomic energy, the United States, after the Atomic Energy Act was signed in 1946, established the Atomic Energy Commission (AEC). Commercial nuclear power generation then expanded explosively and the rapid increase in installed capacity caused difficulties for the AEC in handling both promotion and regulation of the industry. In 1975, the Nuclear Regulatory Commission (NRC) was established [9,10].

The huge technological infrastructure in energy before World War II was fossil fuel power, and after the War, it has been nuclear power generation. Through the development phases of huge-sized technology, accidents and obstacles always occur. If the technology is promising, that which surrounds

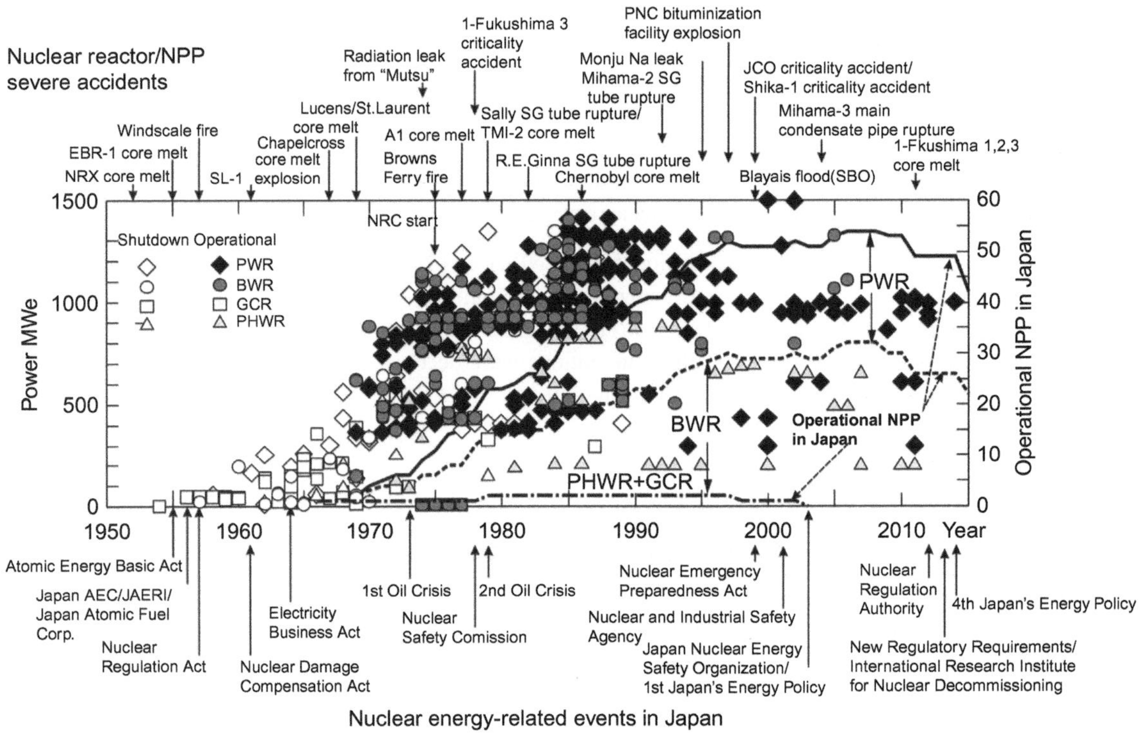

FIG. 17.1 Nuclear power accidents in the world and the history of nuclear power development in Japan. *(Based on data in WNA (World Nuclear Association), Rector database. http://world-nuclear.org/NuclearDatabase/Default.aspx?id=27232 (Accessed 18 June 2015).)*

it grows with it, including material, process engineering, control, and social systems, and then accidents and obstacles are reduced in number. Large-scale fossil-fuel power generation has been around for over 130 years, since the start of electric power transmission in the 1880s, and it had to overcome many technical and social issues to reach the point where it stands now. Nuclear power generation is no exception; however, with its number of installations far fewer compared to those of fossil fuel power stations, it has more to learn from past experiences.

Through the history of nuclear technology development, reference [11] lists accidents with core meltdown. The 1979 Three-Mile Island Unit-2 (TMI-2) accident was a serious one. Japan never had a core meltdown accident until Fukushima, but it had had serious accidents, e.g., broken steam generator tubes or erosion and rupture of condensate piping. There had been station blackouts (SBO) at overseas plants, but the Japanese government or engineers in the nuclear industry never seriously studied these events.

The part count of a system is an indicator of its complexity. The 1970 Apollo 13 with 5.5 million parts was the most advanced system back then. The spacecraft suffered from a relatively simple accident of an explosion of its oxygen tank, but survived the crisis with creative ideas, teamwork of the astronauts and US National Aeronautics and Space Administration (NASA) workers, and the leadership of the flight director. The technically further advanced Space Shuttle Challenger in 1986, despite warnings from the production field, experienced an O-ring failure and exploded 73 seconds into its flight. Business priorities interfered with NASA's judgment in that case [12]. This accident told us that the advancement of a system does not necessarily mean proportional enhancement of safety and redundancy. It may be an indication that a problem with what seems to be a trivial part can often trigger a fatal situation for the entire system.

The 1979 TMI-2 accident started with a minor event of a pump-trip in the condensate polishing system. The series of events that followed was: primary cooling water leakage, operator misinterpreting water level, insufficient coolant supply by the Emergency Core Cooling System (ECCS), fuel rods exposure, and fuel rod cladding and water reaction that eventually led to partial meltdown of the core. This is roughly what happened with TMI-2 [13,14]. This accident revealed that, despite the primary system of defense in depth, the ECCS, being available, human factors magnified the seriousness of the accident. Reports [15,16] of the TMI-2 accident investigation pointed out problems with NRC in its management structure and nuclear plant licensing process. It also blamed the plant in its technical and management capacities, quality of equipment design evaluation, quantitative risk assessment methods, safety awareness in health management for civilians and site workers upon radioactivity release, emergency plans and actions, and opening of information. The US government and the NRC had to take a number of measures to counter these problems. In 1979, the Institute of Nuclear Power Operation (INPO) was established as a private organization for plant assessment, operator training and accreditation, event evaluation and information exchange. The core thermal power density of commercial BWRs is 50 MW/m^3 and for PWRs it is 100 MW/m^3. A large boiler for fossil-fuel power generation of the same scale had a combustion changer thermal power density of only about 0.2 MW/m^3 [17]. For a more familiar example, a 1000 cm^3 home electric kettle has a thermal power density of 1.5 MW/m^3, i.e., much more than a large boiler. This electric kettle takes about 250 seconds to boil its water when filled to its capacity. The thermal power density of a PWR is 67 times that of this electric kettle and it will boil the same amount of water in 3.5 seconds. This simple calculation clearly tells us the importance of maintaining proper cooling in operating a nuclear core with extremely high thermal power density. This fact led to safety research centered on loss-of-fluid tests (LOFT) in the United States [9] and safety assessment and facilities establishment (SAFE) in Japan [18].

In the 1970s, probabilistic risk assessment (PRA) like the Rasmussen Report [19] was applied for safety assessment of nuclear power plants. PRA expresses the reliability of each component in the event tree with probability numbers to evaluate the probability of accidents. Big variations in reliability numbers for components necessitates uncertainty analysis. Lee and McCormick pointed out that different reliability numbers can manipulate the results; thus, PRA should only be used for finding weaknesses in the design rather than assessing the overall safety [20]. The Rasmussen report compared the accident probabilities of a nuclear plant accident with a meteor strike, and could have contributed to creating the myth that odds for a nuclear plant accident are extremely small. PRA, however, is an effective method for evaluating huge systems and it is widely used for real nuclear power plants in the United States [21,22].

Fig. 17.1 shows the history of nuclear power plant installations in Japan with dotted-and-dashed (PHWR + GCR), dashed (+BWR), and solid (+PWR) lines. Major events are shown along the timeline. Nuclear power development started in 1955 with the three acts of nuclear power: Atomic Energy Basic Act, Act for Establishment of the Atomic Energy Commission, and Act to Amend a Part of Act for Establishment of Prime Minister's Office. The newly established Science and Technology Agency, Atomic Energy Commission, Japan Atomic Energy Research Institute (JAERI, later merged with Japan Nuclear Cycle Development Institute (JNC) into the Japan Atomic Energy Agency (JAEA) in 2005), and Japan Nuclear Fuel Government Corporation took the central roles for promoting the development policies. Then later, the Act on Control of Nuclear Reactors (1957), Act concerning Prevention of Radiation Hazards (1957), and Electricity Business Act (1964) followed. The authors recall that probabilistic safety analysis (PSA) was common instead of PRA, due to hesitation in admitting "risk" since the time that nuclear power plants were started in Japan. Even with PSA, JNES reported a significant threat to BWR in the case of large earthquakes and tsunami [23], but TEPCO hardly looked into the possibility [24].

While Japan pursued R&D for a Japanese nuclear power plant, it also introduced plants and nuclear technology from the United Kingdom and the United States in the 1960s to 1970s. Japan, at the time, did not have the technology and it had to bring the whole package of operation, maintenance, and safety standards from overseas. This matches the overall trend in Japan after the no-growth period of World War II. It was the time to reconstruct social infrastructures and actively introduce technology centered on heavy and chemical industries for economic revival [25].

The Atomic Energy Commission, initially established with roles of both promotion and regulation just like the US commission with the same name, was reorganized in 1978 and the NSC was established as a division for safety regulation. After the TMI-2 accident that broke out in the following year, NSC instructed all domestic nuclear power plants to review their structures for safety management and report the results [26]. It also organized a Special Committee on TMI Accident Study and defined 52 items on safety standards, review, design, operation management, disaster prevention and safety study to reflect the lessons learned from the TMI-2 accident [27]. In Japan, however, the idea that serious safety measures are in place based on the concept of defense in depth, and that severe accidents (SAs) are extremely rare, led to excluding SA evaluation and accident management from regulation and they were left to the discretion of business entities [26,28]. This decision was much different from the practice in Europe and the United States, which mandated PRA of SA events and safety measures of vent and containment vessel reinforcement. We must say that this difference foreshadowed the Fukushima nuclear accident 30 years later. The US NRC had also pointed out this difference in comparing regulation requirements among countries [29].

The 1990s saw the sodium leakage accident at Fast Breeder Reactor Monju, the explosion and fire at Power Reactor and Nuclear Fuel Development Corporation asphalt processing plant, and the JCO criticality accident. These events led METI to establish NIST in one of its affiliates, the Agency for Natural Resources and Energy, as a specialized organization for safety regulatory review for authorizing building of nuclear power facilities based on the Act on Control of Nuclear Reactors and the Electricity Business Act. This move gave the appearance of a double-checking system with NISA and NSC; however, NSC lacked expertise in nuclear power technology and only followed the decisions made by NISA. If, for example, the roles had been split into different evaluation systems, such as NISA for technical evaluation and NSC for evaluation of corporate governance, management and safety culture, a real defensive in-depth structure might have been established. Safety regulation in Japan, although it followed the practice of the US NRC, was merely patched together without an overview of what was needed in terms of regulation for the entire technology and society. When we look back at the history, forming NISA under the Agency for Natural Resources and Energy produced the same conflict that the Atomic Energy Commission had before its reorganization in 1978.

17.4 BASIC CONCEPT OF SAFETY WITH IAEA AND NRC

Safety regulation in Europe and the United States places safety culture at the very foundation [30] and builds social and technical requirements on top of it. The word "culture" has its origin in "cultivating the soil to grow plants and animals." The word also has the active meaning of "train and develop body, soul, and action of human" [31]. The IAEA defined "safety culture" to mean the ultimate goal of protecting people and the environment from radiation damage [6]. We, thus, can understand that the phrase stands for activities and a mindset of related people, corporations, and society placing the highest priority on safety, accident prevention and damage reduction. The corresponding Japanese word "Bunka" has the active meaning of spiritual activity towards a goal, like "culture" [32], but the general understanding of the word is more static, standing for a spiritual state and common social understanding. The Japanese translation "Anzen Bunka" of "safety culture" has a subtle difference in meaning.

IAEA explains the problem of safety with an iceberg analogy [30]. An iceberg floating in the sea has its one-seventh tip above the surface and the remaining six-sevenths are hidden underwater. The problem of safety is the same, with only its tip visible above the surface and most of the basics of norms, attitudes, values, basic assumptions of reality and shared understandings are underwater. In the famous Titanic sinking, the damage occurred not from colliding with the visible part of the iceberg above the surface, but by its collision with the iceberg underwater.

IAEA shows the fundamental safety of nuclear power with a hierarchical structure of Safety Fundamentals, Safety Requirements, Safety Guides, and Safety Reports [6]. Safety culture is at the center of the fundamental principle, to form the overall body of activities and mindset for all organizations and individuals concerned for their prioritized safety and accident prevention. This is a requirement for defense in depth as well. The management system must also have its foundation in safety culture. Examples listed include leadership, management, responsibilities of organizations and individuals, legal and administrative framework for securing safety, independence of regulatory organizations, proper equipment and actions for securing safety, high-level protection systems, control of radiation-caused health hazards, daily accident prevention and mitigation activities, preparation and countermeasures in case of accidents, and so on.

The NRC has similar frameworks [33,34]. The NRC has been continually revising concept of safety culture and fundamental safety principles after the 1989 Peach Bottom and 1996 Millstone nuclear plant accidents. The concepts of safety culture and the fundamental principles do not remain unchanged and they are always under review in the United States as well.

17.5 CHANGES OF NUCLEAR SAFETY REGULATION IN JAPAN

The Interim and Final Reports by the Japanese Governmental Accident Investigation Committee [4,5], as well as other organizations in and out of the country, pointed out the following areas in Japanese nuclear power safety regulation that have problems:

(1) Basic concept of safety regulation and conformance to international standards
(2) Countermeasures for severe accidents
(3) Implementation of new knowledge, technology and risk information
(4) Safety awareness of the electricity business organization
(5) Countermeasures for multiple failures triggered by common causes like natural disasters
(6) Meaningless formality in inspection and review

Many of these points overlap with those pointed out by the Kemeny report [15] and Rogovin report [16] on the TMI-2 accident.

Receiving these comments, the Japanese government, in September of 2012, merged NSC and NISA and established the NRA as an affiliated agency of the Ministry of Environment. Further, in 2014, JNES was also merged into the NRA. This move formed a single regulatory authority for the nuclear power industry of commercial companies, research institutions, universities and so on, a function that used to be scattered among a number of organizations. Establishment of this new organization meant its demerging from the Agency for Natural Resources and Energy, an agency for promotion of nuclear power. The NRA has an affiliated administration office of The Secretariat of the NRA. It is also where the following organizations report: Human Resource Development Center, Reactor Safety Examination Committee, Nuclear Fuel Safety Examination Committee, Radiation Council, and National Research and Development Agency Council. The authority also jointly manages the incorporated administrative agencies of the Japan Atomic Energy Agency and the National Institute of Radiological Sciences [35].

The Secretariat of the NRA has the Secretary-General's Secretariat and Nuclear Regulation Department. The former has an assigned Director-General for Emergency Response, Director-General for Regulatory Standards and Research, and Director-General for Radiation Protection Strategy and Security, and the group is the home for the Regulatory Standard and Research Department (Regulatory Standard and Research Division, and four research divisions for Reactor System Safety, Severe Accident, Nuclear Fuel Cycle and Radioactive Waste, and Earthquake and Tsunami) and the

Radiation Protection Department (Emergency Preparedness/Response and Nuclear Security Division, Radiation Monitoring Division, and Radiation Protection and Safeguards Division). The latter Nuclear Regulation Department has a Nuclear Regulation Policy Planning Division and seven Regulation Divisions of BWRs, PWRs, Inspections of Nuclear Reactor Facilities, Advanced Reactors, Research Reactors, Decommissioning, Nuclear Fuel Facilities and Use of Nuclear Material, Radioactive Waste, Storage and Transport, and Earthquake and Tsunami [36]. The Nuclear Regulation Policy Planning Division is in charge of review and inspection of power generation reactors, and the Regulation Divisions manage the actual execution of regulation such as facility inspection. The Regulatory Standard and Research Division is in charge of setting the regulation standards. Regional organizations of the Secretariat of the NRA include 22 NRA Regional Offices, Local Radiation Monitoring Officers, Regional Administrators, Rokkasho Safeguards Center, and the Yokosuka Nuclear Vessel Monitoring Center. Regional administrators provide information to local governments and related organizations.

As we explained previously, divisions that carry out regulatory work are under the regulating agency. In contrast, the NRC [37] has its operations department directly reporting to the commission. The NRC regulatory operations are headed by the Executive Director of Operations who directly reports to the commission Chairman, and has 16 offices: New Reactors, Nuclear Material Safety and Safeguards, Nuclear Reactor Regulation, Nuclear Regulatory Research, Nuclear Security and Incident Response, Information, Human Capital, Small Business and Civil Rights, and so on.

When the NRA was established, a series of laws on nuclear power were also amended. The Atomic Energy Basic Act added a new Article 2, the basic policy on the use of atomic energy, which reads: "The safe use of atomic power is aimed at contributing to the protection of the people's lives, health and property, environmental conservation and national security." The Act on Control of Nuclear Reactors had fundamental enhancements to its regulation standards of preparation and facilities against unexpected events, and accident management by business organizations was lifted from voluntary actions to a mandatory requirement. Furthermore, total risk assessment was turned into a requirement, and the following aspects were pursued: building an organization to implement the most up-to-date knowledge in existing facilities, introducing plant life extension measures up to 40 years, clarifying responsibilities of business entities, unifying nuclear power safety regulations, and separating from the Electricity Business Act [35]. Actual implementations included enhancements in resistance to external events, emergency power sources and core cooling systems, containment vessel damage prevention, radiation spread control, and extended loss of power avoidance. These standards may represent the most stringent requirements, as the NRA insists. However, these new implementations do not clearly show a response to overall management or leadership based on the safety culture that the IAEA and NRC establish as the fundamental basis.

The NRA on September 10, 2014, based on these new regulatory standards, gave approval to Sendai Nuclear Power Station's application for Permission for Changes in Reactor Installation of Units 1 and 2 [38]. More approvals followed with Takahama Units 3 and 4 (February 12, 2015) [39] and Ikata Unit 3 (July 15) [40].

On the other hand, residents of four prefectures including Fukui and Osaka filed an application for provisional disposition to stop the restart of Takahama Nuclear Power Station, and Fukui District Court ruled against the power plant, saying that "The new regulation standards are insufficient and meeting them does not secure safety." A similar application was filed against Sendai Nuclear Power Station, but in this case, Kagoshima District Court, in April 2015, rejected the application for provisional disposition, saying that the new regulation standards are reasonable. Both reactors are PWRs with 30 years of operation with electrical output of 870–890 MWe. What were the reasons for these completely opposite rulings on the applications of provisional dispositions to stop the operations?

When we trace back the history, we find that the Supreme Court ruled in 1992 that the approval of the installation of Ikata Nuclear Power Station was a reasonable judgment by the AEC with its technical expertise as the regulating organization [41]. The Supreme Court made another ruling in 2005 that the AEC technical assessment to approve the FBR "Monjyu" was reasonable [42]. The ruling by Fukui District Court in 2015 was completely different from the past Supreme Court rulings in terms of making a rational judgment. It is questionable whether highly technical discussions blend well with discussions in the courtroom. The distance between the two rulings is hard for the general public to understand and it raises concerns about the future of the NRA. That is, based on the concept of safety culture that values accountability, the NRA has the responsibility to society to explain its basis and its process in reaching its judgments, whether the assessment is an approval or a denial.

17.6 AN IDEAL REGULATORY BODY

Keeping in mind the discussions up to the previous section, we would like to point out what an ideal regulatory organization should look like.

First, the rules of safety regulation are now in the Atomic Energy Basic Act, but society demands that the standards of the NRA be clearly defined, announced, and continually be discussed and revised. Such actions would be in sync with the safety culture followed by the IAEA, NRC and other overseas organizations.

Secondly, the NRA has an organizational problem in that the operational divisions of Nuclear Reactor Safety, Regulation, Standards, Public Affairs, and Inspection are all under the administrative function of The Secretariat of NRA. How the authority plans to fill in the gap of technical expertise between the operational groups from the high-level committees and councils is a big concern. Also communications, leadership and management of the NRA over the operational groups remain as concerns.

The third point is that the purpose of nuclear power safety is to secure safety for the citizens. Whether we continue or discontinue nuclear power generation, we need a framework for discussion involving all the players, i.e., the central and local governments, the business organizations, manufacturers, and the citizens. Unfortunately, however, it seems that the people of Japan have not developed the qualities of thinking and discussing sincerely without arguing from emotional opinions about securing safe energy. The discussions are not necessarily limited to nuclear power. We have hopes for the NRA, with its high level of independence, to start constructing such a framework.

Fourth, we need an accident research division that, in addition to investigating the Fukushima accident, can study possible future accidents independently. As the final report by the Governmental Accident Investigation Committee [5] pointed out, we should understand that the final report was really the starting point of the accident investigation. The Fukushima Nuclear Power Plant accident was one of the worst accidents in the world. Our nation has the responsibility to continue a detailed and thorough study of the events. And publicizing the results will lead to enhancing the safety of nuclear power plants all over the world (e.g., [14]). The NRA should be aware that regulation standards always become obsolete and they should establish an independent research organization dedicated to safety research on nuclear power. This organization should have plenty of opportunities to exchange opinions with academia, manufacturers, and electricity companies in and out of the country. It should be expected to carry out comprehensive research on the safety of nuclear power, including developing new technologies, especially for decommissioning and disposition of high-level nuclear waste, as well as sociotechnological aspects and human factors. We should stop importing overseas standards and start building our own regulation standards that conform to international regulations. Such a research organization would allow training human resources for accident investigation and safety inspections. The Fukushima accident did not expose all possible problems of huge, complex systems. We have to continue planning the actions we must take for our own future.

The fifth point concerns the NRA's scope. The NRA currently is busy with its "regulation" work; however, as the Atomic Energy Basic Act states, there is no question that its primary function is securing the safety of the citizens. Its leadership is needed in cooperating with the local and central governments in solving problems of resident evacuation and health management against radiation. These are problems that will always accompany nuclear accidents.

Sixth is risk assessment. The NRA requires PRA and uncertainty evaluation in its regulations [35]. Even though it contains uncertainty, the reliability of PRA itself relies heavily on the structure of the event tree for a complex system like a nuclear power plant. Even the event tree is a hypothesis made by the plant engineers. Under the assumption that events beyond expectation take place, PRA can effectively expose weaknesses of the system and taking measures against them are important actions for securing safety. This idea must be shared with the whole society, including the media.

The seventh point: we need a private institution for a third-party accreditation of nuclear plant inspectors and operators, like INPO in the United States. If we look into the boiler industry, private organizations, like the general incorporated association Japan Boiler Association, train and accredit inspectors and conduct inspections as well. For the same industry in the United States, the National Board of Boiler and Pressure Vessel Inspectors (NBBI) accredit inspectors [43] and the American Society of Mechanical Engineers (ASME) sets the regulation standards. This structure shows that nongovernmental organizations of an academic society and an association can function properly as third-party authorities.

Lastly, the NRA itself has to act and regulate in sync with safety culture. Activities of the NRA have to undergo an audit function. The NRA must determine if a technology is state-of-the-art and if it is feasible. At the same time, it has the responsibility to explain it to the people of Japan. Its proper leadership should manage the safety of nuclear power generation. The primary party of responsibility is, of course, the owner electricity company. However, the NRA has to ready itself to take the commanding role at the time of an accident, such as TMI-2, and it also must have a spokesperson ready.

These are the responsibilities of the NRA as the central party for securing nuclear plant safety for our society.

17.7 AT THE END

Engineers in the nuclear industry have often said that nuclear power technology is a "consolidated technology." All fields of technology in general exist within society and they can never be separated from it. Thus, "consolidated technology" is not unique to nuclear power. It appears that by making the excuse that nuclear power was "consolidated," those in the field avoided interference and criticism from the outside.

The society we live in has a number of dangerous and risky events. Nuclear power generation handles special material that is radioactive and it must strictly control this material for an extremely long time. Whether our nation abandons nuclear power generation or continues to promote it, we cannot avoid the problems of decommissioning plants and dispositioning high- and low-level radioactive waste. We need to develop the technology for the future and inevitably need to train human resources that can seriously tackle such research topics. At the same time, we have to receive social support through discussions among all the players, including the citizens, local and central governments, equipment manufacturers, electricity companies, and regulatory bodies. We all need to rationally and sincerely discuss the future of energy, the environment and the economy in our nation. Otherwise, the problem of nuclear power will never be solved.

REFERENCES

[1] United Nations Scientific Committee on the Effect of Atomic Radiation, Source and Effects of Ionizing Radiation, vol. I, United Nations, New York, 2010.

[2] United Nations Scientific Committee on the Effect of Atomic Radiation, Source and Effects of Ionizing Radiation, vol. II, United Nations, New York, 2011.

[3] Fukushima Prefecture Office, The state of damage in the Tohoku District–Off the Pacific Ocean Earthquake, No. 1532, 2015. http://www.pref.fukushima.lg.jp/site/portal/shinsai-higaijokyo.html (Accessed 10 October 2015) (in Japanese).

[4] Investigation Committee on the Accident at the Fukushima Nuclear Power Stations of Tokyo Electric Power Co. (shortened as "Governmental Accident Investigation Committee") interim report, Media Land Co., Ltd., Tokyo, 2011. English version is available at http://www.cas.go.jp/jp/seisaku/icanps/eng/interim-report.html (Accessed 23 October 2015) (in Japanese).

[5] Investigation Committee on the Accident at the Fukushima Nuclear Power Stations of Tokyo Electric Power Co. (shortened as "Governmental Accident Investigation Committee") final report, Media Land Co., Ltd., Tokyo, 2012. English version is available at http://www.cas.go.jp/jp/seisaku/icanps/eng/final-report.html (Accessed 23 October 2015).

[6] IAEA, Fundamental safety principles—safety fundamentals, IAEA safety standards series No. SF-1, Vienna, 2006.

[7] National Diet of Japan Fukushima, Nuclear Accident Independent Investigation Commission, Tokuma Shoten Publishing, Tokyo, 2012.
[8] WNA (World Nuclear Association), Rector database. http://world-nuclear.org/NuclearDatabase/Default.aspx?id=27232 (Accessed 18 June 2015).
[9] E.S. Fisher, Nuclear Regulation in the U.S.—A Short History, Nova Science Publishers, New York, NY, 2012.
[10] S. Walker, T.R. Wellock, A Short History of Nuclear Regulation 1946–2009: NUREG/BR-0175, Rev. 2, U.S. NRC, Washington, DC, 2010.
[11] Japan Electric Association Newspaper (every annual edition) Division Pocketbook of Nuclear Power, Tokyo, The Denki Shinbun (in Japanese).
[12] W.P. Rogers, Report of the presidential commission on the space shuttle challenger accident, White House, Washington, DC, 1986.
[13] J.G. Collier, G.F. Hewitt, Introduction to Nuclear Power, Hemisphere Pub., New York, NY, 1987
[14] M. Ishikawa, A Study of the Fukushima Daiichi Nuclear Accident Process—What caused the core melt and hydrogen explosion? Springer, Tokyo, 2015. pp. 3–33 (TMI-2 Accident), pp. 35–132 (Fukushima Daiichi Accident).
[15] J.G. Kemeny, Report of the President's Commission on the Accident at Three Mile Island, The Need for Change: The Legacy of TMI, Pergamon Press, New York, NY, 1979.
[16] M. Rogovin, Three Mile Island—A Report to the Commissioners and to the Public, U.S. NRC, Washington, DC, 1980.
[17] E. Nishikawa, General planning of the boiler gas-side heat transfer surface, in: S. Ishigai (Ed.), Steam Power Engineering—Thermal and Hydraulic Design Principles, Cambridge University Press, New York, NY, 1999.
[18] H. Nariai, Nuclear power plants and heat transfer, J. Heat Transfer Soc. Jpn. Tokyo 50 (213) (2011) 28–35 (in Japanese).
[19] N.C. Rasmussen, Reactor Safety Study—An Assessment of Accident Risks in U.S. Commercial Nuclear Power Plants, U.S. NRC, Washington, DC, 1975. WASH-1400, NUREG-75/014.
[20] J.C. Lee, N.J. McCormick, Risk and Safety Analysis of Nuclear Systems, John Wiley & Sons, Hoboken, NJ, 2011.
[21] U.S. NRC, Severe Accident Risks: An Assessment for Five U.S. Nuclear Power Plants, U.S. NRC, Washington, DC, 1990. NUREG-1150.
[22] U.S. NRC, Perspectives on Reactor Safety, NUREG/CR-6042, SAND93-0971, Revision 2, U.S. NRC, Washington, DC, 2002.
[23] Japan Nuclear Energy Safety, Probabilistic safety analysis of BWR plants against earthquake level 2, JNES/NSAG10- 0003, 2010 (in Japanese).
[24] The Yomiuri Shinbun, Tokyo, 4 Apr. 2011 (in Japanese).
[25] The Heavy & Chemical Industries News Agency, Survey of Introduction of Foreign Techniques, The Heavy & Chemical Industries News Agency, Tokyo, 1965 (in Japanese).
[26] H. Yamana, Development of the Japanese Nuclear Power Generation and its historical background, J. Heat Transfer Soc. Jpn. Tokyo 52 (219) (2013) 7–14 (in Japanese).
[27] Nuclear Safety Commission, Lesson learned from TMI accident and response to the Japan's NPP Safety Measure, White paper on nuclear safety, 1981 (Chapter 3.1) (in Japanese).

[28] Nuclear Safety Commission, Accident management against severe accidents at light water nuclear power plants, Decision on 28 May 1992, 1992. http://www.mext.go.jp/b_menu/hakusho/nc/t19920528001/t19920528001.html (Accessed 2 July 2015) (in Japanese).
[29] U.S. NRC, Report—a comparison of U.S. and Japanese Regulatory Requirements in effect at the time of the Fukushima Accident, 2013.
[30] M. Haage, IAEA safety standards and guidance on safety culture in the pre-operational phases, 2014/2015. https://www.iaea.org/NuclearPower/Downloadable/Meetings/2014/2015-02-03-02-06/D4_S10_IAEA_Haage.pdf (Accessed 1 June 2015); https://pbadupws.nrc.gov/docs/ML1332/ML13326A991.pdf (Accessed 10 July 2015).
[31] J.A. Simpson, E.S.C. Weiner, The Compact Oxford English Dictionary, second ed., Clarendon Press, Oxford, 1991.
[32] I. Shinmura, Japanese Dictionary Kojien, sixth ed., Iwanami Shoten Publishers, Tokyo, 2008 (in Japanese).
[33] U.S. NRC, Safety Culture Common Language, NUREG-2165, U.S. NRC, Washington, DC, 2014.
[34] U.S. NRC, Safety Culture Policy Statement, NUREG/BR-0500, Revision 3, U.S. NRC, Washington, DC, 2015.
[35] Nuclear Regulation Authority, Japan, New regulatory requirements for light-water nuclear power plants—outline, 2013. https://www.nsr.go.jp/data/000067212.pdf (Accessed 3 July 2015).
[36] About NRA, https://www.nsr.go.jp/english/e_nra/index.html (Accessed 28 July 2015).
[37] The Commission, http://www.nrc.gov/about-nrc/organization/commfuncdesc.html (Accessed 28 July 2015).
[38] Nuclear Regulation Authority, Japan, Permission for changes in reactor installation of units 1 and 2, Sendai Nuclear Power Station, Kyushu Electric Power Co., Inc., 2014. For English news release of NRA, see https://www.nsr.go.jp/data/000067246.pdf (Accessed 28 July 2015). Detailed data is available at https://www.nsr.go.jp/data/000029964.pdf (in Japanese).
[39] Nuclear Regulation Authority, Japan, Permission for changes in reactor installation of units 3 and 4, Takahama Nuclear Power Station, Kansai Electric Power Co., Inc., 2015. For English news release of NRA, see https://www.nsr.go.jp/data/000096401.pdf (Accessed 28 July 2015). Detailed data is also available at https://www.nsr.go.jp/data/000096119.pdf (in Japanese).
[40] Nuclear Regulation Authority, Japan, Permission for changes in reactor installation of Ikata Nuclear Power Station Unit 3, Shikoku Electric Power Co., Inc., 2015. For English news release of NRA, see https://www.nsr.go.jp/data/000115068.pdf (Accessed 23 October 2015). Detailed data is also available at https://www.nsr.go.jp/data/000115031.pdf (in Japanese).
[41] Supreme Court of Japan, Judgment upon the method for examination and estimation in the suit for cancellation of the permission for installation of a nuclear reactor; Ikata Nuclear Power Plant case (Case number 1985 (Gyo-Tsu) 133), Minshu 46 (7) (1992). English version is Available from: http://www.courts.go.jp/app/hanrei_en/detail?id=107 (Accessed 23 October 2015).

[42] Supreme Court of Japan, Judgment concerning what matters are included in the scope of matters concerning safety of the basic design of a nuclear plant which should be subjected to the safety examination for granting permission for the establishment of a nuclear reactor (Case number 2003 (Gyo-Hi) 108), Minshu 59 (4) (2005). English version is Available from: http://www.courts.go.jp/app/hanrei_en/detail?id=749 (Accessed 23 October 2015).
[43] The National Board of Boiler and Pressure Vessel Inspectors. https://www.nationalboard.org/Default.aspx (Accessed 28 July 2015).

FURTHER READING

[1] Reconstruction Agency, Current status of the number of the Great East Japan Earthquake Related Death, Report on 31 Mar. 2015. http://www.reconstruction.go.jp/topics/main-cat2/sub-cat2-6/20150630_kanrenshi.pdf (Accessed 20 September 2015) (in Japanese).
[2] National Police Agency of Japan, Damage and police responses in the Tohoku District–off the Pacific Ocean Earthquake, 2013. Available from http://www/npa.go.jp/archive/keibi/biki/higaijokyo.pdf (Accessed 10 October 2015) (in Japanese).

Index

Note: Page numbers followed by "*f*" indicate figures, "*t*" indicate tables, and "*np*" indicate footnotes.

O

P

R